Charles C. Abbott

Primitive Industry

Salzwasser

Charles C. Abbott

Primitive Industry

1. Auflage | ISBN: 978-3-84604-894-8

Erscheinungsort: Frankfurt, Deutschland

Erscheinungsjahr: 2020

Salzwasser Verlag GmbH

Reprint of the original, first published in 1881.

PRIMITIVE INDUSTRY.

PRIMITIVE INDUSTRY:

OR

ILLUSTRATIONS OF THE HANDIWORK,

IN STONE, BONE AND CLAY,

OF THE

NATIVE RACES

OF

THE NORTHERN ATLANTIC SEABOARD OF AMERICA.

BY CHARLES C. ABBOTT, M. D.,

Cor. Member Boston Society of Natural History; Fellow Royal Society of Antiquaries of the North Copenhagen, etc., etc.

SALEM, MASS.:

GEORGE A. BATES.

1881.

PREFACE.

Little need be added by way of a preface to the present volume.

The work may be said to be the natural outcome of the fact that the author lives in a neighborhood once densely populated by the Indians, as attested by the thousands of stone implements that are scattered over the fields.

As in the course of my rambles in search of relics, I have always met with kindness and aid from all who could assist me, it is scarcely practicable to mention every one who kindly offered assistance, gave me specimens, and in other ways furthered the undertaking. Many, however, have been named in the text, in connection with the specimens they have furnished, or the information derived from them.

Still I feel it is incumbent upon me to acknowledge my indebtedness to Prof. F. W. Putnam, Curator of the Museum of Archæology, at Cambridge, Mass., and to Lucien Carr, Esq., assistant curator of the same institution. Without the aid so kindly rendered by them, it is doubtful if the volume could have been completed.

I am also indebted to the late Prof. S. S. Haldeman for the kindly interest he took in all that pertained to the preparation of the work. The treasures of his cabinet were always at my disposal, and I have drawn from it and from his correspondence, almost all the matter relating to the archæology of the valley of the Susquehanna river.

To Henry Carvill Lewis, of Philadelphia, my grateful acknowledgment is also due for assistance in many ways; and particularly for the chapter containing the results of his latest investigations into the geology of the Delaware river valley, which he has contributed to the work. By so doing, he has added greatly to the scientific value of that portion relating to the evidences of Palæolithic Man on the Atlantic seaboard of North America.

It is a pleasant duty, furthermore, to acknowledge, at this time, my great indebtedness to my father, Timothy Abbott, Esq., of Trenton, N. J., whose unfailing interest in my labors, and substantial encouragement in meeting many of the expenses of the undertaking, enabled me to overcome many difficulties that seemed insurmountable.

To the Hon. Abram S. Hewitt, of New York, I am greatly indebted for valuable aid, here, most gratefully acknowledged.

In conclusion, I would add that the series of illustrations of objects from California, which have been used for purposes of comparison, are copied from the Seventh Volume of the U. S. Geographical Surveys West of 100th Meridian; Capt. G. M. Wheeler in Charge; and their use in advance of the appearance of that volume arises from the fact that the author was associated with Prof. Putnam, its principal author, in its preparation.

Realizing how much is yet to be done in the field of investigation treated in this volume, the work has been written, not so much, indeed, with the hope that it may add materially to our common stock of knowledge, as that it may induce others to explore such localities as they have the opportunity of doing, and to preserve such traces of early man as they may find *by placing them in public museums.* If, in so doing, my readers shall find this volume a worthy guide, my pleasant labors during the many years in which I have been a collector will ever remain a happy recollection.

C. C. A.

Prospect Hill: Trenton, N. J.
July 1, 1881.

TABLE OF CONTENTS.

CHAPTER I.

INTRODUCTORY.

While the early inhabitants of the continent of North America are collectively known as " Indians," it has been long ascertained that they present distinctions which widely separate them, and possibly point to several and diverse origins. A study, therefore, of the handiwork in stone, bone and clay, of the former occupants of any one portion of the country is not, of itself, sufficient to give an accurate knowledge of the wonderful variety of forms, and skill exhibited, in fashioning the articles which their needs demanded. Such, however, is the similarity existing among the objects generally known as " Indian relics," wheresoever found, that we are led to conclude that, to a limited extent, through a system of barter or the vicissitudes of warfare, the distinctive weapons and implements of one people became mingled with the home productions of their neighbors. It is evident, therefore, that, in treating of the implements, weapons and ornaments found in any one locality, we cannot be, at all times, positive that any given specimen is the production of the tribe of Indians known to have inhabited the country where it was found. The influence of the introduction of objects of European manufacture must also be taken into consideration, as in the century or more that elapsed between the visits of the first explorers and the arrival of the colonists at Plymouth and at Jamestown, and later, at Philadelphia, the introduction of a few foreign trinkets and metal cooking vessels greatly influenced the home productions of the various Indian nations. Many of the later articles made by them were doubtlessly modelled from similar objects of European origin. From this relinquishment of the customs established by them, during their happy ignorance of European civilization, may be dated the termination of the career of these people in manufactures of a purely American character. As one of their own people has well said:[1] "we lived before the English came among us, as well or better, if we may believe what our forefathers have told us. We had then room enough, and plenty of deer which was easily caught, and though we had not knives, hatchets nor guns, such as we have now, yet we had knives of stone, and hatchets of stone, and bows and arrows, and these served our uses as well then as the English ones do now."

As the one general locality, extensive as it is, treated of in the present volume, does not include any territory known to have been permanently occupied by the so-called Moundbuilders, the relationship of that people to the supposed ruder hunting tribes of the Atlantic seaboard will not be discussed, although the belief may be here expressed that they are not necessarily older than the earliest occupants of the Eastern coast.

On the other hand, the somewhat vague historical references to the presence of the Eskimo as permanent occupants of the coasts of the New England States, and even farther south, will receive brief attention in connection with a class of objects which were probably the handiwork of that people.

It has been necessary to adopt some method of classification in studying the great number of distinct patterns of stone and bone implements described in the succeeding pages. So far as possible this has been done by following the references of the early writers, to the utensils and weapons of the Indians. When there were to be found no such guiding references, the suggested purposes are necessarily conjectural; and if the suggestions are, in a few instances, made in a somewhat *ex cathedra* manner, it is because the author is fully convinced of the *probability* of the suggestion, and not because it could be demonstrated as true. In the more general classification, however, implying a greater antiquity of a given class of objects over those of another, and in separating the traces of the pre-European

---

[1] Acrelius. Memoirs Penna. Historical Society, Vol. XI, p. 52. (Speech of Canassatego.)

occupants of the eastern coast of North America into three divisions, each antedating the other, it is fully believed that the facts justify the implied relative antiquity and chronological sequence. Exception therefore is firmly taken to the view expressed by Prof. Whitney,[2] that "it is evident that there has been no unfolding of the intellectual faculties of the human race on this continent which can be parallelized with that which has taken place in Central Europe. We can recognize no palæolithic, neolithic, bronze or iron ages. Over most of the continent, as it seems to the writer, man cannot be considered as having made any essential progress towards civilization." The careful and systematic examination of the surface geology of New Jersey, of itself, it is believed shows as abundant and unmistakable evidence of the transition from a true palæolithic to a neolithic condition, as is exhibited in the traces of human handiwork found in the valley of any European river. The proofs of this earlier than an Indian occupancy of the Atlantic coast of North America, and of the intermediate period that connects this earlier with the true Indian age, will constitute the second part of the volume.

As the greater number, by far, of the various implements and other objects of stone, bone and clay that are here described, are unquestionably the handiwork of those tribes which were in peaceful possession of the Atlantic coast when first visited by European adventurers, a brief reference to them collectively seems necessary; although the scope of the present volume does not include a discussion of the ethnic relationship of the various native races of the Atlantic coast, but merely an illustration of how far their ingenuity had expressed itself in utilizing stone to supply their several wants. These so-called Indians, which have figured so largely, in the earlier historic times, as well as filled a large place in the prehistoric annals of America, are known to ethnologists as the Algonkins and Iroquois.

---

[2] Memoirs of the Museum of Comp. Zoology, Vol. VI (1st Part): The Auriferous Gravels of the Sierra Nevada of California. By J. D. Whitney, Cambridge, Mass., 1879. (Chapter III, Sec. V, Human Remains and Works of Art in the Gravel Series, p. 287.)

These are[3] "peoples of wholly diverse descent and language, who, at the time of the discovery, were the sole possessors of the region now embraced by Canada and the eastern United States north of the thirty-fifth parallel. The latter * * * * occupied much of the soil from the St. Lawrence and Lake Ontario to the Roanoke. * * * * They were a race of warriors, courageous, cruel, unimaginative, but of rare political sagacity. They are more like ancient Romans than Indians, and are leading figures in the colonial wars.

The Algonkins surrounded them on every side, occupying the rest of the region mentioned, and running westward to the base of the Rocky Mountains, where one of their famous bands, the Blackfeet, still hunts over the valley of the Saskatchewan. They were more genial than the Iroquois, of milder manners and more vivid fancy, and were regarded by these with a curious mixture of respect and contempt. Some writer has connected this difference with their preference for the open prairie country in contrast to the endless and sombre forests where were the homes of the Iroquois. Their history abounds in great men, whose ambitious plans were foiled by the levity of their allies and their want of persistence. They it was who under King Philip fought the Puritan fathers; who at the instigation of Pontiac doomed to death every white trespasser on their soil; who, led by Tecumseh and Black Hawk, gathered the clans of the forest and mountain for the last pitched battle of the races in the Mississippi valley. To them belong the mild mannered Lenni-Lenapé, who little foreboded the hand of iron that grasped their own so softly under the elm-tree of Shackamaxon; to them the restless Shawnee, the gypsy of the wilderness; the Chippeways of Lake Superior, and also to them the Indian girl Pocahontas, who in the legend averted from the head of the white man the blow, which, rebounding, swept away her father and all his tribe."

---

[3] Brinton. Myths of the New World, 2nd ed. p. 26. New York, 1876.

CHAPTER II.

GROOVED STONE AXES.

JUDGING from a series of over two hundred examples of grooved axes gathered from nearly every county of the state of New Jersey, and the full series of axes from New England, in the Peabody Museum of Archæology and Ethnology at Cambridge, it is evident that no one pattern of this form of stone implement is peculiar to any neighborhood, and no form occurs in the south or west, in any way differing from those occurring on the Atlantic seaboard.

The material of which they are made varies almost as much as does that of the arrowheads, although it is rarely that we meet with axes either of quartz or jasper, yet such are not wholly wanting. In most cases, a close-grained, heavy mineral, susceptible of polish was chosen, although the grinding of any part of the surface, except at the edge, was mostly omitted.

In glancing over any considerable series of axes, we find the same variation in the degree of finish, that we see in every form of weapon or domestic implement; but perhaps the most noticeable difference is in the size. Many of these axes appear to be too large to be readily wielded for any purpose; while the smallest doubtlessly were toys.

As so large a number of these implements have had their shape determined by the contour of the pebbles from which they were fashioned, it is difficult to determine what is to be considered a typical axe, if indeed, there is one. The use of a water-worn pebble as a hammer, simply held in the hand, was among the first acts of primitive man, and it was not long before the advantage of a handle to such a stone was recognized. Once hafted and in use, either as a hammer or a weapon, if the ends at all differed in sharpness of point or thinness of edge, the desirability of increasing the effectiveness of these features would quickly suggest itself. In this way we can see how readily a water-worn pebble would be converted into what we have here designated as an axe, and see also, how indefinite must have been the shapes of those first in use.

If we endeavor to trace any development of the more specialized forms of stone implements, it will prove a puzzling problem to connect the pecked and polished grooved stone axes of neolithic times, with the chipped implements of the river drift. No two forms could be more dissimilar, yet the hatchet of the one period and axe of the other are supposed to have had similar uses. For use as a cutting implement, one might well hesitate between the two forms, for rude as the better made chipped implements of the drift appear to be, they are quite as available for most of the purposes for which the grooved stone axe was designed. The people that used the one certainly never developed them into the other; nor, would the chipped hatchets of the better class always be discarded for the axe. It is, in truth, inconceivable that these latter should be the production of the same people, who used the chipped implements; and if they originated among the descendants of the palæolithic folks of the same river valleys, they are of quite recent times comparatively, and came gradually into use, as a specialized implement, intended for but few purposes, just as, at an apparently later date, we find the gouge, hoe and celt. There is no relationship to be traced between the two forms, or evidence that the makers of the one were at all related to the inventors of the other.

In examining any large series of grooved stone axes, the fact that the shape was of less importance than some other of its features becomes very evident. Occasionally, we meet with an axe that has no trace of work upon it, other than the groove; the edge being a natural feature of the pebble; others again have only some slight inequality pecked away; and from such an one, we have a regular gradation in degree of alteration of the original surface, to such as are wholly artificial in shape, of perfect symmetry, and exquisite polish.

Of those that are shaped without reference to the previous size or shape of the selected stone, there is not a very extended range of patterns. Apparently, every Indian being his own axe-maker, if not content with the simpler form of a slightly modified pebble, fashioned *de novo* such an implement as pleased him, and whether the edge was very broad and the axe short, or just the reverse, was a mere whim of the maker. One feature, in this respect, may be held as true, that the depth of the groove, symmetry of the blade, and degree of polish, are all more marked in those specimens that are wholly of artificial design, and have been shaped from a mass of stone that originally bore no resemblance to the finished implement.

The maximum size of stone axes may be stated to be twelve inches in length, and six to eight inches in width. Those of this maximum length are very rare, while those measuring one-half, and two-thirds of that size, are extremely common. The weight, of course, varies with the size and the density of the mineral used. Some axes, of diorite, or of porphyry of small size, are as heavy as others that are much larger, of sandstone. The smallest axe I have ever found or seen, from New Jersey, measures but two and three-fourths inches in length. It is well shaped, has the groove extending entirely around it, and has had an excellent edge. As the surface is now so weathered, and as the mineral is not very compact, it is impossible to determine whether it has been polished or not.

The most noticeable feature of grooved stone axes is the groove that either partially or wholly encircles the implement. While in many cases, it is merely a shallow depression roughly pecked away, in others it is very deep and occasionally highly polished; but why it should be so, considering the one object of the groove, it is difficult to conjecture. Certainly the material of which the handle was made, whether hide or wood, would not polish it, even if it were wrapped tightly about the axe, as was necessary, in order to secure it effectively. As

I have already mentioned, this groove is occasionally protected by prominent ridges on each side, which also deepen it considerably. The position of the groove varies from very near the centre of the axe to a point so near the head, that but a narrow rim protects it. On measuring nearly one thousand examples, I find that the groove in most cases is about one-third of the total length distant from the head, and is at right angles to the upper margin of the implement. In many specimens, where the groove is quite in the middle of the axe, it is evident that the blade has been repeatedly shortened by grinding the edge anew, and so, in some instances, fully one-third of the implement has been worn away. In the valley of the Susquehanna river, in Pennsylvania, the majority of the grooved axes have had an oblique direction given to the groove, which of course would alter the position of the handle, from that of right angles to the axe. What object there was in this peculiarity, it is difficult to determine. This pattern is not unknown, but is very rarely met with in the valley of the Delaware river, or elsewhere in New Jersey. They are of more frequent occurrence in New York, but probably do not occur in New England. There is no example of this pattern in the large series of New England axes in the Museum at Cambridge.

A large proportion, possibly two-thirds of the stone axes found in New Jersey, have the groove extending along the sides and across one margin; the opposite margin, in these cases, being flat or slightly concave. This same pattern is common also to New England, but not to such an extent; one-half, probably, of the axes found in Connecticut and northward having the groove entirely encircling the stone.

As the implements under consideration are called "axes," it is natural to infer that the edge, which in many specimens is quite sharp, should be adapted to cutting. If not, the term axe is a misnomer. It is generally conceded that, with these implements, standing timber could not be cut. This is not true of all axes, however, and one of unusually sharp edge was found, by experiment, sufficiently sharp to enable me to cut—not bruise—a small tree, by bringing the edge in contact with the tree, at an angle, say of forty-five degrees. But the labor was such as would have disheartened an Indian, and the task could scarcely be successfully repeated on trees of larger girth. No stone axe, that I have ever seen, would be available for felling trees of even a foot in diameter in a reasonable length of time.

There is such a difference in the finish of the edge of any ordinary series of these axes, that it is probable this feature, in connection with the size, determined the particular use of each; and there was that range in use, as is apparent in the case of celts which, varying in length from fifteen inches to one inch, could not have been intended for one and the same purpose.

The thickness of the blade varies to a considerable degree, and many of the broadest examples are quite short. The slope to the edge in such cases is very pronounced, and however sharp the edge may be, its cutting power is necessarily much reduced. Such short and thick axes were even better weapons than those that are thinner.

From the great number of stone axes already gathered, and that remain to be gathered from the area of the state of New Jersey, it is clear that this form of weapon or implement, as the case may be, was in constant and universal use among the Delaware Indians. In some localities, of several square miles in extent, there have been found from three to five axes in every one hundred acres, and still others are occasionally brought to light by the plough. Allowing but one-half the smaller number to have been left lying in every one hundred acres of the state's area, when abandoned by the Indians, there would remain, for the benefit of archæologists, the enormous number of one hundred and twenty-five thousand stone axes. If these axes are as abundant elsewhere, as they are in Mercer and Burlington counties, New Jersey, it is quite within reason to believe that one-half that number were left by the resident Indians, when they relinquished their territory to the founder of Philadelphia.

However incorrect the above estimate of the abundance of axes in New Jersey may be, it is certain that there have been many hundreds gathered in the past, without any apparent diminution of their numbers. Yearly the plough upturns as many as in previous years, and the thoughtful observer, who chances to seek for these scattered relics, is amazed at their frequent occurrence.

Supposing that these grooved axes, except such small examples as were probably toys, were only used and owned by men, does this great abundance of them indicate a larger population, than is generally supposed existed at any one time, or may we take it as indicative of a smaller community, whose occupancy extended over an immense lapse of time? So far as axes not in use, or those deposited in graves affect their numbers, it may be remarked, that we find that several early authors mention the fact of the handing down from father to son, of the cherished stone axe, to fashion which, "the life of a savage is often insufficient."[4] This shows that the custom of depositing them in graves was only occasional; and therefore, it may be considered, that such setting aside of a certain number would not materially affect any calculations based upon their numbers, as now found scattered broadcast over the entire area of the state.

In New England, grooved stone axes are by no means abundant. They are a well known form and a dozen or more may be collected in the course of a summer's work, but in no portion of this extended area, do they appear to have been in such common use, as in the more southern states. Prof. George H. Perkins[5], in treating of the archæology of the Champlain valley, remarks that "grooved axes are not common, though some very fine specimens have been found, but I have seen none that would compare favorably with the finest western specimens either in size or elegance of form. The largest I have seen is nine inches long and four and one-half inches wide, but most of the grooved axes are much smaller. Such specimens of axes as have been collected have been obtained, one here and one there, singly; nowhere in such groups as some collectors describe. I presume that all the grooved axes ever found in the Champlain valley, unless many were destroyed before collectors began to save them,

---

[4] Mœurs des Suavages Ameriq., Vol. I, p. 110, Paris, 1724. Quoted by C. C. Jones, jr.

[5] American Naturalist, Vol. XIII, No. 12, p. 738, 1879.

would not amount to so large a number as Dr. Abbott mentions from a single small excavation made in digging a cellar in Trenton, N. J."

In Massachusetts, there is a greater abundance of these implements, than would appear to be the case in western Vermont, as is shown by the large series in the Museum at Cambridge, and in the collection of the Peabody Academy of Science, in Salem, Mass.; but while relatively more abundant in eastern Massachusetts, than in the Champlain valley, they are scarce as compared with the numbers found in New York and New Jersey. Nor is this abundance at all confined to the middle states. Southward of Pennsylvania, they are abundant; and along the greater part, if not the entire extent, of the Atlantic seaboard, the statement that in Georgia they "are frequently met with in the sepulchred tumuli, upon the sites of old villages, in relic-beds, and in cultivated fields,"[6] is equally applicable.

As found in New England, axes are essentially a "surface" find, no reference to them being made by any of the archæologists who have so carefully examined the graves of the Indians of Massachusetts. Celts and ornaments, as will be mentioned in a subsequent chapter, are frequently found, but never the grooved axe. In New Jersey, a small proportion of the graves of Indians contain, among other patterns of stone implements, a small grooved axe, such as, from the size and finish, may be considered the prototype of the more modern iron tomahawk. It is doubtful if the largest of these implements were used as weapons, as they do not appear ever to have been placed among the "treasures" of any warrior, who was fortunate enough to secure a careful burial at the hands of his friends.

Fig. 1 is a very good example of the most common type of grooved stone axes. This specimen measures eleven and one-half inches in total length, and is but four inches wide at the broadest portion, the ridge immediately in front of the groove. The groove itself is but seven-eighths of an inch in width, and the head, or that portion posterior to the groove, varies from one and one-half inches

[6] Antiquities of Southern Indians, p. 274. New York, 1873.

to one inch in length. The cutting-edge is but two and one-fourth inches in extent, and is still moderately sharp and well preserved. Although the specimen still shows the marks of the hammer, yet it might also be placed under the head of polished axes, as the weapon has been so carefully smoothed down that the slight inequalities and shallow indentations can scarcely be felt by the hand. Its weight is seven and one-half pounds. With the handle placed where the groove is, it must have produced a great strain upon the wrist. Axes, with the groove so far removed from the middle of the implement, as in this instance, are not often met with, except such as are more nearly of equal length and breadth.

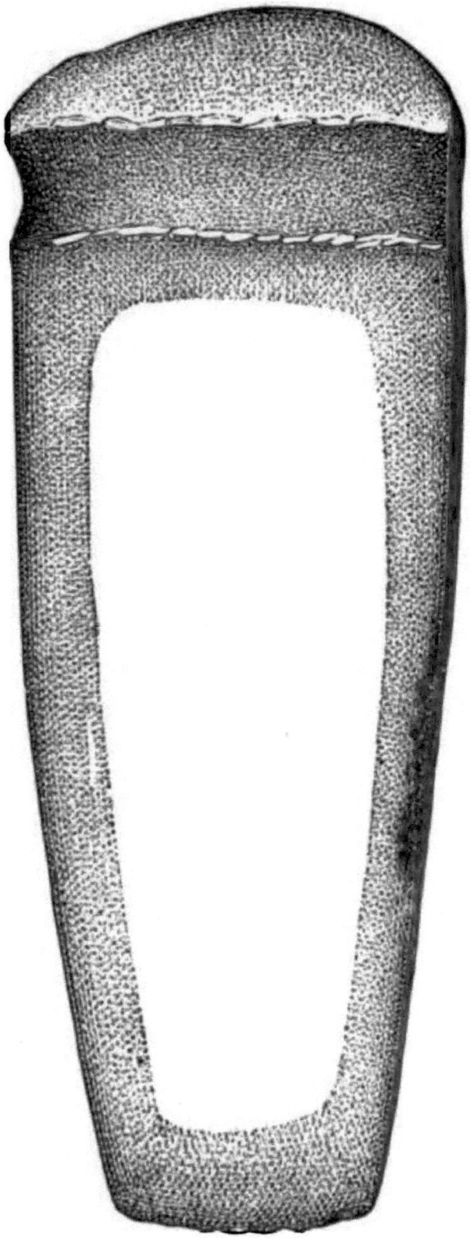

FIG. 1. — New Jersey. $\frac{1}{2}$.

It is remarked by Loskiel,[7] of the stone axes of the Delaware or Lenni Lenapè Indians, that "they were not used to fell trees, but only to peel them, or to kill their enemies." Cer-

[7] Hist. of Mission of United Brethren, p. 54, London, 1794.

tainly the axe before us might have been used for either purpose; but in an attempt at classification, I should refer the smaller specimens to the category of weapons or "tomahawks;" to which class fig. 1, judging from its weight and size, could never have belonged. Few Indians of to-day, certainly, would care to place this axe among their side-arms, or be compelled to use it in a hand-to-hand conflict.

This specimen was found on a small gravelly island in the Delaware

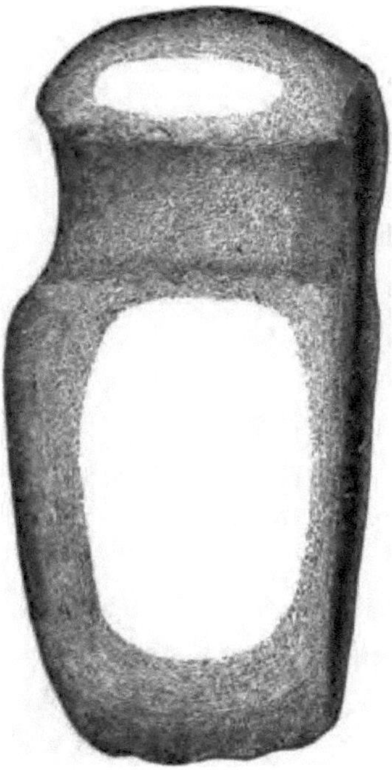

FIG. 2.—New Jersey. ½.

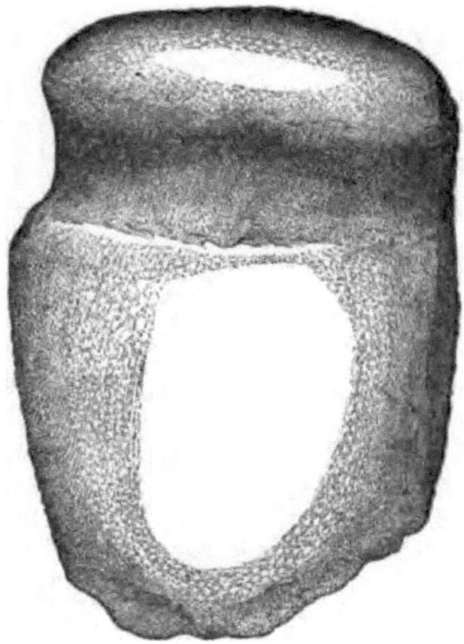

FIG. 3.—New Jersey. ½.

river, and was presented to the author by his friend Mr. William Dean, of Lambertville, N. J.

Fig. 2 represents an average example of this pattern of grooved axe, which is found in greater or less abundance throughout the United States, east of the Mississippi river. Such axes generally are about seven inches in length, the extremes being from two and one-half inches to eleven and one-half, as in the preceding illustration. It is very seldom that one is found that measures less than four inches, and but few are seen in any collections so small as the minimum here mentioned.

When of such diminutive size, these objects become toys rather than weapons, although if mounted on a slender, flexible handle, their value as a weapon would not be inconsiderable. Figs. 3 and 4 represent excellent average examples of the smallest size of grooved axes. Whether toys or weapons, as much care has been expended upon them, as upon the largest, and the battering to which the edge of No. 3 has been exposed shows that, if a toy, the play must have been rough and very realistic, that so completely demolished the cutting edge. In the smaller of the two specimens, fig. 4, it will be noticed that the edge, which is still preserved, is very smoothly polished. It is the more probable that these small axes were made for children's uses as we find not only all weapons reproduced in miniature, but even such prosaic articles as mortars, and cooking vessels. The same is true of the pottery; especially of the kind found in such great abundance in the western and southwestern states; where besides clay images,—the counterparts of modern children's dolls,—miniature vessels of most of the patterns in vogue at that time are common.

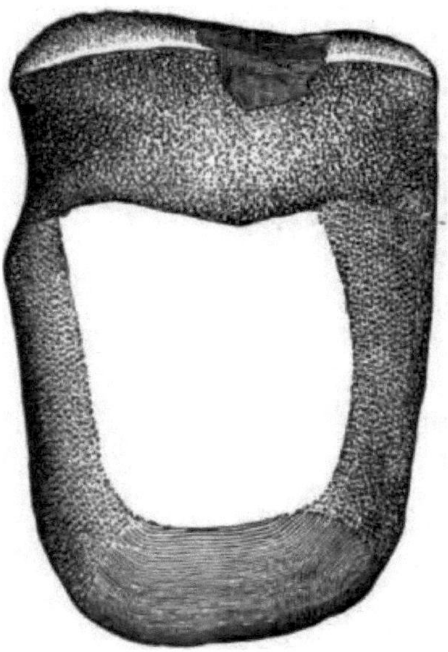

FIG. 4.—New Jersey. ½.

Fig. 5 is an admirable example of an axe made of a porphyry pebble of this pattern, worn down by continual resharpening. The specimen now measures four inches in length by three and three-eighths in width, and is two and one-half inches across the head or back. It has a well-defined groove running along one margin, a feature common to this pattern of stone axes. Into this marginal groove a wedge is supposed to have been driven in order to tighten the handle. This is possibly the true explanation.

Several specimens of axes have been collected quite recently, however, which have this marginal groove duplicated, each being quite deep, and the separating ridge as high as the exterior edges of the grooves. Were the object of the single depression the insertion of a wedge, as suggested, these doubly grooved axes would need two such tightening wedges, which is not probable. If, however, the end of the handle was placed against this margin of the axe, and so notched as to fit closely the single or double groove, as the case might be, then the binding withe or sinew wrapped about the implement would closely adhere to the transverse groove at every part, and need no wedges to secure the attachment of the axe firmly to its handle.

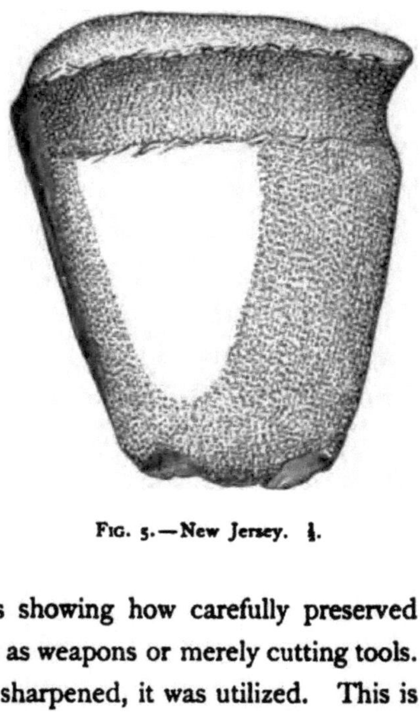

FIG. 5.—New Jersey. ½.

In many axes of this pattern the groove is but very slightly defined, and in none from the Atlantic seaboard is it so deeply cut, as in a number of those found west of the Alleghany Mountains.

Fig. 5 is of especial interest as showing how carefully preserved were these stone axes, whether used as weapons or merely cutting tools. So long as any blade was left to be sharpened, it was utilized. This is more marked in the above than in any other axe in the collection of the Museum at Cambridge.

This variety of axe is usually of sandstone, and the ordinary cobblestones, or water-worn pebbles of the adjacent river-beds. At and above Trenton, N. J., the bed of the Delaware river is wholly composed of loose stones of various sizes, with here and there an outcropping of rock. These loose pebbles are frequently found on examination to bear considerable general resemblance to finished axes, and to need little work upon them other than making the groove and rubbing one end down until a cutting-edge is produced. So abundant are the well-adapted stones, in shape and size, that we wonder why so frequently one meets with stone axes that have been carefully pecked over the whole surface to bring them down to the proper shape. This may be explained, perhaps, by the suggestion that many axes were made where suitable stones were difficult to obtain, and that the frequent wars or wanderings of a community and bartering may have resulted in the commingling of the axes of a multitude of localities, many of them miles distant from each other. It is known, too, that tribes came from long distances to make autumnal visits to our seacoast, and, of course, on such journeys they would always be provided with, and frequently lose, as they passed through the state, many specimens of both weapons and domestic implements.

The routes taken by the Indians who annually crossed New Jersey, from their homes in the mountains of Pennsylvania, on their autumnal visit to the seacoast, were well known to the early surveyors of the state; and several of the principal thoroughfares, extending from the Delaware river eastward, are the sites of those trails over which the Indians had been accustomed to pass, for unknown centuries.

It is in the immediate vicinity of these trails that we still find a great number of the various patterns of stone implements and fragments of pottery, which are largely the traces of those inland communities which passed yearly, by the same path, to their chosen locality on the coast. Year after year, they camped at the same spot, while en route, and left imperishable traces of their sojourn by the sea, in the well-known Indian shell-heaps.

It has been thought by many, and confidently asserted by a few, that this particular form of stone axe was peculiarly a moundbuilder's weapon or implement. Certainly many of the finest examples of this form have been picked up in the immediate vicinity of mounds; though it is well known that axes of any description have been but rarely found in the mounds; moreover those of the maximum size and highest finish are not confined to the vicinities of mounds.

A very fine specimen of a large axe is in the cabinet of Rutgers College Museum, at New Brunswick, N. J. It was found within the limits of that town, on the banks of the Raritan river, which was probably a favorite locality with the aborigines, on account of the native copper that was formerly found there, and which they highly prized for a variety of purposes, especially ornamental. The axe above referred to is of identical pattern with that figured by Squier and Davis in Smithsonian Contributions, vol. 1, p. 216, fig. 108 (Anc. Mon. Miss. Valley), but is somewhat larger and heavier. The former measures nine inches in length by six inches in width, and weighs an ounce or two over nine pounds. The western specimen "is made of very compact greenstone, and measures eight inches in length by five inches and a half in its greatest breadth, and weighs eight pounds." Squier and Davis further state that this "is regarded as a genuine relic of the moundbuilders. *Its form is almost identical with that of the forest-axe of the present day.*" The result of investigations up to the present time, in and about these same mounds, renders it almost certain, that the above mentioned axe was itself a "forest-axe."

Fig. 6 represents one of the finest specimens of a large stone axe that we have ever met with. Very many that we have seen have been as large; a number have been of more finished workmanship, but no one has as many features of interest as this. This specimen measures eleven inches in length. The conical head is three inches long, the groove and ridges together two and one-quarter inches, and the blade within a small fraction of five and three-quarter inches. The conical head does not appear to have met with any very hard usage, and was probably intended for ornament. It would seem as though the ridges, at each margin of the groove, would be of great advantage in fastening the handle to the axe, inasmuch as it secures greater depth to the groove without cutting too deeply into the body of the implement itself; but such plausible reasoning somewhat vanishes when we come to compare weights and find that this specimen (fig. 6) weighs but six pounds, whereas fig. 1, with one pound and a half greater weight, has a groove only one-half the depth and width, placed as near as practicable to one end, while in fig. 6 it approaches more nearly to the middle.

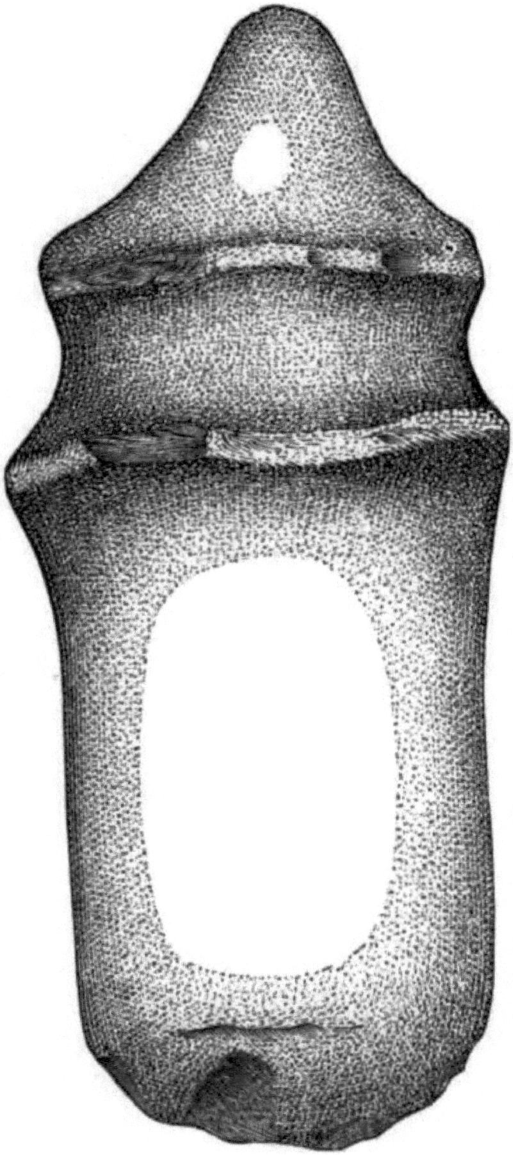

FIG. 6.—New Jersey. ½.

Axes of this pattern, having both the projecting and protecting ridges at the groove and a conical head, are common within limited localities in New Jersey. Thus, in Gloucester county, in a series of one hundred specimens, about twenty were of this pattern. All were similar to the specimen here figured though not so large, and in no instance was the conical head so symmetrical.

The material of which this axe is made is very compact and unyielding; and considering the amount of work upon it, one can well realize that much time was consumed ere it was completed.

Axes of this form do not appear to be widely distributed, as judged by the examination of large series from various states. In some collections, there was but a single example, whilst in others, nothing at all similar to it was seen.

Fig. 6 was found on the shore of the Delaware river, near Trenton, N. J., and was presented to the writer by Dr. J. W. Ward of that city.

Fig. 7 represents an axe of somewhat similar outline to the preceding, having the ridges that are on the margin of the groove very well defined, but the tapering, conical head is by no means as artistically finished as in the former instance. As the illustration shows, this specimen has been pecked over its whole surface, and is a good example of the perseverance and patience of the primitive folk who accounted such weapons among the chiefest of their worldly goods. Axes of this shape and pattern occur in many parts of New Jersey, but are less common in New England. No better opportunity for contrasting the rude with the elaborate specimens of a similar implement is furnished, than by a study of this axe and the preceding. In the specimen before us, we have the same high protecting ridges to the groove, as in fig. 6, but without a trace of the care and workmanship there shown. The conical head is symmetrical, and has been brought by pecking to its present shape, but it is not much altered, in comparison with the pointed back or head of the preceding specimen.

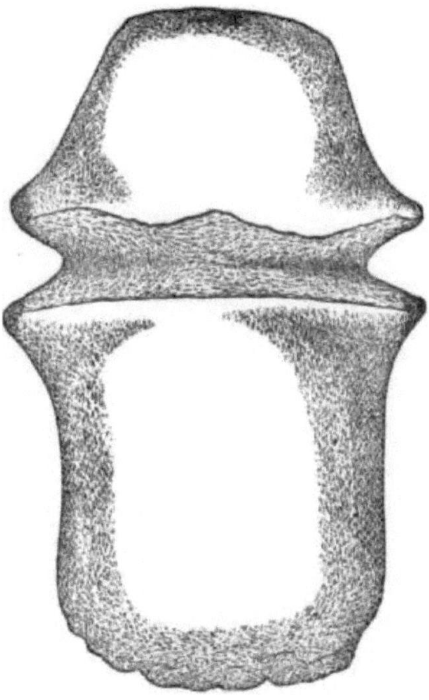

FIG. 7.—New Jersey. ½.

Much has been said, by early writers, of the use of these axes as wedges for splitting wood; and as this implies the use of a maul of some sort, there is an apparent explanation of the frequent fractures of the head of the axes, as now found. Still, it is doubtful if grooved axes were customarily used in this way, as the labor of making them was far too great to warrant their being subjected to blows which would very quickly destroy them. Certainly axes like the present and preceding examples, which show no marks of violence, could not have been so used.

Fig. 8 represents a remarkably fine example of a polished grooved axe. The illustration gives a better idea of the specimen than can any description. Suffice it to say, that the whole surface has been beautifully polished, and the edge, still perfect, is as sharp as it can be made, and describes nearly an accurate arc of a circle. It will be noticed that this axe has two grooves, one of them shallower and much less well defined than the posterior and deeper one.

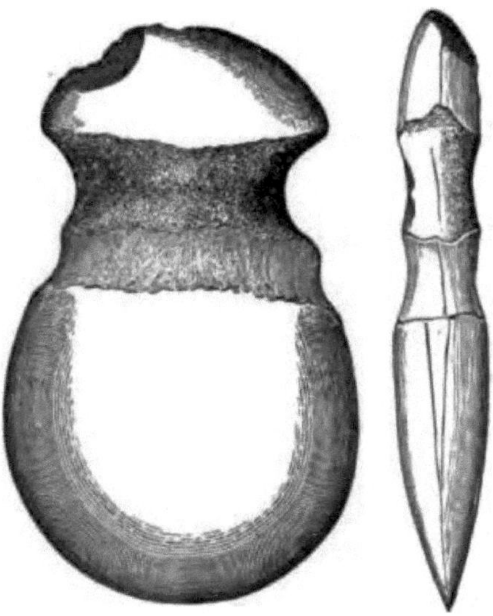

FIG. 8.—New Jersey. ½.

The object of the double groove, which is but seldom met with on the Atlantic seaboard, is by no means clear. No similar example is found in either of the large New England collections at Cambridge, or at Salem, Mass., nor do such axes occur in any numbers, apparently, except in the immediate vicinity of where fig. 8 was found. Polished axes of somewhat similar form, but without such a well defined cutting edge, are met with in New Mexico.[8] Specimens have been found in the ancient pueblos, with the two grooves, very well defined.

[8] U. S. Geog. Survey of Territories, west of 100th merid. Vol. VII, Pl. xvii, xviii, xix, p. 377.

These doubly grooved axes, or axe-shaped implements possess several marked peculiarities. Some of those described in the volume referred to, "seem to have had cutting edges, but they now so blunted that they appear to have been used more for giving blows than for cutting purposes. They are of small size and may have been mounted for use as implements of war, similar to the iron tomahawk of a later date. * * * Others ought rather to be called axe-shaped hammers than axes, for they evidently never were furnished with anything approaching a cutting edge, and their rounded and fractured ends show that they have long been used as hammers. * * * These pueblo axes are highly polished."

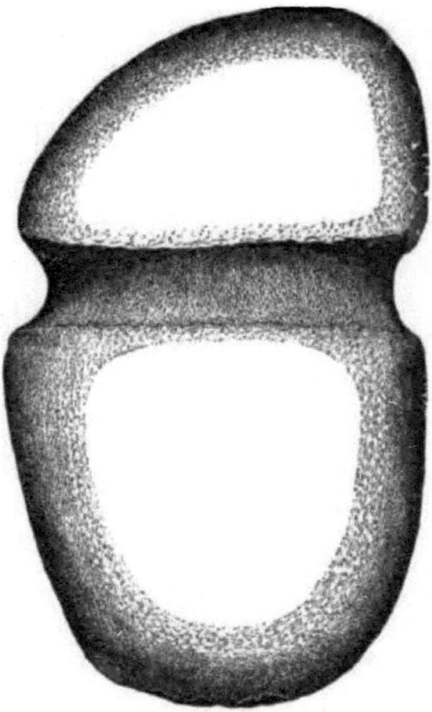

FIG. 9.—New Jersey. ½.

While no doubly-grooved axes have been found on the Atlantic coast that have so completely lost their cutting edges as have those described by Professor Putnam, it is no uncommon occurrence to meet with ordinary axes that have been converted into hammers by the gradual loss of their cutting edges, and the fractured surface worn to a comparatively smooth one either by design or as the result of use as a hammer. It is probable that most of the grooved hammers, other than those cylindrical in shape, were originally axes of the ordinary patterns.

Fig. 9 represents a fair average specimen of a cobble-stone axe, in which the groove extends entirely around the weapon. One feature is particularly noticeable in these axes, viz., that the groove is more nearly in the middle of the specimen. I have examined a large series of axes, and find that the following characteristic is common to all the examples that have come under my notice, viz.: that when the groove extends entirely around the axe, it is in advance of the usual position of the grooves that do not meet above, or on the upper margin, as in fig. 1. There was something in the method of using these implements that is yet to be learned, before an explanation can be given of this curious feature of the varying position of the groove. Certainly, the original shape of the selected pebble had nothing, or very little, to do with determining the location. This specimen (fig. 9) is about the average size of any ordinary collection of these stone axes as gathered from any one neighborhood. They range from four to eight inches in length, seldom exceeding this limit; and the number of instances of axes less than four inches in length is comparatively few. As a class, the completely grooved axes do not appear to be as well finished as the preceding style; and being usually of "crooked" or irregularly-shaped stones, when a number are together, there appears to be but little in common except such features as pronounce them all "axes."

Such an axe as fig. 9 is the simplest, if it may not be considered the primitive form of this implement. This is true, if the minimum amount of labor expended upon them is indicative of the first steps towards the production of such an axe, as the one immediately preceding. There is little, however, to warrant such an opinion and it would rather seem that if axes were the outgrowth of some simpler form, they originated beyond our boundaries. It has occurred to the writer, that as axes were made not by professionals, as was largely true in the case of arrowheads, but by each man for himself, they are really something of an indication of the great individual differences that existed among the Indians; a rude axe being the workmanship of a lazy man, and a symmetrical one, the result of long, patient labor on the part of a person who had the taste to design and the pluck to accomplish what, to them, was a serious undertaking, and involved, as we have seen, a great expenditure of time.

Fig. 10 represents an exceedingly crude axe, that, when figured, was the very "plainest" specimen I had ever met with. Since then, however, others, even more primitive, and yet unquestionably "grooved stone axes," have been collected. The specimen here figured has the groove on each side and above and below of a uni-

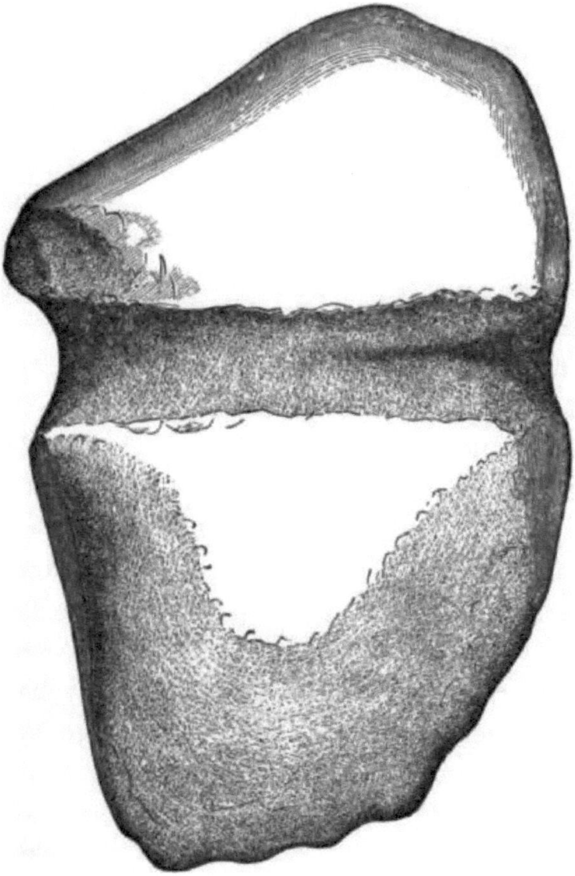

FIG. 10.—New Jersey. ½.

form depth, and is well defined throughout, as the illustration indicates; but, in the still plainer specimen, the groove consists of a faint roughening, that seems of little use, being scarcely uneven enough to prevent the fastening from slipping; but, like fig. 10 the groove at the top and bottom is practically deepened by a projecting knob of the stone, at which points all the strain of the fastening of the handle must have come. In the specimen figured (fig. 10), the sides of the blade of the axe have been dressed down with a hammer-stone to a pretty well defined edge; but in the still plainer specimen before us there is found that upon one side a few chips only have been struck off, and on the other two great portions have been artistically knocked away, and the then roughly-prepared blade has been rubbed with a polishing-stone until a small but highly-polished edge has been produced. I cannot imagine any more difficult task than *really* cutting or splitting wood with such a weapon as this, and would, therefore, restrict its use to bruising the bark of trees. Judging from their present appearances, the edges only of these axes have been used; the back, which is very uneven in each case, does not show any trace of having ever been struck with a hammer; and I find in many of the axes, especially in the pattern of figure 1, that they were so struck, thus converting the axe for a time into a wedge. Loskiel[9] says: "Their hatchets were wedges, made of hard stones, six or eight inches long, sharpened at the end and fastened to a wooden handle."

Occasionally we meet with crooked or bent axes, which have, however, more method in their irregularities than has fig. 10. Several such specimens have been collected in New Jersey, which were originally nothing but ordinary cobble-stones slightly bent or bow-shaped. Stones of this shape were frequently chosen, as I have seen a number of specimens from widely distant localities. The best illustration of such bent axes is one measuring nine inches in length by three and one-half inches in width. The head alone is the natural surface of the stone, except a narrow strip immediately in front of the groove; all the rest has been carefully worked but never polished. These so-called bent axes are attractive in appearance, but do not seem to possess any especial advantage over other forms.

Fig. 11 represents an irregularly shaped cobble-stone with three uncommon features: the near approach of the groove to the middle

---

[9] Mission among N. A. Indians (Delawares), page 54.

of the specimen; the almost flat surface of one side of the implement; and the *intended* double edge. It may be that the shorter end has had a cutting-edge, although there is now no trace of it left, or the broken condition of this end may be the preparatory chipping, to have it ready for grinding to a cutting-edge at any time it might be desirable to do so. This specimen measures six inches in length by two and one-half inches in breadth, except at the projection immediately in front of the groove on the upper margin, which projection is about one-half an inch in length. What might be the object in having one side flat, or nearly so, does not appear; but it will be seen that this peculiarity is not confined to this axe, or to a few axes as a class by themselves, but occurs in weapons and implements of very different character.

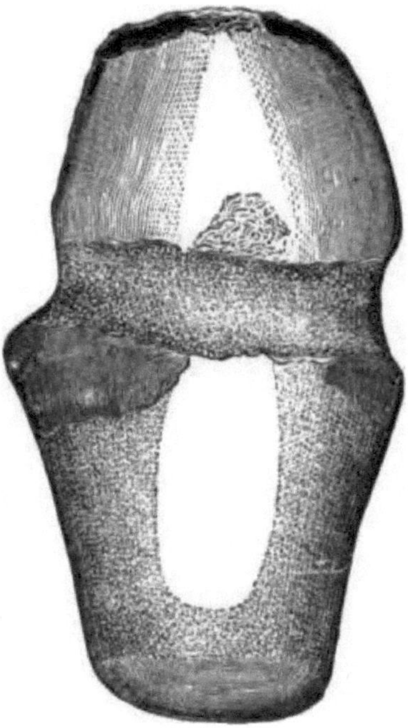

FIG. 11. — New Jersey. ½.

If the specimen here considered is, or was, a double-edged axe, it is the only example, so far met with. In describing axes found in the southern states, Col. C. C. Jones[10] remarks, that while "in most cases the groove is near the head of the axe; occasionally this transverse furrow runs across the central portion, thus affording an opportunity for a double edge. Specimens of this latter variety, so far as our observation extends, are carelessly made, and of soft material. They could have been used for little else than offensive purposes."

The above does not apply to the axes found either in New England or the middle states. Properly speaking there are no double-edged axes

---

[10] Antiquities of Southern Indians, p. 276, New York, 1873.

in tne Museum at Cambridge, or elsewhere in the United States, so far as I have seen, and the impression among many people, that such are occasionally found, arises from the habit of applying the term "axe" to the small ceremonial objects which, whatever their significance, were certainly never used as weapons, or implements of any kind.

Mention has already been made of grooved axes with the furrow extending obliquely across the implement. Such axes are characteristic of, but not peculiar to, the valley of the Susquehanna river, Pa. In a great majority of these the groove does not encircle the axe. A few specimens have been noticed that were simply water-worn pebbles, such as fig. 9. An estimate based upon the examination of several hundred specimens from that valley, and which tallied with the information received from correspondents, who had collected at different and widely separated points, leads me to infer that in about sixty per cent. of the axes found there the groove was oblique. In New Jersey, not more than three per cent. are of this pattern.

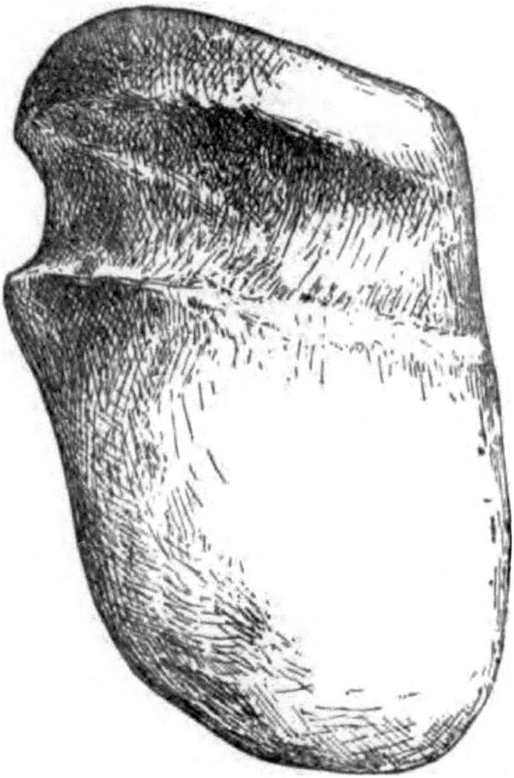

FIG. 12.—Pennsylvania. ½.

Of the stone implements generally, as gathered in the valleys of the Delaware and the Susquehanna, there is little to be said of the one series that does apply to the other. The same people were at the same time the sole occupants of the two localities, and it is a fact of peculiar interest that so marked a difference should occur, in so prominent an implement of daily use, as the grooved axe. It is in vain to attempt to offer any reason for the difference. The bare fact is all that we can ever know, and we are only left to wonder why the oblique groove, apparently much less desirable for every purpose, should have been preferred by a people to whom the straight and more desirable method must have been familiar.

Fig. 12 represents an average example of this pattern, and well exhibits the peculiarity of the oblique groove, referred to in the preceding paragraphs. It is common to the axes found in the Susquehanna valley, whether of the largest or smallest size, and thus, of itself, is not indicative of any particular purpose, as might reasonably be supposed, if it were limited to axes of a certain size.

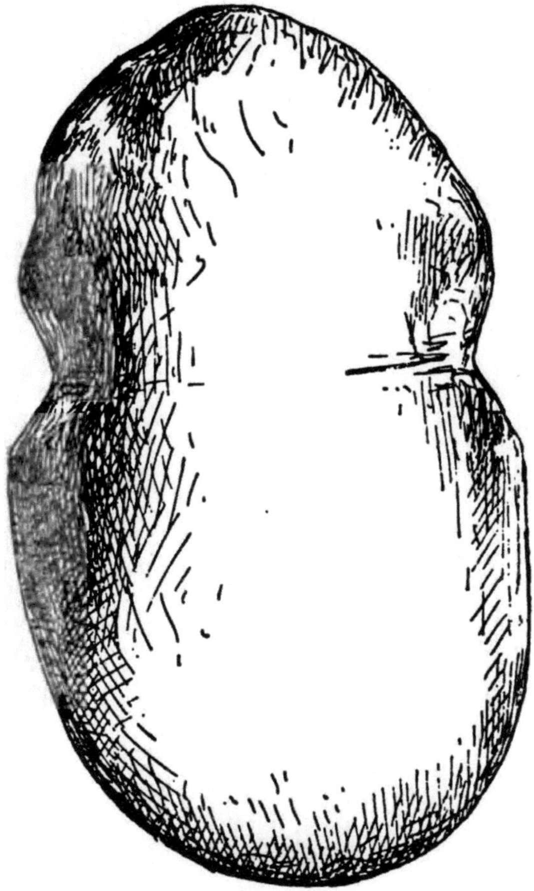

FIG. 13.—New Jersey. ½.

In any considerable series of axes from the New England states, there will be found a certain proportion with the groove running in a more or less oblique direction, but very seldom is it so pronounced, as in fig. 12.

A modification of the grooved axe is to be seen in that pattern where the groove is reduced to a deep marginal notch, as in fig. 13. This illustration is an excellent example of this form of stone axe. It is four inches in length, by two and one-half in width. It is of uniform thickness, the edge being quite abrupt, and hence of very little cutting power. The surface of the implement has been polished except at the head, which has been pecked to the desired shape.

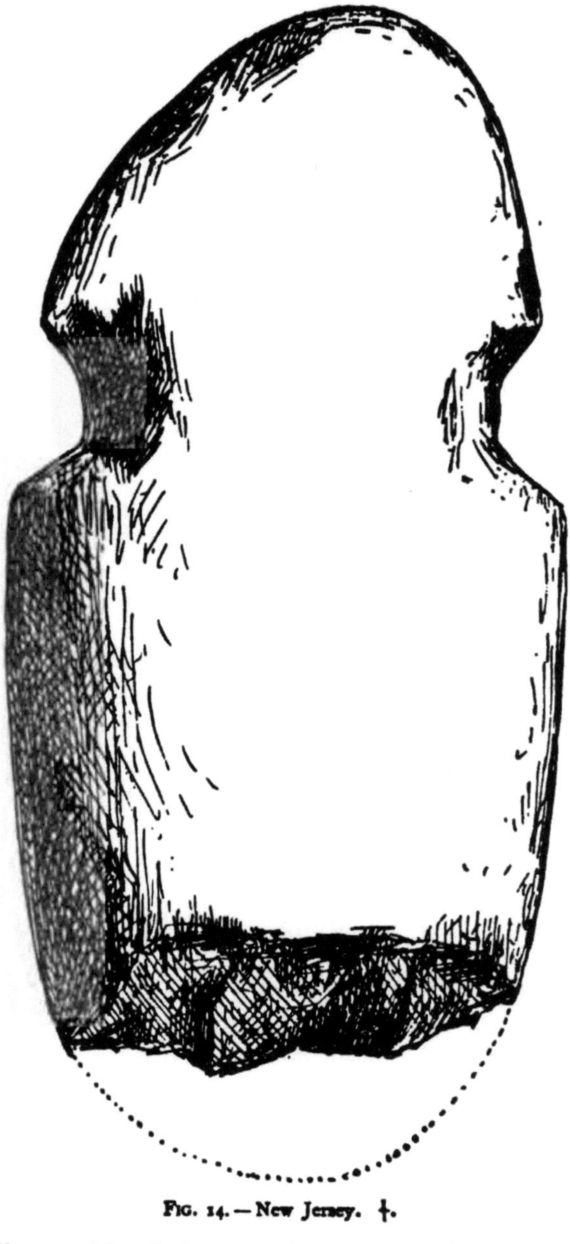

FIG. 14.—New Jersey. ½.

Fig. 14 represents a second example of this form, and differs principally from the preceding, in being an ordinary water-worn pebble, that has been notched and by a little modification brought to its present shape. Except this, all the examples of notched axes have been well polished, and have received that finish over the entire surface which characterizes a polished celt.

Two other examples of this pattern have been collected in Gloucester Co., N. J., one of which shows a trace of wear upon one side, extending from notch to notch, though it does not amount to a groove. This scratched and worn surface was caused, probably, by the band that passed around the implement to secure the handle; although how the handle was attached, if differently from the method followed in hafting the ordinary grooved axes, is not made clear, by the mere presence of the notches, unless they can be considered as replacing the groove which is hardly possible.

Notched axes, such as the above, while not common in any part of North America, are of comparative frequency in South America. In the cabinet of the late Prof. Haldeman there is a beautiful example from Demerara; in the Museum at Cambridge, there are others from Brazil and the West India Islands, and in the Archives of the National Museum of Brazil,[11] a specimen very similar to fig. 14 is represented.

Fig. 15 represents a highly finished example of a grooved implement, which may originally have been an axe. At present, the point that replaces the edge makes it difficult of classification, as it is quite unlike any other specimen from the localities, whence are derived the material upon which this volume is based. It is not, however, by any means a unique specimen. Axes with an edge so narrow, as to be almost pointed, are occasionally found in Pennsylvania; and in Ohio, specimens that are distinctly pointed, and not simply with a narrowed edge, are even more common. They do not occur in New England, at least nothing of this character is found in the collections of axes in the museums there.

The upper margin and a portion of each side of the specimen here figured, from near the middle to the extreme point, are quite highly polished. The under surface and head of the implement are pecked

---

[11] Archivos do Mus. Nac. do Rio de Janeiro, Vol. 1, Trim 1º. Est. 1, Fig. 2, 1876.

and smooth, but not polished. This polishing of one surface only is indicative, probably, of the fact that the specimen was originally an ordinary axe, with an edge equal in length to the width of the blade; and the polished surfaces result from grinding the face down to its present outline, in consequence of some accident by which the edge was destroyed. This, however, is wholly conjectural.

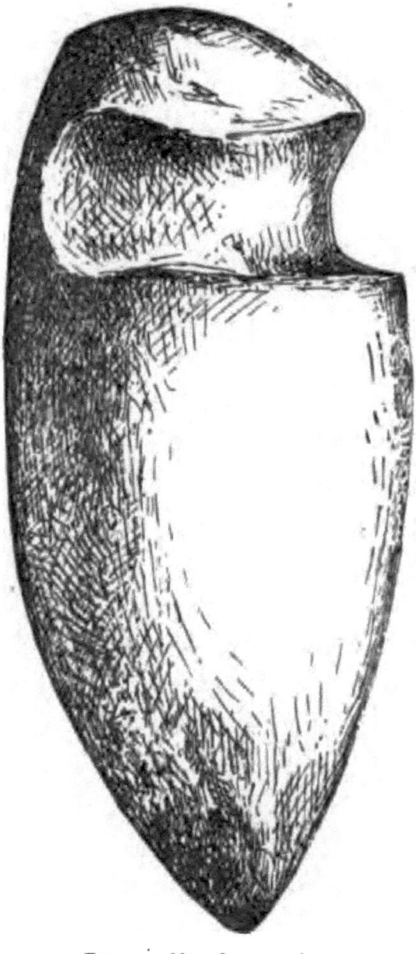

FIG. 15.—New Jersey. ½.

As an instance of the occurrence of a well known form of stone implement far beyond its supposed boundaries, fig. 16 is worthy of attention. Not only does this specimen prove the occurrence of grooved axes on the coast of California, but it is also an instance of the pointed instead of edged axe. This axe, if such it may be called, was found by Mr. Schumacher, at the steatite quarry, on St. Catalina Island, and it shows in the thin coating of steatite dust which is ground into it, that it has been used for the purpose of pecking or hammering out masses of soapstone, for subsequent conversion into the cooking vessels, that are found in that neighborhood in such great abundance.

While in the immediate vicinity of the locality, where the preceding example of a pointed axe from New Jersey was found, there is an extensive soapstone quarry, that was well known to the Indians, it is not probable that this specimen was used in working it, as it is free from such scratches and slight fractures as would result from the constant hammering upon even so yielding a material as steatite. In all probability, the pointed axe, fig. 15, was a weapon, and the somewhat similar specimen from California was a tool.

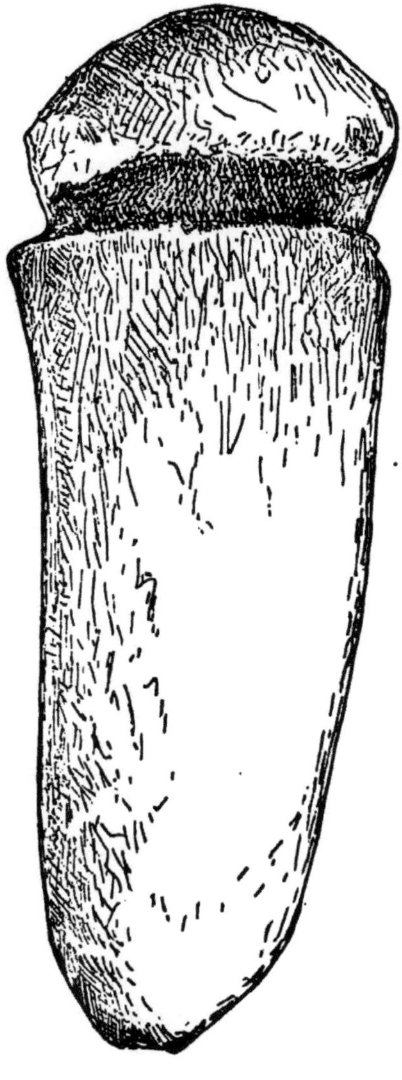

FIG. 16. — California. ½.

Dr. Daniel Wilson, in "Prehistoric Man," page 412, gives an illustration and comments on an "inscribed axe" that was found in New Jersey, and so claims a notice here. We quote in full: "In 1859, Dr. John C. Evans, of Pemberton, N. J., communicated to the American Ethnological Society an account of a stone axe inscribed in similar [that is, to the 'Yarmouth Bay Stone,'] unknown characters, which had been recently ploughed up on a neighboring farm. The axe, which measures about six inches long by three and a half broad, is engraved from a drawing furnished to me by Dr. Evans. Dr. E. H. Davis, after carefully examining the original, informs me that, though the graven characters have been partially retouched in the process of cleaning it, their edges present an appearance of age consistent with the idea of their genuineness, and the circumstances attending its production furnish no grounds for doubting its authenticity. Two of the characters are placed on one side in the groove for the handle; the others apparently form a continuous line, running round both sides of the axe-blade."

An example of an inscribed axe is now in the collection of the Museum at Cambridge, from New Jersey, which certainly cannot be considered as of modern origin, nor is the "inscription" of an "alphabetical" character. Yet there can be no doubt that it is a carefully carved inscription, whatever its meaning may be. The figures or marks are as follows: "1110." They are deeply cut, and are about as closely placed, as they would be, were it intended to write that date.

Such an instance as this cannot be attributed to plough marks or any accidental occurrence of that character; and if it has any meaning such as ownership, it is of great interest, and may be allied to the peculiar notchings on many ornamental objects, which are of such a character as to lead to the supposition, they are not simply attempts at ornamentation.

The Pemberton axe, referred to above by Dr. Wilson, has at times, excited considerable discussion and is a fraud, so far as the "inscription" goes. It should in all such cases be remembered that the discovery of such specimens, even when made by people of known integrity, does not by any means, settle the question of the genuineness of the inscription. Those who unfortunately have so little to do, that they can find time, and are malicious enough to perpetrate such hoaxes, desire, above all things, that their handiwork should fall into just such hands as did the Pemberton axe. When such remarkable objects are found by men of respectability, they are received with that attention and interest, which the fabricators could not hope to gain, were they to present the discovery as their own. Fortunately the poor fool, who carved a few meaningless lines on an axe, overshot the mark, and in making them too much like known characters, he failed to mystify the honest workers in archæology.

And here, also, it may be well to caution archæologists against frauds of a different character, which are even more likely to deceive the unwary. These are imitations in soapstone of well known patterns of implements, rare or even wholly wanting in some cases in the localities from which they are said to come. Fraudulent axes of polished steatite, beautiful in form and finish, do not unfrequently find their way into the cabinets of private collectors, where, however, they accomplish but little harm.

It would be easy to go on for an indefinite time, and point out peculiar features in the multitudes of stone axes that are to be found in every museum, and scattered throughout the country, but it is unnecessary to give additional examples.

Whatever may be thought of the scientific value of single specimens of these axes, or of other relics found lying upon the surface of the ground, that value is enhanced perhaps, or at least interest is attached to the specimens, when we occasionally have the good fortune to unearth a so-called "deposit" of these specimens, sometimes numbering several hundred.

In one case, in digging a cellar in Trenton, N. J., one hundred and twenty were found, "all closely huddled up together," as described by the man who found them. They were about three feet below the surface, and a "foot deep" in the gravel underlying the soil. They were surrounded by, and entirely covered with, a bright brick-red powder. Again, in digging the receiving-vault of the Riverview cemetery, near Trenton, N. J., "a bushel-basketful of these axes was found, packed closely together, six feet deep in the ground." In the neighborhood of the bluff fronting the Delaware river, about three miles below Trenton, N. J., several such instances have come to the notice of the writer. In the first two instances, the specimens were all grooved cobble-stone axes. In another case, fifty porphyry celts were found. These appeared to have been carefully deposited, and not thrown pell-mell into the hole dug to contain them.

In all such cases of "deposits" of either axes or celts, there has been no commingling of a number of forms of implements, nor any trace of fire. The inference, judging from the conditions under which those in New Jersey were found, is that they have been buried for the purpose of temporarily concealing them.

But one single fact has come under notice that in any way bears upon the subject of the age of these relics.

The instance referred to was as follows: on the 3d of July, 1869, a large white oak, measuring twenty-seven feet in circumference at three feet from the ground, during a high gale of wind, was uprooted. A short time afterward the immense stump was removed preparatory to levelling the ground. The hole that the extracted roots left measured seven feet in depth and thirty-three in circumference. Four feet below the bottom of this hole, or eleven feet from the surface of the ground, was found a very rude stone axe, entangled in a mass of fibrous roots that had been cut off from the main roots of the tree. In this case the axe must have been buried in the earth before this old tree was an acorn. Now, as to the age of the tree. There were not less than five hundred rings clearly to be traced on a section of the tree afterward made; and a large portion of the centre and another portion about the circumference could not be determined accurately, but which, on comparison with so much of the tree as retained the rings sufficiently distinct to be counted, might safely be estimated at as many more circles. Without allowing for any time to have elapsed from the time the axe fell to the ground, or was intentionally buried, we have here with considerable certainty, indications of the long stretch of one thousand years that this axe has been quietly resting in the ground.

CHAPTER III.

CELTS, CHISELS AND GOUGES.

THE term "celt," from the Latin *Celtis*, a chisel, has been universally applied to certain polished cutting implements of stone, which may be considered as the typical form of implement of the polished stone, or neolithic period.[12] The term, as ordinarily used, however, includes such a considerable range of forms, that some limited subdivision seems desirable to avoid confusion.

The series of polished and worked stone implements that are collectively described in the present chapter, are, in accordance with this plan, subdivided into three classes: the celts proper, with broad, convex cutting edges; the chisels or narrow celts with straight edges, and the gouges or celts with concave blades and curved cutting edges.

As in all cases, where an attempt is made to classify a large series of stone implements, so it will be found here, that many objects are so closely related to each other, that places are assigned to them according to the fancy of the collector.

In considering the various forms that have been gathered, the grouping has been made about the most pronounced examples, and the boundary line in most cases has been obliterated by the many intermediate forms.

Of the celts proper, much has been written, yet little positively ascertained as to their uses. If there were no grooved axes found along

---

[12] Not all implements of this, the later division of the Stone Age, are *polished*, but many are simply pecked and hammered into the desired shapes. Such implements are strictly "neolithic" in age.

our Atlantic seaboard, the larger celts might be considered as axes; but under the circumstances it cannot be shown that they were merely a simpler form of that implement. The smaller celts have been supposed to be used as knives for skinning animals, yet no savage was ever seen to skin an animal with one of them. On the contrary, stone knives of a very different pattern are used for this purpose.

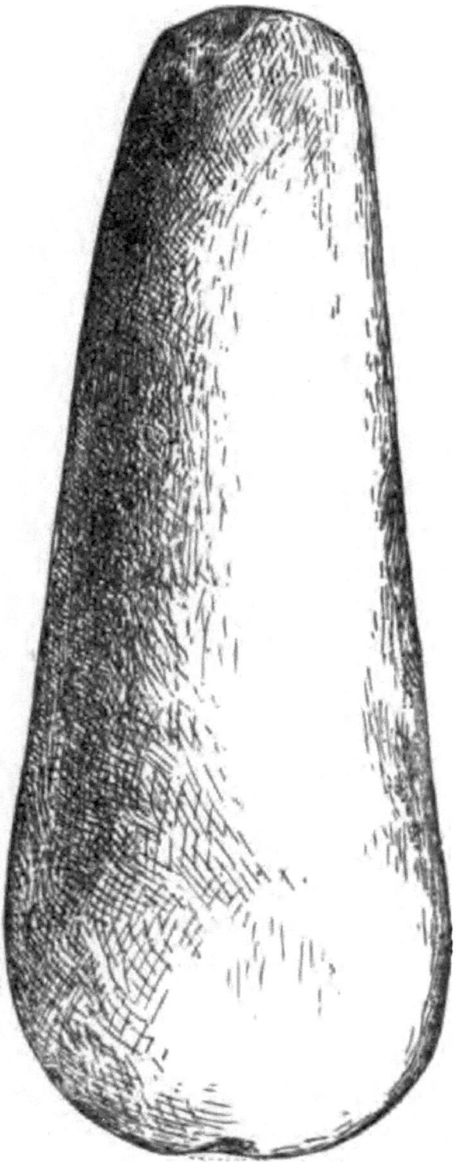

FIG. 17.—New Jersey. ½.

The typical chisels were doubtless used in much the same manner as the steel chisels of to-day: the wood having been previously charred, so as to make these primitive tools available. Such of these as are of very small size are a puzzle to the archæologist; and they are called celts or chisels from their general resemblance to other and larger objects, of a similar form, the use of which is indicated by their size and shape.

There exists less doubt in the mind of the collector, as to the gouges, than as to any other form of stone implement. That they could be used in any other manner, than as their name indicates, is inconceivable; nevertheless, we are here confronted by the difficulty that besets the satisfactory classification of other forms, that there are numbers of these gouges too small to be of any practical use. In this case, they cannot be disposed of as "toys for children," and hence, as far as they are concerned, we are left in the dark.

Fig. 17 represents what may be called the typical celt or axe without a groove. This specimen measures eleven inches in length by four in greatest width, and has a beautiful, even, sharp cutting edge, of a semicircular outline, which is perfectly true in every detail. The entire surface is evenly polished, but the material—a very compact fine grained sandstone—has not that glossiness of surface found on celts, made of porphyry or diorite.

Admirably fashioned and well adapted as it is even to cutting wood, it is difficult to conjecture any use for such an implement, except it be securely hafted; and that it was used in connection with a handle, there can be but little doubt. If such celts as these, which are very abundant along our northern Atlantic coast, had been usually hafted in stag-horn, as are many of those found in Swiss lakes, it is scarcely possible that all trace of such *horn* handles should have disappeared, especially as other implements of antler are exceedingly common; and in one case a finely polished celt of that very material has been taken from a burial mound in Arkansas.

In the collections of the American Museum of Natural History, at Central Park, New York, there is preserved a beautiful example of a polished celt of this pattern, still retained in its wooden handle. This handle is made of some hard wood, apparently black walnut, worked to a nearly cylindrical shape, and about fifteen inches in length. The handle has been perforated for the insertion of the implement, in the same manner as the wooden and horn handles found in the Swiss lakes. In the New York specimen, the celt was hafted so that the implement projected equally from both sides of the handle, and thus rendered available for use, the pointed end, as well as the cutting edge of the tool. This hafted celt was found near Lake Luzerne, N. Y.

If such implements as fig. 17 could be referred to the simple class of chisels, which is scarcely warranted, and considered only in connection with a hammer, as implements for working in wood, as canoe-making, then, indeed, there would be no reason for supposing that they were ever hafted, but the specimen found with its handle, to which reference has been made, shows that, at least to some extent, these larger celts were attached to handles.

Mr. Morgan[13] has remarked of the Iroquois, "for cutting trees, and excavating canoes and corn-mortars, in a word, for those necessary purposes, for which the axe would seem to be indispensable, the Iroquois used the stone chisel Uh′-ga-o-gwät-hă. In cutting trees, fire was applied at the foot and the chisel used to clear away the coal. By a repetition of the process, trees were felled and cut to pieces. Wooden vessels were hollowed out by the same means. Fire and the chisel were the substitutes for the axe. The chisel was usually about six inches long, three wide and two thick; the lower end being fashioned like the edge of an axe. Stone gouges in the form of a convex chisel were also used when a more regular concavity of the vessel was desired."

If it be proper to call all such polished stone implements, chisels, and consider them only as a carpenter's tool; it is evident from the fact of their very frequent occurrence in graves, that they possessed a high value in the estimation of the natives, which is somewhat improbable, if they were never put to other uses than boat-building and the felling of trees. Fig. 17, associated with another celt of equal size and scarcely inferior finish, was ploughed up in a field known to have been an Indian burial place. When discovered, the two were lying side by side, in actual contact.

As there is a well-marked class of implements found in the same localities where celts occur, which are true chisels, it is not without reason that such large celts as fig. 17 should be considered weapons.

In figs. 18 and 19 we have examples of common celts. These are

---

[13] League of the Iroquois, p. 358. New York, 1849.

the ordinary forms as gathered in the ploughed fields, found in graves, or unearthed in digging about village sites. Although not so abundant, they are almost as well known as arrowheads, or grooved axes. In the two specimens here figured, there will be noticed one marked difference. Fig. 18 is acutely pointed at the upper end; while fig. 19 is as markedly blunt, and further shows that it has been subjected to hard usage, as from blows from a stone hammer. Does this difference between the acutely pointed and the blunt head indicate that the

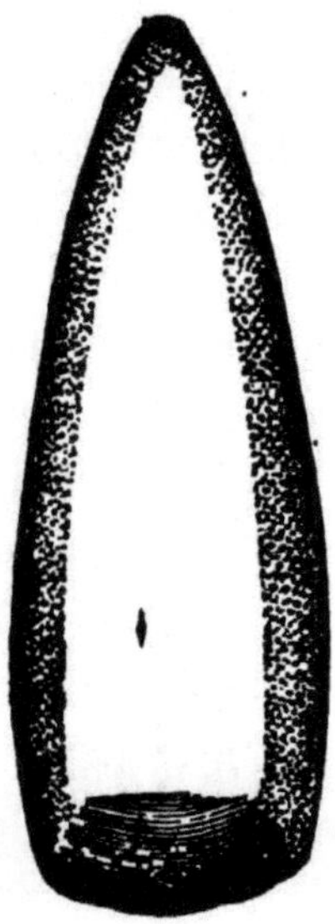

FIG. 18.—New Jersey. ½.

FIG. 19.—New Jersey. ½.

latter was used as a chisel? It would indeed be difficult to strike an effective blow, with certainty, on the pointed head of such a chisel as fig. 18.

This pattern is found in every part of the globe where polished stone implements occur, showing that it best met the common wants of mankind, everywhere; and possibly, if we could determine one use to which such axes were adapted, of a strictly universal nature, it would be safe to apply a name suggested by such use to this form, now known by the somewhat objectionable term of "celt."

Sir John Lubbock[14] figures a celt, similar to fig. 18, from Ireland; Nilsson[15] figures them from Scandinavia; and the pattern is nearly approached in axes from Accra, West Africa, figured by Sir J. Lubbock.[16]

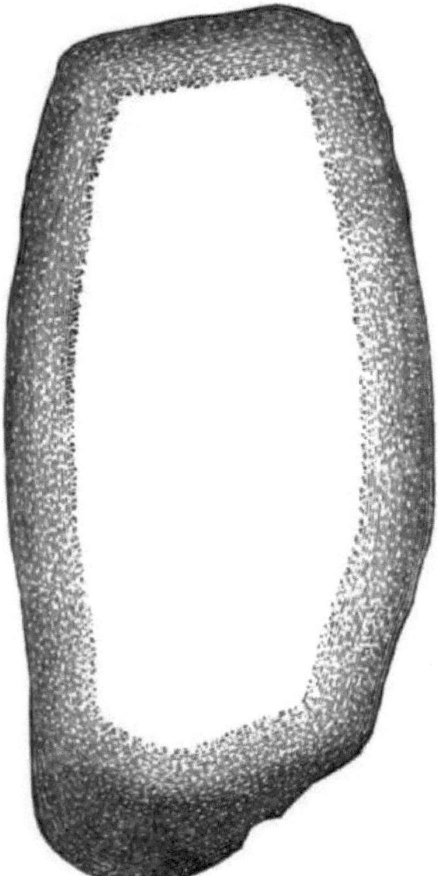

FIG. 20.—New Jersey. ½.

Fig. 20 represents a specimen of the larger ungrooved celts, that show but little trace of human workmanship, other than the finely-wrought edge, and a limited polished surface on the upper and lower margins. It measures eight and one-quarter inches in length, by four inches in width, at a point a little in advance of the middle. It is of ordinary sandstone, and originally was very nearly of its present shape. One side is flatter than the other, and appears to have been pecked and then somewhat polished. The margins have been polished for a short distance from the edge, and, on the lower margin, there is a very smooth surface, a little over an inch in extent either way, that appears to be such "peculiar polished space, which has been produced by the friction of the wood," described by Lubbock as exhibited in

[14] Prehistoric Times, 2d ed., figs. 97–93, page 88.

[15] Stone Age in Scandinavia, plate vii, figs. 151 and 162.

[16] Journ. Anthrop. Inst., London, vol. 1, page xcv (Proc. Eth. Soc.).

some specimens found in Europe. There does not appear to have been any hard hammering upon the head of this celt.

Rude celts of this very primitive pattern are by no means common in New Jersey, nor are they any more so in the New England states. Possibly, to some extent, they may have been overlooked, for such a celt as fig. 20 would scarcely be recognized if found lying in a stony field. Pebbles thus sharpened at one end may have been used for a short time only and then thrown aside, and under most circumstances, such discarded implements would have the edge broken by violent contact with other stones.

The puzzling feature of such rude implements as the above is, that one fails to comprehend why such should ever have been used, when there was apparently such an abundance of better ones, and when the edge is too limited in extent apparently to be of any use for cutting purposes.

FIG. 21.—New Jersey. ½.

Fig. 21 represents a comparatively common style of celt made from a piece of serpentine, pecked to a blunt point at the back. From about the middle of the implement to the edge it is very smoothly polished. This specimen measures a little less than three inches in length, and two inches in width along the cutting-edge, and is a very good average example of this class of implements.

Objects of this character, made, not only of serpentine, but of much denser mineral are very common, wherever stone implements of any pattern are found. This particular shape is even more abundant than the modification of it, which has the head nearly or quite as broad as the edge.

In New England, though not rare, these small celts are less frequently found than throughout New York and New Jersey. This comparative scarcity throughout New England is the more noticeable from the fact, that there does not appear to be any common form of stone implement found there which might readily replace it. As a rule, these small celts are still found with entire or but slightly fractured edges, which is proof of the fact that they were not exposed to violent contact with substances harder than wood. Indeed, their size and shape show that they were not mounted in handles, as were grooved axes, however it may have been with such long and slender forms as fig. 17, but may have been hafted in deers' horn, as were all the similar celts found in the Swiss lakes.

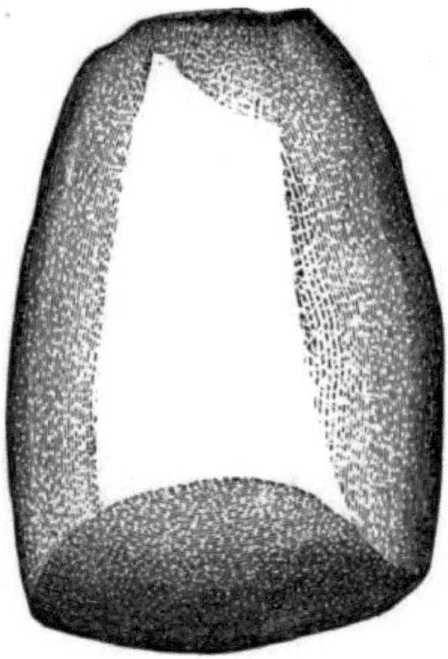

FIG. 22.—New Jersey. ½.

In the magnificent series of small celts, contained in the Clement collection of the Archæological Museum at Cambridge, there are hundreds which are mounted in horn or bone, by being inserted into the handle in such a manner that only a small portion of the blade projects. The use to which these diminutive implements were put, is, of course, a matter of conjecture. They may have been weapons, and possibly were also used for splitting the long bones of those animals which were used for food. The identical character of the implement as found in America, does not, however, indicate necessarily that they were used in this country in the same manner as were those found in the Swiss lakes.

Fig. 22 represents a slightly different pattern, being nearly square in outline. This blunt head, however, is carefully smoothed, and shows no trace of hammering, as might be looked for, had the implement been used as a chisel. This difference in outline does not indicate any difference of purpose, probably, and no use could be made of this celt which could not be made of the one immediately preceding.

As yet, the collections of stone implements made do not supply us with any examples of pebbles of this form, with simply an edge worked upon one end, as we have seen is true of such larger celts as fig. 20. Small celts, like figs. 21 and 22, and, as will be seen later on, fig. 25, have had their entire surfaces carefully worked, and it is not unlikely that, in many cases, these smaller specimens were originally much longer, and constant resharpening has reduced their original length, possibly one-half. Grooved axes, we have seen, have been reground until their length was quite disproportioned to the breadth and thickness of the blade, and such, too, was the case with many of the grooved hoes that have been collected.

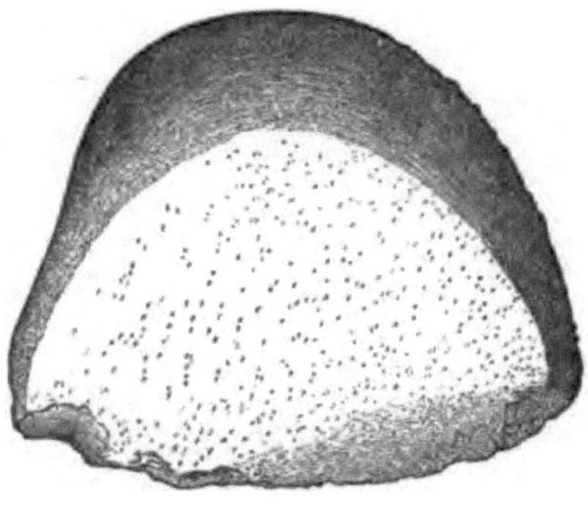

FIG. 23.—New Jersey. 1/1.

Fig. 23 represents an excellent example of these small celts greatly reduced in length by successive sharpenings. This specimen is a porphyry pebble, originally globular in form, although it may have been oval, and hence the original grinding away of so much as was necessary to produce an edge was not as great as now appears. It, indeed, seems incredible that a pebble of one inch and three-eighths in thickness, and less than two inches in length, should be chosen for making one of these small implements.

Quite small celts, with convex sides, curved edges, and of a thickness usually equal to one-half the width, are almost as abundant as those I have called the typical, medium sized celt, such as fig. 18. It is difficult to determine the relative abundance of various patterns, as we are still, of course, ignorant of how great a number of these same forms are left in the soil, for future explorers to collect; but basing calculations on such large series as have been gathered, it may be stated, that celts are abundant in proportion to their approach to such sizes as those represented in figs. 18 and 19; and that larger and smaller specimens are rare in proportion as they exceed in size, or are smaller than, figs. 17 and 21.

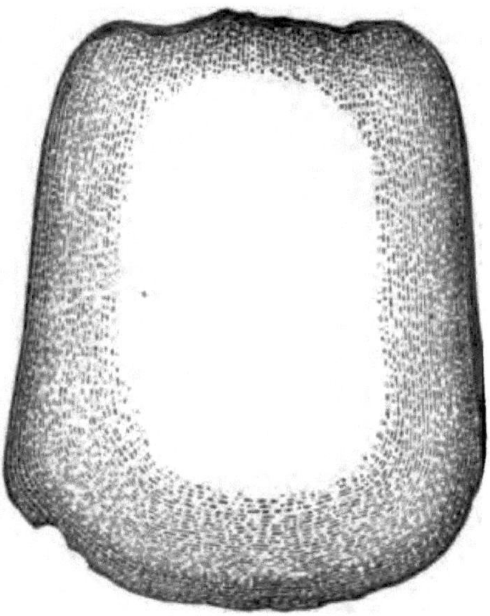

FIG. 24.—New Jersey. ½.

Fig. 24 represents an example of a thin pebble, quite similar to fig. 22. This little celt is made of very compact stone, and has not only a carefully worked edge, but is evenly polished over the entire surface. Such thin, square celts as this are frequently found in New Jersey, and occasionally in New England. In Ohio and westward, they are in greatest abundance, and taken collectively, those found in that region exhibit the maximum degree of skill in shaping and finishing.

Fig. 25 varies considerably from the preceding, in that the surfaces are not made to blend into each other, but are ground so as to make the celt angular. This is noticeable not only at the edge, but at the margins also, which are smooth and almost at right angles to the broad sides. This specimen differs in these respects from the average implement of this pattern, found in New Jersey, and is similar to the majority of those found in Ohio and Indiana.

Fig. 26 may properly be placed in the same "class" with the preceding. Although a much less finished specimen, it was unquestionably put to the same uses. It is made of a fine-grained porphyritic stone, and has been polished over its entire surface. This little "celt" measures two and one-eighth inches in length by one and three-quarters in width. The cutting edge was originally good. The back

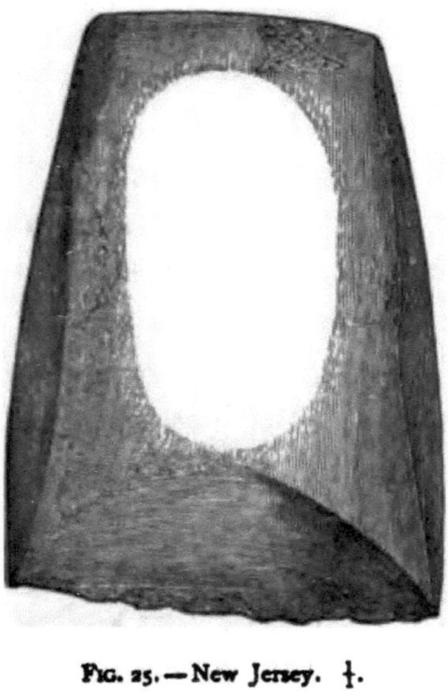

FIG. 25.—New Jersey. 1/1.

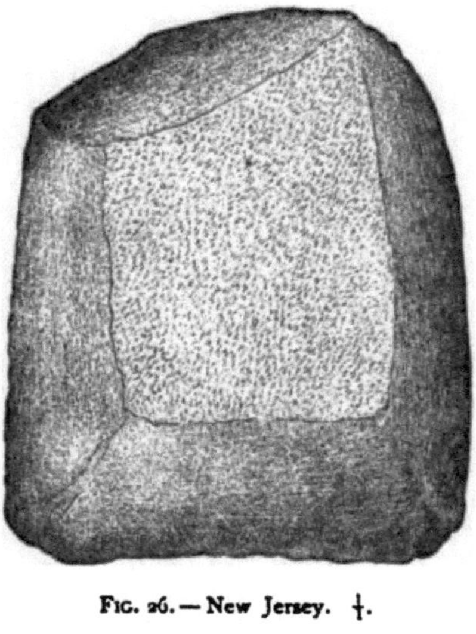

FIG. 26.—New Jersey. 1/1.

has a ridge running obliquely across it, from which the surfaces slope at an angle of forty-five degrees.

Sir John Lubbock,[17] in some "Notes on Stone Implements from Africa and Syria," gives figures of actual size of stone axes, which certainly are identical in shape, and have been used, no doubt, in an identical manner. The author says, with reference to them: "Some of the West African axes, as will be seen by the figures (plate ii, figs. 1 and 2), closely resemble some of the smaller axes so common in

[17] Journ. Anthrop. Inst., London, vol. 1, page xcii, plate 11, figs. 1 and 2 (Eth. Soc. Proc.).

Western Europe;" and adds, as has already been observed of the preceding pattern, "Indeed, this type may be said to be cosmopolitan, and needs no description."

Fig. 27 represents one of those very diminutive celts that are not uncommon in almost any considerable collection of Indian stone implements. This example is a serpentine pebble carefully rubbed down, until brought to a convenient shape, and then given an excellent cutting edge. Except the still smaller hematite celts, of which but very few examples have been found in New Jersey, and none in New England, this is about the smallest of this class of polished stone implements. For what purpose they were made, it is in vain to conjecture; but it must not be inferred because this specimen, measuring less than two inches in length is called by the same name as fig. 17, which lacks little of being twelve inches in length, that they were regarded as identical implements by the people who fashioned them.

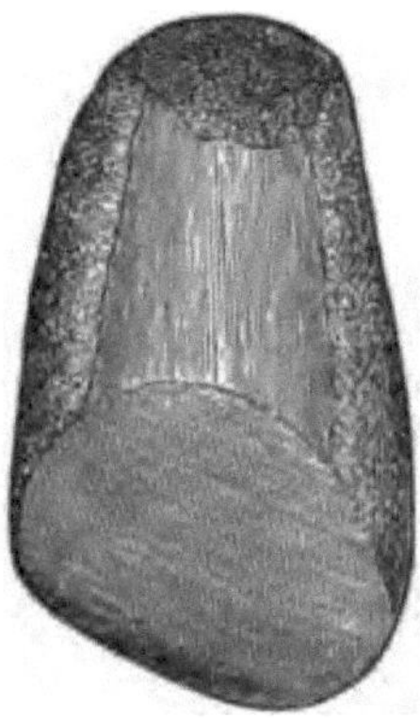

FIG. 27.—New Jersey. ½.

The hematite celts, to which reference has been made, are seldom smaller than fig. 27, but are thinner and have had every surface carefully worked down to a polished state. The edges of these celts are well preserved, and while usually sharper than even the best specimens made of hornstone or porphyry, there is not that difference which would seem to warrant the extra labor of reducing hematite to a suitable shape.

It is an interesting fact with reference to the use of this mineral, by the Indians, for making either implements or ornaments, that in Hunterdon and Warren counties, New Jersey, where hematite occurs in the greatest abundance, that so few specimens of implements made of it should be found. Ordinary stone implements, of every pattern occur in the greatest profusion, but of thousands of specimens examined, less than a dozen were made of hematite. It is not improbable that the few that have been found were brought from a distance, and that the resident tribes were not accustomed to make use of this material.

Fig. 28 is a remarkably pretty example of a celt of totally different shape and character, being a long, slender stone, edged at one end, instead of on the margin of one of its longer sides. The illustration will convey a better idea of the specimen itself than can any description. The specimen is a hornstone pebble, beautifully polished over the greater portion of its surface. One end is blunt, as though abruptly broken off, but is now as well polished as any of the other parts. From this blunt end, the width of the specimen gradually increases, with about a corresponding decrease in the breadth or

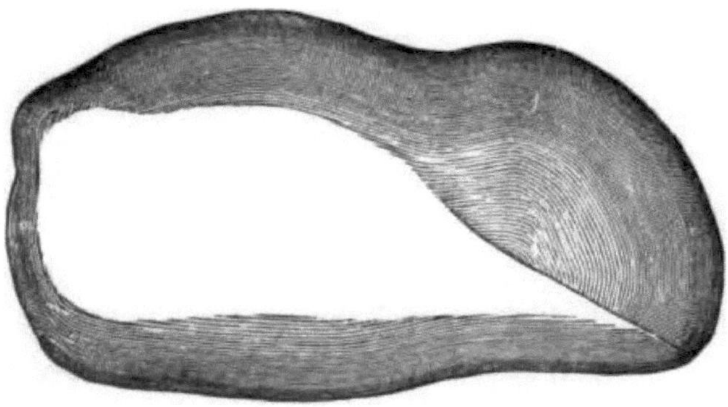

FIG. 28.—New Jersey. ½.

thickness for the distance of an inch, when the width decreases by a beautiful curve more marked upon the upper margin, which margin becomes the edge at the descent of the curve, and continues so until it joins the straighter portion of the lower outline of the specimen. The blade, or edged end, is slightly bent, or, at least, has that appearance, from the edge not being in a line with the middle of the thickest portion of the implement. If the specimen is held with the straighter side (*lower side*, in our figure) up, then the blade is bent to the right and has just the proper "twist" to separate the skin from the muscles most readily, if the implement is, as was probable, a knife for skinning animals. Certainly to such a purpose, it is in every way well adapted.

Still other forms presenting slight variations might be readily given, and an unbroken series from one extreme to the other might be provided from the hundreds of these celts that have already been gathered; but a sufficient number have been noticed in detail to enable any one, who happily may chance upon one of these implements, in the course of a summer ramble, to recognize it without difficulty, even if our studies of hundreds of specimens do not throw any light upon the part they played in the daily lives of the savage people who made them.

FIG. 29. — New Jersey. ½.

In the preceding illustrations of this chapter, the cutting edge has been produced by carefully grinding the stone from each side, so that the sloping should be equal, and the edge in the middle of the natural margin of the pebble or slab of stone used. Furthermore, all these celts have a distinctly convex edge, as has been already mentioned. Fig. 29, while in many respects quite similar to the celts proper that have been described, varies in having a straight edge, in being but little altered over much of its surface, and in having a large portion of the pebble split off, thus giving the blade of the chisel an upper surface that is flat. Were there no other examples of this form in the collections made in New Jersey, this might be considered a chance occurrence, and had resulted from finding a pebble with a convenient fracture, which had been utilized by being ground down to a cutting edge at one end. A sufficient number of this pattern have been found, however, to show that it is evident that the partial splitting off of one surface of the pebble was intentional. It is not of the least interest in this connection to determine how this was accomplished; for the evidence favors the conclusion that the specimens were shaped by this process, and were not brought to the general pattern of fig. 29 by pecking, or grinding. Chisels of this pattern are not abundant in New Jersey.

Fig. 30 represents a polished hornstone pebble, perfectly flat upon one side, and convex upon the other. There is a symmetrical conical head, below which is a contraction in the width of the blade, producing shallow notches upon the margins, but which do not meet so as to form a groove. The cutting edge is very sharp, and has been quite straight. At present the corners are worn away. This implement, supposed to be a chisel, is essentially the same as the preceding, and measures seven and one-half inches in length, by two and three-eighths in greatest width. In speaking of a chisel, we are apt to associate it with the idea of a hammer, as it is of but little use, except a blow, as from a hammer, be given it. This is true of the modern steel chisel, and is the more so with an implement made of stone, such as fig. 29, even if charred wood be the principal substance cut with it. In this instance the narrow conical head seems to offer a serious obstacle to the use of a hammer; and moreover, the polished condition of the head clearly indicates that it has never been exposed to any violent usage.

FIG. 30.—New Jersey. ½.

It has been suggested that this implement might have been used in detaching bark from trees, either for canoe purposes or for coverings for huts; that by placing the flat side down or upon the body of the tree, and pushing the implement forward, the bark would be detached from the trunk of the tree without danger of cutting it, as the edge of the implement pressed upon the wood, and the curved back of the blade lifted the bark up as the blade moved forward.

Other chisels, or chisel-like modifications of the better known celt, consist of cylindrical and quadrangular implements, usually of serpentine, measuring from three to five inches in length and often less than an inch in width if square, or half an inch in diameter if cylindrical. These very small chisels have an edge varying from one-fourth to three-fourths of an inch in extent. But very few have been found in New Jersey, and they are still more rarely met with in New England; but as we pass westward of the Alleghany mountains, they are of more frequent occurrence, and are fairly abundant throughout Ohio and Indiana.

The purpose of these diminutive chisels is difficult to determine; and we can only safely place them among that indefinite class of stone implements which, it is evident, were used as tools for making other objects, and were in no way ever brought into play as weapons. This is, indeed, a somewhat unsatisfactory method of disposing of a vast number of the most interesting forms of stone implements; but as the early writers either did not see them in actual use, or else have neglected to mention it, we are left in ignorance of the part they were intended to play in the daily life of the native races, or are forced to take refuge in conjecture, as to the purposes of a considerable proportion of these objects, which is, at best, a hazardous experiment.

Stone gouges, such as that represented in fig. 31, are very abundant in New England, and comparatively rare in New York and New Jersey. In eastern Massachusetts, and particularly in Essex Co., they are even more abundant than stone axes in central New Jersey, and are as characteristic of every series of stone implements from that vicinity, as are the axes from the former locality. Still, they do not in any way take the place of the axes, for these are fairly abundant in New England, although nowhere of as frequent occurrence as in more southern portions of the Atlantic seaboard.

Fig. 31 is made of a compact, granitic rock, and has been carefully wrought by pecking, until brought to its present symmetrical shape. It is nowhere polished or smoothed, but the surface is, nevertheless, quite even, and has a finished appearance, notwithstanding the slight inequalities. This gouge, which is not, however, of the maximum length, measures eight and three-fourths long, by about three inches in greatest width. Unlike many, the curved edge is of limited extent, while the hollowed portion of the implement itself is nearly equal to the greatest width. A second peculiar feature, which, with various slight modifications, is common to many of the New England patterns of stone gouges, is the transverse dorsal ridge, which in this case, is duplicated. The object of these ridges is quite clear, when we consider the implement as a gouge for working in charred wood, as in hollowing out a log canoe; for, as such, it required a handle to which it was attached in much the same manner as a New Zealand adze was hafted. Except in the one feature of a more or less degree of curvature of the cutting edge, the New England gouges are the American representatives of the Pacific Island adzes referred to. This, of course, does not apply to all of them, as many are too small to have been used as cutting tools, unless in cases where the material worked upon was very yielding, and the implement was held in the hand, or possibly used in connection with a hammer. While many gouges have been collected in New England considerably longer than this specimen, it is seldom that they are wider or even as wide, and therefore this example may be fairly considered as about of the max-

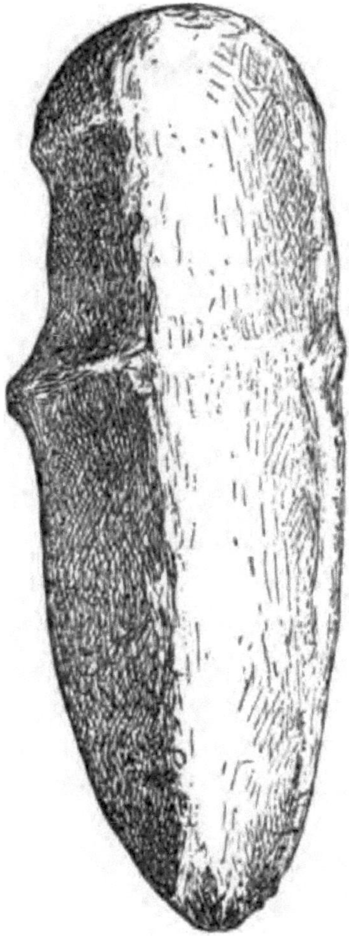

FIG. 31. — Massachusetts. ½.

imum size. As has already been mentioned, these larger gouges, particularly, vary greatly in minor details, and the same slight differences are noticeable that have already been pointed out as common to the grooved stone axes. As in that case, so it may be with gouges, every Indian was his own implement maker, and made the gouge of the pattern he thought would best serve his purpose.

Of the gouges of the Champlain valley, it has been remarked, that,[18] "though not among our most abundant specimens, they are yet relatively quite common, and of an almost endless variety of form. That all of these grooved implements were used as gouges is very doubtful. Of some of the specimens I have found it impossible to do more than conjecture the use. None of the objects found are more carefully formed and finely finished than some of these 'gouges,' and most of them are far more carefully made than the 'chisels' or celts. Some of the larger specimens are a foot in length, of basalt or other hard stone, but yet are made with a degree of skill, as exhibited in the symmetry of form and smoothness of surface, that excite great admiration. In some the groove is deep and wide, and reaches from end to end, each end, in some cases, being ground to an edge, in others it is short. Some are flat on both sides, others on one, others convex on both. Some have one edge finished like a gouge and the opposite like a chisel, and in these the gouge end is flatter than usual and the excavated portion but little concave. Some are of such soft material, steatite, that it is difficult to see how they could have been of much service as implements, but most are of hard stone. Several long 'gouges' have been found which are somewhat peculiar in form. All of these are very finely made; in cross section they are shaped like a narrow Gothic arch, the point coming opposite the groove, or, in such as have only a short groove, the portion above this may be nearly cylindrical." In this communication reference also is made to two gouges, measuring, respectively, eighteen and nineteen inches.

Fig. 32 represents a second example of New England stone gouge,

---

[18] Perkins, Amer. Naturalist, Vol. xiii, p. 744.

which has two very marked differences from the preceding. In this case, instead of the two transverse ridges near the upper end of the implement, there is a flattened oblique stem or projection forming the head of the gouge, and the blade is wider at the broad cutting edge with which it terminates, than at any other part. The curvature of this cutting edge is very pronounced, and it is sufficiently sharp, even now, to make the implement a very good tool for the purposes for which it was intended. The projection at the head does not afford a surface of sufficient width to enable one to use a hammer; and it is safe to suppose that, when made after this pattern, they were intended for removing such yielding material as charred wood, and not for cutting wood in its natural state. Whilst there is a great variation in the finish of the heads of these gouges, it is seldom that we meet with one of this peculiar pattern. Indeed, the form is not particularly desira-

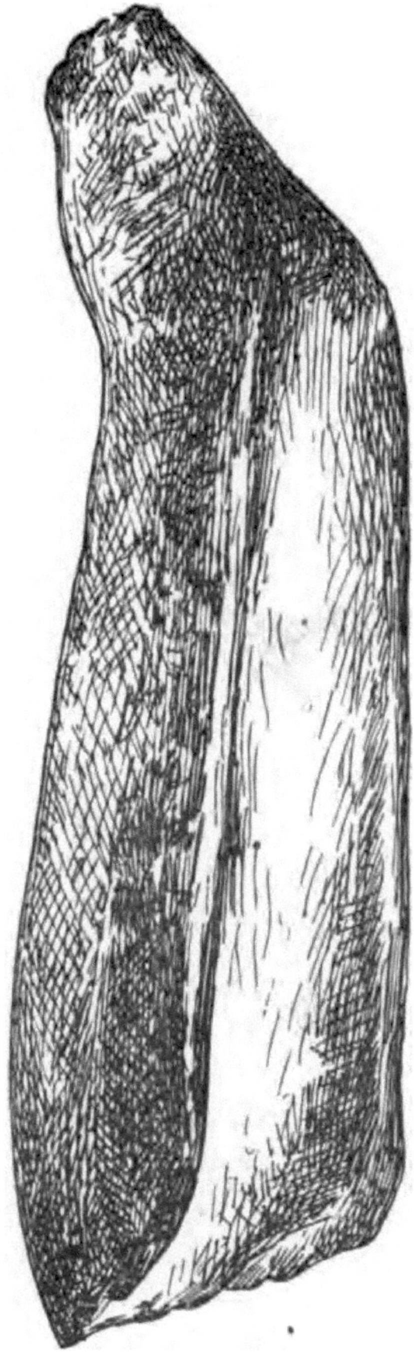

Fig. 32.—Massachusetts. 1/1.

ble, in view of the use to which is put the carpenters' gouge of to-day.

Fig. 33 represents an excellent example of the simpler form of gouge, such as is more or less common along the entire Atlantic coast. They are quite common in New York; and in some portions of Pennsylvania they are more numerous than the straight-edged celts, or chisels proper. In New Jersey, this form of implement is not common, in comparison with ordinary celts, but it is found in such numbers, as to be generally represented in all local collections. The specimen here figured is made of very compact, unyielding sandstone, and has been hollowed out as deep as was compatible with the strength of the implement. The back has been quite evenly smoothed, or it may have been worn so by long continued use. At the upper end, a hole has been drilled partly through the stone. This specimen presents the maximum degree of curvature of the cutting edge, as seen in these gouges. Of the very large series from Massachusetts, in the Museum at Cambridge, and at the Academy of Science at Salem, not one exceeded and but very few approached it in this respect. From the worn condition of the cutting

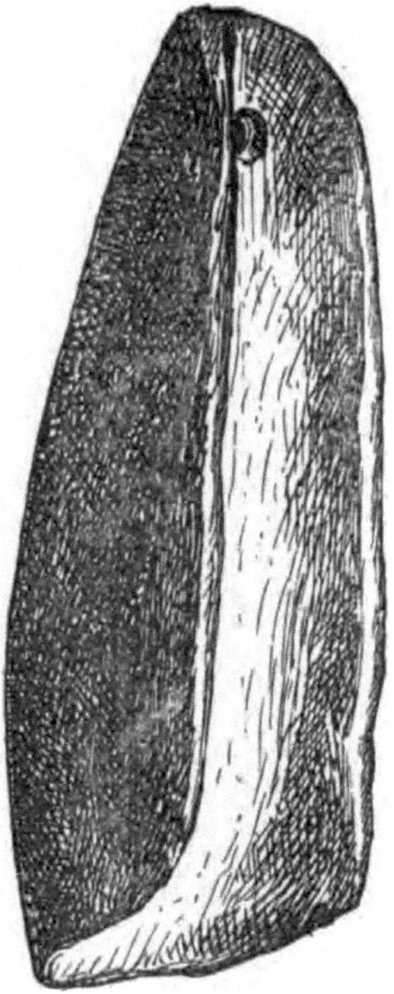

FIG. 33.—Massachusetts. ½.

edge, it is probable that, when made, it projected forward, as is well shown in the following example, from New Jersey.

Fig. 34 represents a similar plain gouge from New Jersey. This specimen differs only in having a projecting or convex edge, which makes the specimen in this respect more like the ordinary celts. It is probable, however, as before suggested, that this was a common form of gouge, and that the edge was gradually worn away by long continued use. Of the series of gouges, from New Jersey, whether plain or with worked knob-like heads, none possessed so marked a degree of curvature of the blade, as does this specimen. It is not without interest to know that of the gouges of this pattern, that have been found in New Jersey, all have the upper end or head badly battered, thus showing that they had been used with a maul or hammer; whilst those with a conical or otherwise designed worked-head, such as the chisel in fig. 30, show no trace of any such usage. Whether this fact indicates a different use of these implements, and hence the inference that they are not gouges in the ordinary acceptation of that term, is left for the reader to determine.

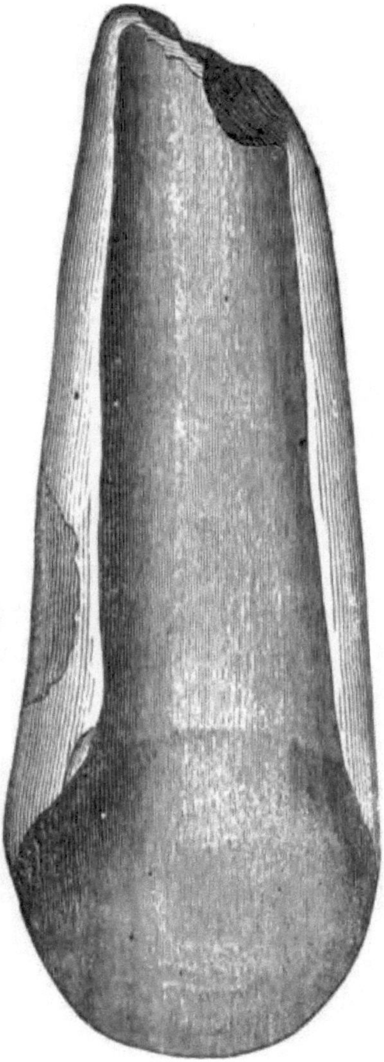

FIG. 34.—New Jersey. ½.

In the northern counties of New Jersey, and about the region of the Delaware Water Gap there is frequently found a pattern of gouge or chisel, as the case may be, which is strictly an intermediate form. It is made of compact stone, is well polished, and has a straight cutting edge which is bounded by a narrow rim upon each side, extending at right angles to the blade. It is not apparent what advantage arises from these narrow boundary ridges, if the implement is a chisel, such as is now used; and the slightly elevated edges of the sides of the blade scarcely convert the implement into a gouge proper.

CHAPTER IV.

GROOVED HAMMERS.

In an attempted classification of stone implements, there is often great danger of making a distinction where no difference exists. This is particularly true of the grooved globular pebbles and small bowlders, that are so common throughout the whole extent of the Atlantic coast. Examples of these, of a small size and carefully worked, have been described as probably "club-head" stones. Others, of ruder make, as the natural pebble with a groove only worked about it, were probably used in the same manner as the notched, flat pebbles. Besides these, there are larger examples of practically the same form, which are of such dimensions, and, in some instances, show such evidence of work, other than the pecking out of an encircling groove, that their use as hammers or mauls seems to be indicated. While none have been found in New Jersey, or throughout New England, as large and as heavy as the largest stone mauls from the Lake Superior copper regions, they do occur of such size and weight as render them available for all ordinary purposes for which a hammer is required. A small series from Gloucester Co., New Jersey, are hard, silicious pebbles or small bowlders, varying from five to seven inches in length, and three and one-half inches to four and one-half in diameter. None ot these are worked otherwise than by being grooved or by having pecked out a shallow channel around them, at or near the middle of the stone. The ends, in two specimens, are slightly battered, as though used for hammering substances equally as hard as the mineral of which they are made. The others show no trace of usage.

Stone hammers of this character, but of beautiful workmanship, having wholly artificial surfaces, and the groove protected by a narrow, slightly elevated ridge have been obtained from Indian graves in Kansas. A series of these are contained in the collections of the Museum of Archæology, at Cambridge, Mass.

Figure 35 represents the more usual shape and size of these stone hammers, as we find them in New Jersey. This specimen is exactly five inches in length. It was originally of the ordinary oval outline so common to the cobble-stones of the river-bed, and afterward pecked at the head to make it flatter. It has a very shallow groove pecked irregularly about it; the dressing down was apparently more with a view to obliterate projecting angles than to secure a depression or groove for the handle-fastenings. At the end or point there is a small pecked

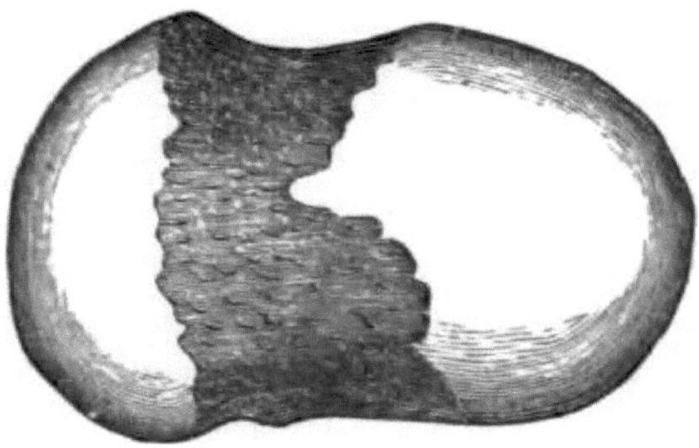

FIG. 35.—New Jersey. ½.

surface which may have been intended to produce a blunter end, or caused by hammering upon other stones, as in using the stone chisel or gouge.

Fig. 36 represents a natural pebble which, being grooved already, was utilized as a hammer. The circumstances under which it was found are the only reasons for placing it with the grooved hammers. It was associated with a number of arrowheads and other objects, and the assumption is, that, like them, it had been placed where found, by the Indians. So admirably shaped is this stone, that, could they be readily obtained, there would be no incentive to make and groove less shapely pebbles, for in no respect is this specimen inferior to the others.

Fig. 37 represents an unaltered pebble, that has been grooved; and thus, it is supposed, converted into a hammer. In the preceding examples, the length and diameter of the pebbles have been more nearly equal than in this instance. Here we have a flatter pebble, and one that needs but a cutting edge at one end, to make it a good example of a common form of axe. The absence of this edge, unless it can be shown that it is an unfinished axe, indicates its use as a hammer, or possibly the head of a club-like weapon. This specimen is seven inches in length, three inches wide in the middle, and tapers quite uniformly to the narrowed ends. The ends are not battered, nor is there any trace of use of any kind to be seen. As a weapon, such an implement would long preserve its natural surface uninjured and

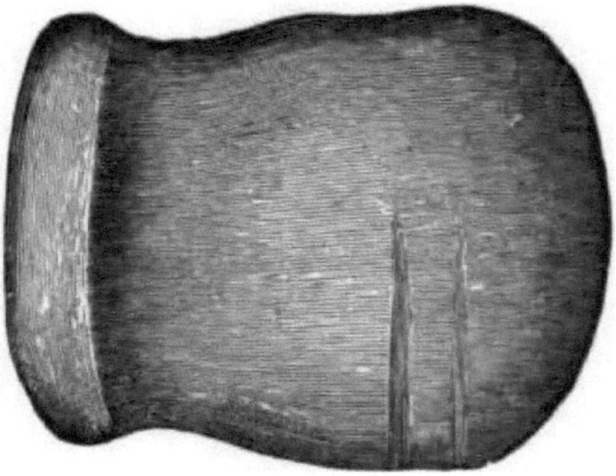

FIG. 36. — New Jersey. ½.

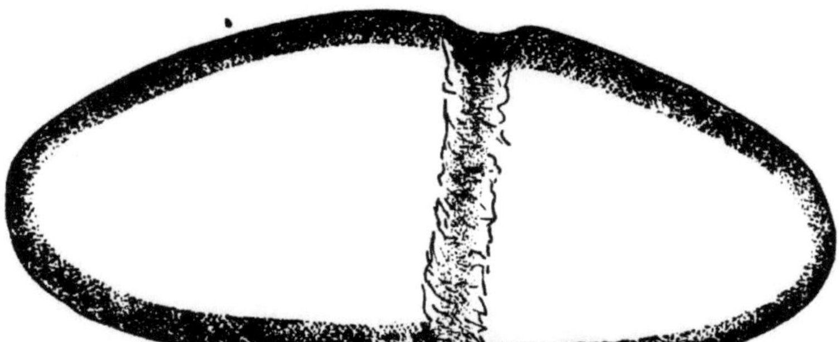

FIG. 37. — New Jersey. ½.

unmarked. If used as a maul, in connection wirh a stone gouge, the surfaces would soon become battered.

Fig. 38 represents a peculiar pattern of stone hammer, in which we have the hammer-head and the handle, in one. This so-called "handle" is a continuation of the head, but has been pecked, ground and chipped, until it is cylindrical. While the outline is that of a hammer, it is not improbable that it was used as a pestle, although the end is not battered or worn in any way.

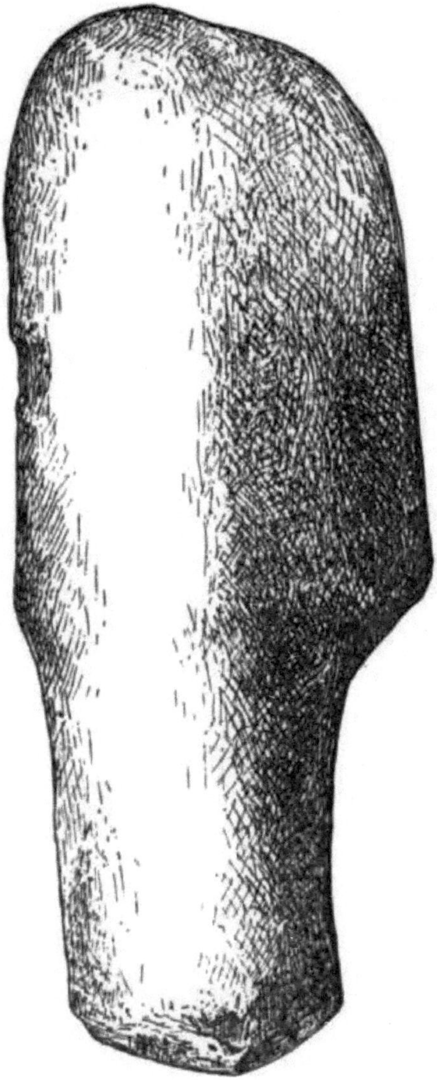

FIG. 38.—New Jersey. ½.

The "head" varies considerably in thickness, and on that side projecting from the handle is quite narrow, and has the appearance of having been chipped; or, possibly, this is the result of long continued usage, in hammering against stone.

Perforated stone hammers, similar to those found in such abundance in northern Europe, are of rare occurrence in North America. I know of but a single specimen of perforated stone implement, which may be considered as probably a hammer. It is in the collection of the late Professor Haldeman. This hammer measures eight and one-half inches in length, by four in width. It is oval in shape, and through the middle is a carefully drilled perforation, four and one-half inches in length, and one inch in diameter. The stone has been pecked over its entire surface, and is a heavy, finely grained sandstone pebble from the bed of the river. It was found near Chickies, Lancaster County, Penna.

Wooden mauls, it would appear, were sometimes used. Mr. E. W. Ellsworth [19] has described the circumstances of the discovery of a wooden maul, in the valley of the Connecticut river. This implement is twenty-one inches in length. "Length of head, eleven and one-fourth inches; length of handle, nine and three-fourths inches; diameter of the head measuring in a plane coincident with the curve of the handle, five inches." This specimen shows two indented and battered surfaces. "Professor Rau has suggested that the mallet was used for driving stone celts. This is probable, both from the form and position of the indentations in the mallet."

---

[19] Ellsworth. Smithsonian Annual Report for 1876, p. 446, fig. 1.

CHAPTER V.

SEMILUNAR SLATE KNIVES.

IT is proposed to consider here, a class of knife-like implements, which, from their shape, have been designated as Semilunar Slate knives. Throughout New England, and as far south as Maryland, these knives occur with more or less frequency, as the localities happen to be rich, or are wanting, in the ordinary patterns of Indian stone implements.

A typical knife of this form may be briefly described as a thin, broad blade, of about six inches in length, and two in greatest breadth, the cutting edge being curved, and extending from end to end of the blade, thus making it semilunar in outline. As a rule there extends, along the back, a narrow, thick ridge of uniform width, which affords a convenient and secure hold, when the knife is taken in hand.

In New Jersey, these knives, now usually much broken, are not uncommon on former Indian village sites; but in other localities, or singly scattered about our fields, they are seldom or never found. Whatever may have been the use to which they were put, their shape certainly indicates that they were a domestic implement, a household knife, for cutting flesh or equally yielding substances; and not such a knife as the men would carry with them.

These knives possess an additional interest from the fact that they are a well-known form of cutting implement of the Eskimo and Alaskan Indians. In the archæological collections of the museum at Cambridge, are two specimens (P. M. Nos. 2053–54) of this form of knife, each with a blade of slate inserted in a wooden handle. Except that the handle is of another material, they differ in no respect.

Knives of this pattern, made of iron and ivory by the Eskimos of Cumberland Sound, have been described by Ludwig Kumlein, in the Bulletin (No. 15) of the U. S. National Museum. This author says: "The favorite and principal tool of the women is a knife, shaped like an ordinary mincing knife. Nearly all the Cumberland Eskimo have now procured iron enough from some source or other so that they can have an iron knife of this pattern. Before they could procure enough iron they made the knife of ivory, and merely sank flakes or pieces of iron into the edge, in the same manner as the natives of North Greenland do at the present time. This same practice of sinking iron flakes into the edge was also used on their large skinning-knives, which were made from a walrus tusk, and much after the pattern of an ordinary steel butcher-knife. Some of these ivory knives have no iron in them; but at the present time they are used principally, if not entirely, for cutting snow and removing ice from their kyacks.

"The women seldom use any other kind of knife than such as just described. With them they remove the blubber from the skin, split skins, cut up meat, and when sewing this instrument is used instead of scissors. They begin a garment by sewing together two pieces of skin and shaping them as they go along by means of the knife, cutting for an inch or two and then sewing. They always *push* the knife *from* them when working it."

As these semilunar knives are more abundant in New England than in the middle states, and do not appear to have been in use among the southern coast tribes, it is probable that the pattern was derived from the Eskimo with whom the northern Algonkins were frequently in contact.

Fig. 39 represents a fine example of these slate knives, such as are found in the localities mentioned. It was ploughed up in a field bordering on Crosswick's creek, Burlington Co., New Jersey, in which locality not only scores of ordinary axes, knives and arrowheads have been found, but fragments of *at least thirty different specimens of this pattern* of knife. They were about equally divided between such as were perfectly plain, as in this instance, and such as were ornamented with incised lines of various patterns on the sides of the dorsal ridge or back of the knife.

This specimen, fig. 39, is made from a slab of compact, fine-grained yellowish slate, or, more properly, clay-slate, and has been at one time, highly polished. Portions of this polished surface are still to be seen in one or two places. Fig. 39 measures six inches in length at the back, and the blade, which is one-fourth of an inch thick, is an inch and a half wide at the widest portion. Thence, towards each end, it describes a curve and decreases in width.

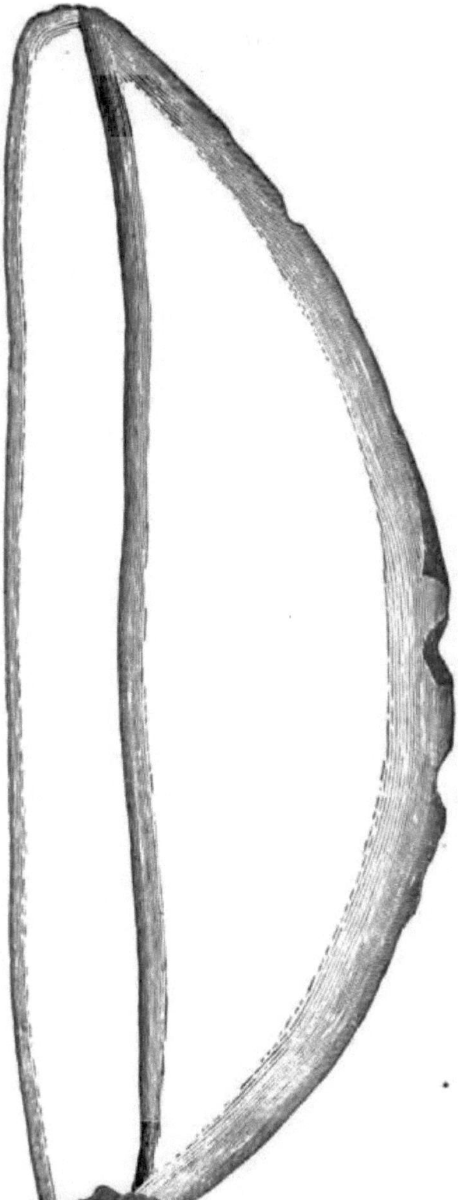

Fig. 39.—New Jersey. ½.

These knives, of which so many are found in New England, have been so well described in detail by Prof. F. W. Putnam, in the "Bulletin of the Essex Institute" of Salem, Mass., that we quote his remarks in full.

"Many beautiful cutting implements have been found in various countries, especially in North America. Schoolcraft, in his extended work on the Indian tribes, figures several fine specimens, notably the one represented on plate 45, figures 1 to 3 (vol. ii), found at Hartford, Washington county, N. Y., which he states to be carved from a piece of green serpentine. This knife is somewhat sickle-shaped, five and three-quarters inches long, with a curved triangular blade descending from a well formed rounded handle. Schoolcraft also figures (vol. ii, pl. 49, fig. 4) a cutting implement with a blade five and three-quarters inches long by an inch in width. The figure shows a thickened portion answering for a back or handle. This specimen was found in Genesee Co., N. Y. The drawing is, however, very poorly executed and the description is so brief as to leave us in doubt as to the exact character of the implement. The specimen figured on his plate 50, figures 5 and 6 (vol. ii), under the title of 'fragment of a blade of a battle-axe,' and described as made of silicious slate, is far too thin and fragile an implement for a battle-axe, and is more likely another form of slate knife, perhaps having two symmetrical blades, through the centre of which (the figure shows a broken groove, which may represent a hole drilled through the centre of the blades) a wooden handle was inserted.

"Squier and Davis in their work on the 'Ancient Monuments of the Mississippi Valley,' p. 216, give a small woodcut of a semilunar-shaped knife, which they state is a form 'occasionally found in the Eastern states. They are sometimes composed of slate, and are of various sizes, often measuring five or six inches in length. They are well adapted for flaying animals, and for other analogous purposes.' Their figure represents a knife of the same shape as the one here engraved (fig. 40).

"Of these semilunar knives, I have seen quite a number of specimens in various collections, but thus far all, as stated by Squier and Davis, have been from the Eastern states. In the Peabody Museum of Archæology, at Cambridge, there are several of this form, one of which is about eight inches long and is labelled 'Paring Knife, Amoskeag Falls, 1795.'

"The one represented here as fig. 1 (Fig. 40) is beautifully finished and perfect. It was found in Salem and placed in the Museum of the East India Marine Society. It is not quite five inches in length and is a little less than two inches in greatest depth of blade and back. The back is about half an inch in depth and a little over a quarter of an inch in width at the centre, narrowing at the ends, perfectly flat above. The blade is one-fifth of an inch thick along the under side of the thick back; it is gradually thinned out to the cutting edge all round, which is only one-tenth of an inch thick about one-fifth of an inch from its outer margin, which is evenly and nicely brought to a sharp cutting edge. The engraving shows the shape of the knife better than words will describe it. It will be noticed that the blade is slightly more pointed at one extreme than at the other. The material is a gray slate having several fine veins of a harder substance (quartz?) as shown in the engraving: it is quite ornamented with several dark wavy lines, light streaks and bands, and a number of irregular wavy lines of a red mineral running in all directions over the surface, but not indicated in the engraving.

FIG. 40.—Massachusetts. 1/1.

"Another specimen, also received by the Academy from the East India Marine Society Museum, was found on the farm in Danvers formerly owned by Governor Endicott. This specimen consists of about one-half of the knife, and was evidently, when perfect, about six inches long and two and a quarter deep. It was made of a slate very much like the Salem specimen, but without the dark and red veins and mottlings."

Although these knives are made of a material readily obtained and easily worked, they were not always discarded, when they chanced to get broken in halves. Many of them had the broken end ground down to a smooth blunt edge, and the cutting edge slightly ground away at the same end, and thus a new knife was made out of half of an old one, which was almost as good as the original, for the purposes for which such knives were made. It is evident, however, that these knives were thrown aside, when others of metal were obtained, as the appearances of the fractured edges of such as we now find indicate that they have been broken or crushed, as by the tread of a horse, in comparatively recent times.

Fig. 41 represents "a knife of dark talcose slate which is unlike any other that I have seen. It was found near the church in Putnamville (Danvers), Mass. and is thus of marked interest to us as a relic from Essex county. It is slightly over five inches in length, and about one and one-half inches in depth at its centre. It is worked to a rounded point at each end, as shown by the engraving, and the smooth cutting edge is from point to point. The greatest thickness of the blade is one-fifth of an inch. The back of the knife is ground off to quite a thin edge, but evidently was never sharpened to form a cutting edge, though the back is so thin as to render its being held in the hand an uncomfortable matter while using the knife in this way; and the three holes that have been rudely cut, apparently by scraping backwards and forwards with a pointed stone, on both sides, until a hole was made, are evidence that the knife was mounted on a handle by passing bands through the holes and around the handle, which was probably grooved along its under side to fit over the sharp back of the knife. In common with the other slate knives, this specimen was finished with care and is perfectly smooth and well sharpened along its cutting edge.

"Evans, in his instructive work on the 'Ancient Stonè Implements of Great Britain,' mentions (p. 311) that in some Esquimaux knives the blade is tied to a wooden back by a cord which *passes through a hole in the blade*.

It would thus seem that our New England Indians, for to them I think we must look as the makers and owners of the knives I have specially described, were not satisfied with using simple flakes of stone and broken arrow and spearheads for knives, but that with them as with us to-day there were many, and often elaborate, styles of this most useful implement, and who can say that to possess a good knife was not as much the ambition of the men of the departed race as it is with those who have succeeded them?"

As will be noticed, by referring to the introductory remarks on these knives, as a class, I do not agree with the conclusion in the above quotation. The New England Indians, as well as those in New Jersey and elsewhere, had better knives for ordinary purposes, than these of slate.

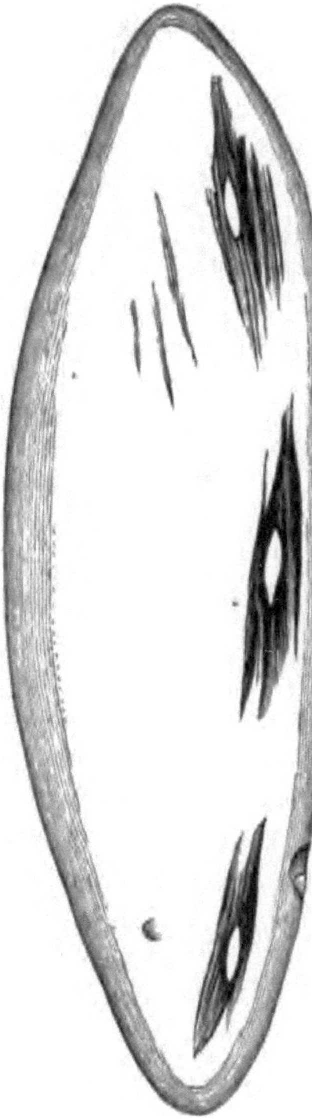

FIG. 41 —Massachusetts. ½.

Another specimen, fig. 42, "is nearly perfect, one end only being broken off, as shown in the figure.

Allowing for this missing fragment the knife was about seven inches long: the back is three-quarters of an inch deep and the blade about one and one-half inches. The thickness of the blade in the centre is about three-tenths of an inch. The peculiar workmanship of the back, as shown in the section, in the form of a series of uneven knobs, was probably intended to give firmness to the hold when grasped by the hand. This specimen was found in a sand deposit near Kingston Falls, Kingston, New Hampshire."

A finish very similar to that on the back of these slate knives is seen in some of the fragmentary specimens from New Jersey, now in the archæological Museum, at Cambridge, Mass. Upon one of them is a series of lines similar to those upon the illustration of the New Hampshire knife. These etched or graved lines are, in some instances, upon

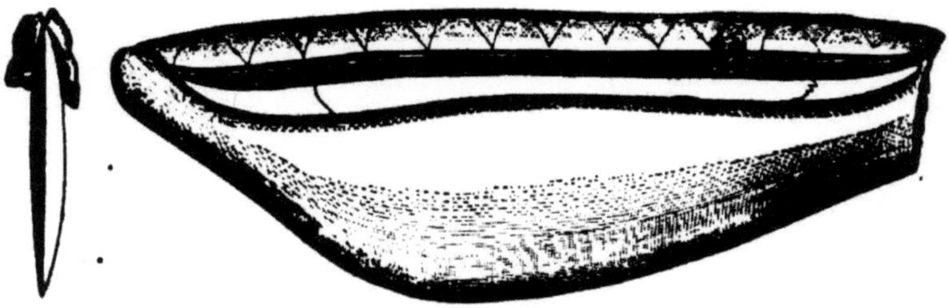

FIG. 42.—New Hampshire. ½.

so limited a portion of the back, that they can scarcely be considered as ornamental; and it seems not unreasonable to consider them as marks indicative of ownership.

Fig. 43 represents a most remarkable form of these slate knives, which, although of more than ordinary rude finish, has certain peculiarities which are of great interest. These striking features consist of a series of etchings and deeply incised lines of perhaps no meaning. Taken in order, it will be noticed that at the back of the knife are four short lines at uniform distances apart, and a fifth, near the end of the implement. Besides these are fifteen shorter parallel lines, near the broader end of the knife and about the middle of the blade. A series of five zigzag lines are also cut on the opposite end of the blade. Is it possible that such simple markings can have been intended as mere ornamentation? However "primitive" the mind of the red

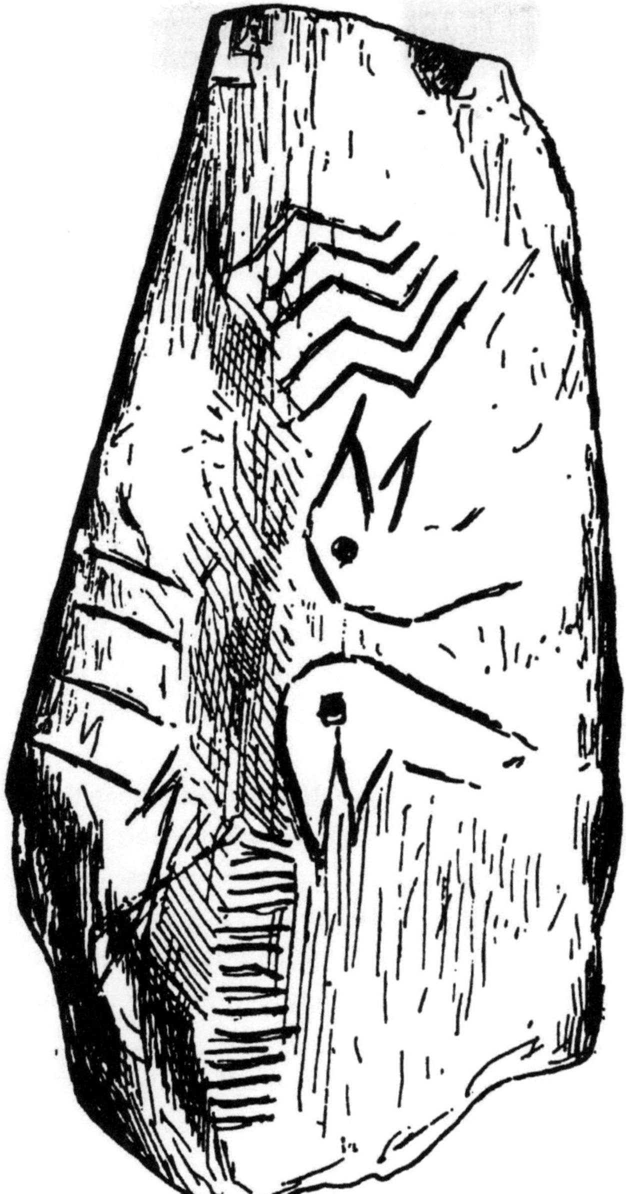

FIG. 43. — New Jersey. ½.

men of North America may have been, in times prior to the advent of

Europeans on this continent, it is difficult to imagine that their ideas of ornament and decoration should have been so crude as to be satisfied with slight and inconspicuous scratches. Nor does it accord with their evident appreciations of symmetry and elements of beauty, which we see in various specimens of their handiwork. Hands skilled in the production of faultlessly designed celts, and of animal sculpture, such as the bird-shaped objects figured in another chapter, could never have intended such simple devices as those upon this knife blade as an attempt at ornamentation. Setting aside, then, this explanation of their meaning, it may reasonably be asked, have they any other meaning? That they have is more than probable, though what that meaning is remains to be determined. More prominent than the numerous lines to which reference has been made, are the clearly defined, unmistakable birds' heads, placed midway between the two series of lines. What they indicate is of course a mystery; but it is not a little curious that there were, among the Lenni Lenapè or Delaware Indians, three grand clans or principal divisions of the nation. These were, respectively, the Wolves, the Turkeys and the Turtles. In the mountainous region of the head-waters of the Delaware river, and southward to the Water Gap, the Wolves or Minsi were in undisturbed possession of the land. Southward, and westward to the Susquehanna, were the Turtles, or Unâmis; and along the coast, the Turkeys or Unalachtgo. If we can, by any allowable use of the imagination, see in the crude attempts at pictorial representation, heads of turkeys, a bird once as abundant in a wild state, as it is now under domestication, the inference is legitimate that the pictures on the knife may have some reference to the people, whose "totem" was the bird in question.

As bearing directly upon the question of the significance of these representations of heads of birds, attention is here called to a shell disk from Tennessee, fig. 44, upon which are four figures similar to the two on the knife from New Jersey. Such carved shells are quite common in Tennessee and southward, and have been usually taken from graves. Can they be regarded as totems? It has been stated of the Virginia Indians,[20] "of this shell [cunk] they also make round tablets of about four inches in diameter, which they polish as smooth as the other [shell beads, etc.], and sometimes they etch or grave thereon circles, stars, a half-moon, or *any other* figure suitable to their fancy. These they wear instead of medals before or behind their neck, etc." Here we see a reference apparently to just such shell disks as fig. 44; and the interest in the reproduction of the same figures on other objects, found in New Jersey, lies in the probable indication that there is, in the latter, a trace, at least, of tribal relation-

FIG. 44.—Tennessee. ½.

ship with the southern Indians. Did we not learn from the writings of Heckwelder, that the Lenapè had "the turkey totem," we might suppose that this drawing of such bird heads originated with the intrusive southern Shawnees, who, at one time, occupied lands in the Delaware valley, and who are supposed by some writers to have been closely related to the earliest inhabitants of the southern and southwestern states. Inasmuch as we shall find that, not only on this slate

---

[20] Beverly: "History and Present State of Virginia," Book III, Chapter xii, p. 58. London, 1705. Quoted by C. C. Jones, Jr., in "Antiquities of Southern Indians."

knife, but upon a bone implement also, similar heads of birds are engraved, it is probable that the identity of the design is not a mere coincidence, but that it must be explained either in accordance with the statements of Heckwelder, or be considered as the work of southern Shawnees, after their arrival in New Jersey. In the latter event, the theory that these disks were the work of a people different from and anterior to the Indians found in the Cumberland valley, at the time of the discovery of that region by the whites is, apparently, not sustained by the facts.

CHAPTER VI.

CHIPPED FLINT KNIVES.

On the same fields, and along the same valleys of our creeks and rivers, wherever we chance upon spearpoints and arrowheads, there are found certain other chipped implements, equally variable in size and perhaps, also, in shape.

When the cutting edge is the one prominent feature, these objects are readily recognized and confidently classed as "chipped knives;" but many others are found so like ordinary spearpoints, that it is not always practicable to determine to which of the two forms they really belong.

So readily can an effective knife be made, by striking from a pebble a splinter of stone, that we wonder why such great pains should have been taken with many of the jasper knives that we find, especially since much of the extra labor upon them was not for the improvement of the cutting edge. It has been remarked of primitive stone knives, that "flint, chert, obsidian, any stone which will chip easily to a sharp edge, will constitute man's first cutting implement, and is much better adapted to this use than we are wont to suppose. Some of the modern California Indians make use of the most primitive form of these implements. Miners of 1849 have described to me practices occurring under their own observation, illustrating the mode of manufacture and the efficiency of the crudest cutting tools. The Indian, without a knife, would skin and dress a deer almost as quickly as his white brother hunter armed with his hunting knife. Picking up the first thin stone he could find, that, under sharp blows with another stone, would flake to an edge, with a few blows he would bring it to the desired form, and, commencing his work with a drawing motion, would open the skin with great nicety."[21]

Implements that were unquestionably used as cutting tools are nowhere as abundant as arrowheads, but they seem to bear such a relationship, in point of numbers, to arrowheads, as would be expected from the use of the two forms. The one not being liable to be lost, and lasting perhaps a lifetime, is much more scarce than the other which was certain to be quickly lost, and of which a vast number were necessary even if they were not used in warfare, as we well know was the case. On the other hand, if we call every object, that seems badly designed for a spear or an arrow, a knife, we have then a vast increase in this class of objects and almost as many specimens are found of one form as of the other. Unless, however, it is supposed that knives were made, used and thrown aside, on the spot, it is manifestly wrong to consider every small spear or large arrowhead as a knife. To do so presupposes a very dense population, or a long occupancy. If we call all these objects knives, we lessen materially the number of arrowheads, which latter were, necessarily, vastly more abundant, than any other implement the Indians possessed.

While a few spearlike forms have been classed as knives, for reasons mentioned, it is by no means certain that they were not quite as much in use as spears, as they were as knives. If we endeavor to picture the ordinary avocations of the natives of our coast, especially before the arrival of European settlers, we shall find, that hunting necessarily occupied much of their time. The skins of animals were the dependence for clothing, and their flesh with that of fishes was an important element in their food-supply. So varied in size, and in habits, were the mammals they hunted, that it can scarcely be doubted, that there was in common use a greater variety of hunting implements, than of weapons of war, if indeed, any distinction was made between them. For hunting certain animals, as well as for spearing certain fish, the Indians doubtlessly made use of particular patterns of stone imple-

---

[21] M. C. Read. Tract of Western Reserve Historical Society. No date. p. 12. Ohio.

ments which have been called spearpoints, arrowheads and knives. So few, comparatively, of the latter, and so many of the larger arrowheads and smallest spears, were required as spears for the smaller mammals, that it is safe to consider, simply as knives, as few objects of chipped flint as possible, and to regard all such as could be so used, as heads of spears and arrows.

The circumstances, too, under which occasional specimens have been found, have a direct bearing upon this matter. Chipped implements that seem ill adapted for use as arrowheads, have been found, in one instance, deeply embedded in a lumbar vertebra of an elk, and a small blunt spear forced into the shoulder-blade of a bear. These instances suggest the use of a bow and arrow rather than a spear. Certainly they do not indicate their use as knives. A third instance is known of an indifferently chipped and scarcely a pointed arrowhead, being found in the very centre of a large white oak. It had evidently pierced the tree, when a sapling; and, remaining in the tree, had been in time enclosed by its growth. When felled, the tree measured nearly five feet in diameter. How unlikely is it that this implement used as a knife had been left sticking in a tree !

Fig. 45.—New Jersey. 1/1.

Fig. 45 represents an excellent example of what may be considered a typical chipped flint knife. That it is a cutting tool alone s evident. Whether held in the hand, by grasping the implement as it is, or whether it had a wooden or bone back, into which the more irregular margin was inserted, is not evident, from any peculiarity of the specimen; but that the hand was protected in some way from the jagged back of the knife is very probable.

Knives of this pattern are of common occurrence in all parts of the country, and bear the same relationship to the common forms of chipped objects that the ground celt does to those that are polished. Such knives, however, are not always as coarsely flaked as in this case. Those that are of a less wavy surface, and crooked edge, have no advantage as mere cutting tools. That they were much used, except for skinning and dismembering animals, is very improbable.

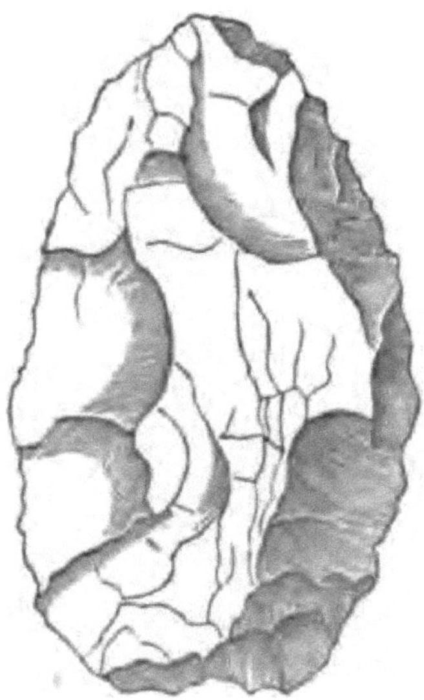

FIG. 46.—New Jersey. ½.

In the limitation made in this chapter, fig. 45 is considered as a knife proper and is of about the maximum size. Those of the size of fig. 46 are much more common. This specimen, which strongly suggests the blocked-out arrowhead, is a most common object on all village sites, and wherever ordinary forms of stone implements are found.

In some localities, as in the immediate vicinity of inland ponds, and for short distances along many small streams, often a hundred or more of these knives will be found scattered over the ground. When a number are thus found together, they are very uniform in chipping, and present a great similarity to figures 46 and 47. When, further, it is noticed that they were all apparently made from the same bowlder, it becomes evident that, for some given purpose, and probably by some one person, a quantity of knives were made, and used by the little community, which often has left scarcely any other trace of its ancient sojourn on the spot.

Fig. 48 represents a common form of knife, that also bears a strong resemblance to an unfinished leaf-shaped arrowhead. Were such implements gathered from the refuse, left where arrowheads have been made, it would be looked upon as simply a discarded specimen; but when we find them singly, whatever may have been their origin, it is presumable that they were used as knives. Many show indications of use by a striation and semi-polish of the surface; when this is found, it is evidence that, to some extent, they were used as knives. The width, in comparison with the length, and the obtuse point, presuming this to be a finished implement, clearly indicate that they could have had no other use than that of knives.

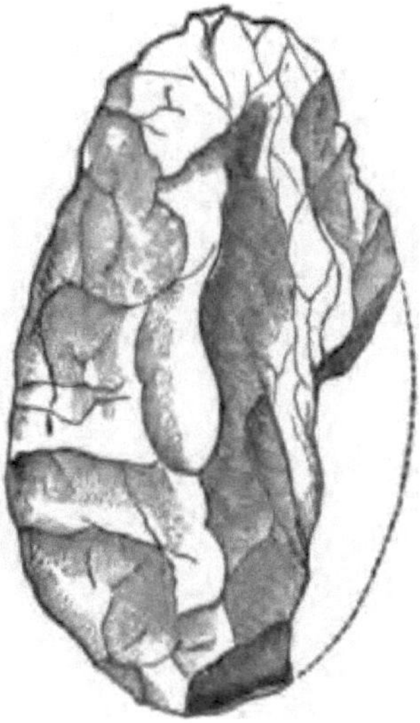

FIG. 47.—New Jersey. ½.

Fig. 49 represents a flint knife from California, to which attention is called, in connection with the descriptions of the preceding examples of chipped implements, figs. 45–48, which have been considered as cutting tools. This California specimen is a flake of striped jasper or hornstone, which has been detached by a single blow. The surface shows no trace of secondary chipping, except at the very edge, which has been slightly modified by the removal of a series of minute chips. The cutting edges are not sharp, nor the point acute. At the base, there still remains much of the asphaltum used in securing a handle to the specimen. These handles were usually of wood. (See Vol. VII, of Geographical Survey of U. S., West of 100th Meridian, from which volume this illustration is taken.)

FIG. 48.—New Jersey. ½.

While asphaltum was not known to the Atlantic coast tribes, they still had an abundance of glue of excellent quality, and it is quite probable that such knives as figures 45 to 48, inclusive, were hafted in much the same manner, as the specimen represented in fig. 49.

In fig. 50 we have the highest type of these oval knives, and an excellent example of the dexterity acquired in chipping flint to any desired form.

FIG. 49.—California. ½.

As compared with such as have just been considered, these knives are not abundant in New Jersey; and they are comparatively rare throughout New England, except in the Connecticut valley where they are quite common. In Ohio, these knives are more abundant than along our seaboard, and there, they are usually made of the bluish hornstone so characteristic of the best examples of flint chipping found in that state

The New Jersey examples of this pattern, which vary much in size, although not generally larger than fig. 49, have been found mostly on the banks of the Passaic and Hackensack rivers, in the eastern part of the state; and, as in many other localities, they have been considered as "fish-knives," for no other reason, apparently, than that they were found in the immediate vicinity of the rivers. There is some danger of being misled, in thus naming an implement from the character of the locality where it was found, because villages were invariably situated near or on the very banks of all our rivers, and especially at such points as afforded favorable opportunities for fishing. In such localities, therefore, all the stone implements used by the Indians are likely to be found, and great care must be exercised in drawing inferences from the mere position of an implement when found, or from the nature of the locality, in which it was discovered. That such knives as fig. 50 are admirably adapted to cutting and scaling fish is unquestionable; but until they are found so associated with other objects, and in kitchen middens where fish-bones characterize the heap, it cannot be shown that they were designed for a particular purpose, but rather were applied to all the ordinary uses of such articles.

FIG. 50.—New Jersey. ½.

Figs. 51 and 52 represent other forms of these broad-bladed knives, which are comparatively common throughout the length and breadth of the land. The fact that the sharp and carefully worked margins are the prominent feature of such implements leads very naturally to the inference that they were used as cutting tools.

Fig. 51 appears to have had a short drill-like projection at the upper end, which has been broken off; but that it was really such a drill is not evident. It bears, however, a close resemblance to certain perforators or combined drills and knives.

Fig. 53 represents a beautiful specimen of the maximum size, of a

FIG. 51.—New Jersey. ½.

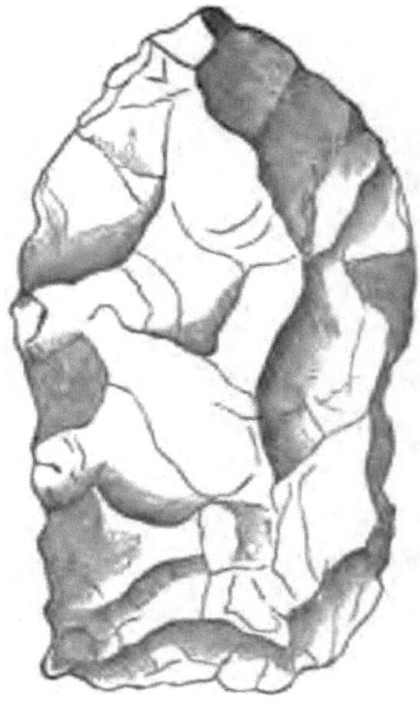

FIG. 52.—New Jersey. ½.

class of chipped implements which is rarely met with in the New England states, but more frequently in New York and New Jersey, particularly on a much smaller scale, and then it is known as the triangular arrowhead.

Like the long oval knives, these triangular knives, if such they are, are chipped from flint or jasper, that is free from foreign mineral, and hence susceptible of being worked to a very sharp, straight edge. The blade itself is as thin, as in any implement of its size, of any pattern.

Were it not that dagger-like implements are almost as abundant as specimens of this character, it would seem proper to refer these to that class of weapons, rather than to consider them as knives, or as plain spearpoints. The absence of the worked base seems to be a sufficient reason for not considering them as spears. This may be an error, but as unquestionable spearpoints are abundant, and implements like fig. 53 comparatively rare, it is quite as probable, that they were used as knives, possibly for some one purpose alone, as that they had any other significance.

FIG. 53.—New Jersey. $\frac{1}{1}$.

Knives of the plainer patterns, here mentioned, are equally abundant in the valley of the Susquehanna river. In the very complete collection of stone implements from this region, made by the late Professor Haldeman, are many chipped knives of the size and shape of figs. 45 and 46. The majority are made of jasper, and are very evenly chipped; but a few are made of the limestone of the vicinity, and are admirably well finished. A few specimens of the highly finished oval, and long triangular knives also, were collected by him, and in every case, they are made of the peculiar bluish hornstone found in Ohio. It is probable, therefore, that these were brought from that state.

This implement, except in being somewhat shorter, is of the same character as the large chipped chert implements from the California coast, figured in the seventh volume of the U. S. Geographical Survey, West of the 100th Meridian. While many of these latter are fully double its length, they do not differ in any other particular. When of such great length, however, they could scarcely have been used as knives. Knives of this pattern have occasionally been found in "deposits" of considerable numbers. In the museum at Cambridge is a series constituting a portion of "a deposit of about two quarts," which was ploughed up near Denmark, Lewis Co., New York. Those of the series sent to the museum are all of a dull bluish color, made of jasper and remarkable for their uniformity of size and finish.

FIG. 54.—New Jersey. $\frac{1}{1}$.

In the American Naturalist, vol. vii, p. 277, a description is given of a series of forty-two specimens of knives, which were found associated with other implements, about a buried fireplace. The knives of this series averaged three and one-half inches in length by one inch in breadth.

Figs. 54 and 55 represent two characteristic examples of a class of chipped implements, which, being evidently finished, may safely be considered as small knives, though their size seems to cast a doubt upon their use as cutting tools. There does not appear to be any evidence that a knife, or saw-like implement, where many of these were placed in a row in a wooden or bone handle like the Aztec "Macehuatl," or the shark's tooth-sword of the Pacific Islanders, was ever used by the Indians of the Atlantic coast. Such implements are found in other localities, but they are usually armed simply with pointed flakes of flint or obsidian. These small chipped knives are all carefully finished, and are not, as now found, sufficiently sharp to have been used, collectively, in the manner described. Like the smallest scrapers, they seem to have had

FIG. 55.—New Jersey. $\frac{1}{1}$.

some special use, but what that was cannot now be determined. They are nowhere very abundant, or else they have been strangely overlooked by collectors. They have been found throughout New Jersey, and in the Connecticut valley, and occasionally specimens have been found in Massachusetts. It is not improbable that many implements of diminutive size are supposed to be of very rare occurrence, simply because they have thus far escaped notice.

Figs. 56 and 57 represent specimens of so-called knives, which are of quite common occurrence in some localities. Those found in

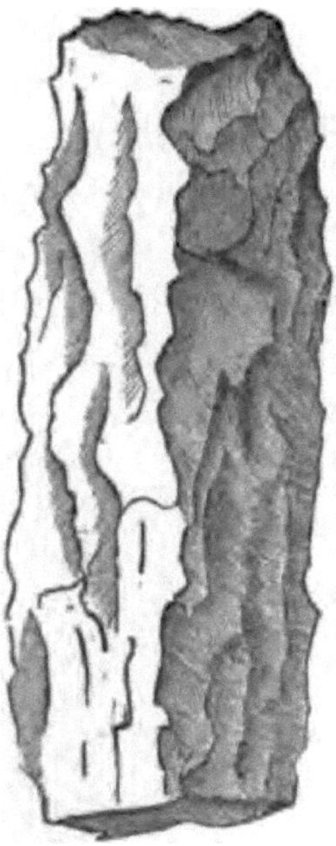

FIG. 56.—New Jersey. 1/1.

FIG. 57.—New Jersey. 1/1.

the valley of the Delaware river have almost invariably been broken at the ends, as shown in the illustrations. This fracture is believed to have been intentional and made when the knives were chipped. While, as a class, they may be considered as double-edged knives, some are much blunter along one margin, than on the cutting edge proper. Whether this thicker edge constituted the back of the implement, or was used for some purpose other than cutting, is uncertain.

Professor Perkins[21] has found many knives of this pattern in the Champlain valley. He remarks, "it is interesting to notice that, on both shores of Lake Champlain, we find the same quadrangular forms, broken across one end, as those figured in the Annual Report of the Smithsonian Institution for 1875, page 301, figs. 111 and 112," and adds, "from the appearance of our specimens I am inclined to agree with the author in believing that the break was not accidental, or rather, I should think it more probable that the broken end is simply the original surface of the block of flint or quartz from which the knife was flaked, and while all the other sides were chipped this was left, perhaps for insertion in some sort of handle of wood or bone.

FIG. 58.—New Jersey. 1/1.

We now come to consider a class of chipped flint knives which are so worked in every detail, that their import cannot be mistaken.

Figs. 58 and 59 represent stemmed knives of jasper, of about the maximum and minimum sizes. In both these examples, the character of the chipping is such, that it is evident that the implement is now in its original condition, and not a re-chipped spearpoint. The cutting edge proper is supposed to be the one on the left, as shown in the illustrations, although there is no difference in the character of the chipping of the two sides.

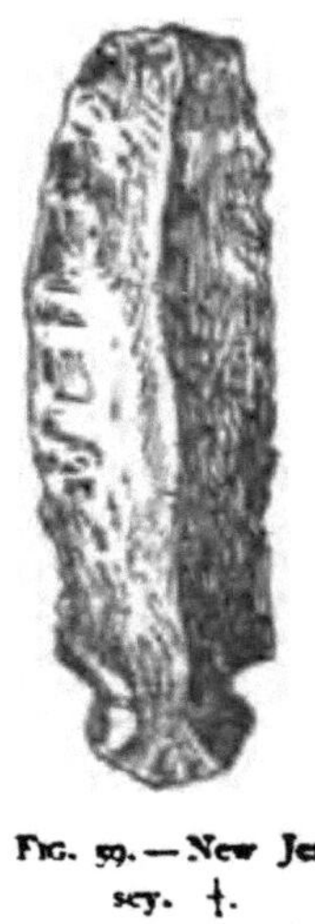

FIG. 59.—New Jersey. 1/1.

[21] Perkins, *l. c.* p. 745.

Fig. 59 differs from the larger specimen in being a flake, chipped only upon the one side. When stemmed knives of the above pattern are as small as fig. 59, they were generally made of flakes; but as yet, no flakes as large as fig. 58 have been met with that were so used. That such will be found is, however, very probable.

Knives, thus fashioned, are common in the Champlain valley, and somewhat abundant along the Connecticut; they are occasionally found in eastern Massachusetts, and are of more or less common occurrence in New Jersey, though they do not anywhere appear to be more numerous than the oval or quadrangular knives already described.

FIG. 60.—New Jersey. ½.

Fig. 60 represents a stemmed knife that strongly suggests the spearpoint, and, indeed, it may possibly have been made from an implement of this kind which had been broken near the point, and was rendered of value again by the substitution of a cutting edge for the point it originally possessed. This knife, thus made from a spearpoint, if it really were so made, does not seem so desirable a cutting implement as the foregoing; nor is the cutting edge as well wrought. Of a series of stemmed knives from New Jersey, four-fifths are of the pattern of figs. 58 and 59, and the fact that so many were made directly from the original mineral, is somewhat indicative of the fact, that those resembling spearpoints have also been made directly from the unworked pebble or bowlder of jasper.

Fig. 61 represents a fourth example of a stemmed knife which, if found but rarely, would probably be classed as an arrowhead. A sufficient number have been collected from one limited locality, to show that the form is not a chance occurrence; and the marked convexity of one side and slight concavity of the other show clearly, that the implement was intended as a cutting instrument, and not as an arrowhead. All these stemmed knives, it is supposed, were mounted in wooden or bone handles.

FIG. 61.—New Jersey. ½.

Figs. 62, 63 and 64, represent chipped implements, which, gathered casually from the fields, as they there occur associated with arrowheads of all patterns, might be classed with them. Yet, it is certain that, even when armed with sharp points, many of them were used as knives.

The archæological explorations in southern California, made under the direction of Capt. G. M. Wheeler, U. S. A.,[22] and the subsequent labors of Schumacher, Powers and others in the same localities, have thrown much light upon the uses of various forms of stone implements. Especially is this true of those large arrowhead-like implements, which seemed too small to be classed as spearpoints, and were undesirable as arrowheads by reason of their size and weight. These are now known to have been used as knives, and it is fair to assume that they were put to the same use when found along our Atlantic seaboard.

FIG. 62.—New Jersey. ½.

In the plate referred to, there is represented a series of knives of flint, four of which have still attached to them the short wooden handles in which they were mounted. Of the series of seven flints, six are acutely pointed; and, if dissociated from their handles, they would be generally con-

[22] U. S. Geog. Survey West 100th Meridian, vol. vii, Archæology, pl. iv, figs. 1–7. Washington, D. C., 1880.

sidered as leaf-shaped arrowheads; but on comparing these Pacific coast specimens with the arrowheads found in the same locality, it will be noticed that true arrowheads, while of the same shape, are much smaller and invariably thinner. This feature of size, and also the relative thinness of specimens, must be always borne in mind, in classifying these objects; for, if thick and heavy, they would require an expenditure of force, to enable them to penetrate the body of an animal, that perhaps, an Indian could not command.

Fig. 65 represents one of these large leaf-shaped knives, from New

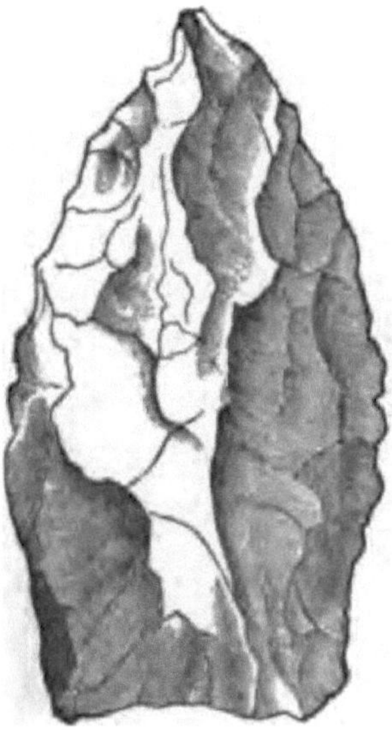

FIG. 63.—New Jersey. 1/1.

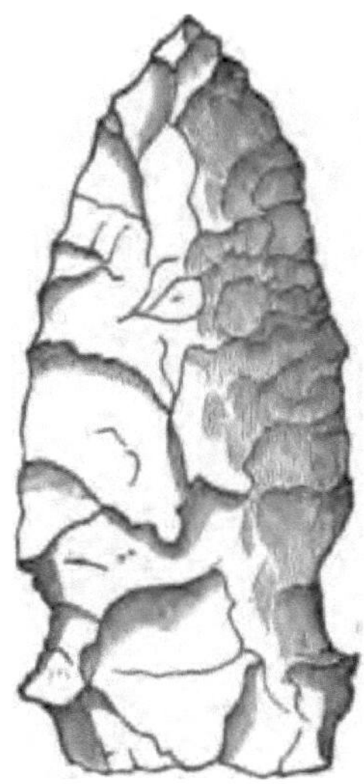

FIG. 64.—New Jersey. 1/1.

Jersey, which was formerly supposed to be an arrowhead. If we compare this implement with the leaf-shaped arrowheads figured in a subsequent chapter, the difference will at once be seen; and in classifying these objects as found on the Atlantic coast, as it is necessary to do, in studying the same forms from California, it is obvious that the distinction should be drawn between leaf-shaped knives and leaf-shaped arrowheads. Of a somewhat similar character, are the arrowheads mounted upon short wooden shafts, found in Peru.

Fig. 66 represents one of these, which, although differently secured to the wood, is otherwise of the same character as those found in California. The Peruvian example may indeed have been intended to be attached to a reed and so used as a spear, but the size and shape of both head and shaft are so similar to those found in California, that it is probable at lèast, that they were also used as knives, even if primarily intended for use as spearpoints. Judging from what we learn of the native races of other countries, it seems safe to conclude, that a proportion of the arrowhead-like implements we gather along the Atlantic seaboard were used as knives, and were hafted in a practically similar manner to that shown in fig. 66.

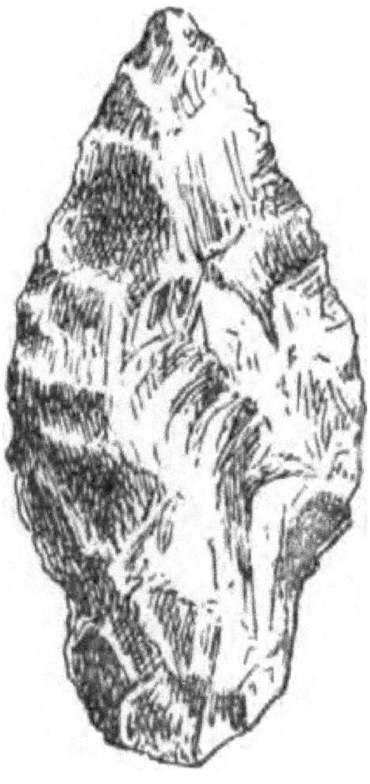

FIG. 65.—New Jersey. $\frac{1}{1}$.

Figs. 67 and 68 represent specimens that are in outline quite similar to ordinary arrowheads, but the width in comparison to the length is such, that even as spearpoints they would be of very little value. Placed at the end of an arrow they would be so clumsy and heavy, that it is questionable if an arrow could be discharged with any accuracy that carried such a load. As a knife this same form, especially when made of jasper, is in every way desirable. Fig. 68 is of slate and much weathered. An example of this kind is of course more open to question as to its use; as, with a moderately long shaft, it might have been used as a spear. The probabilities are, however, that the few slate examples that are found of this pattern were used as knives.

A class of supposed knives, of very different patterns, are represented by the specimen illustrated in fig. 69. So many have been found of this shape and size that it is evident, as in the instance of specimens like fig. 65, they were designed for some particular purpose, and there is little in their shape certainly to suggest the arrowhead, rather than a cutting implement.

Fig. 70 represents a peculiar spearlike implement of slate, found in Vermont, which was first described[23] as a spear-point; but, since then, others have been found varying somewhat in details, and Professor Perkins is inclined to consider them, not as spearpoints, but as knives. Of the series found, he remarks,[24] "of the dozen specimens . . . . no two are alike in form or size, but they all resemble each other in being made of slate, usually red roofing slate, ground, not chipped, and with the stem notched on each side by a series of semicircular depressions. In some specimens the notches are small, in others large; one specimen having but two deep and wide notches on each side of the rather short stem, while another has five smaller ones. In some the surfaces are flat, bevelled only near the edges, which are straight and sharp in all, while in others the whole surface is bevelled from a median line. One or two are so long and narrow that they seem wholly unfit for any use as implements; one especially, made of talcose slate, is nearly nine inches long."

While it is quite possible that these objects may have been used as knives, rather than spearpoints, it is evident that as the former, they would compare very unfavorably, when used in place of many of the chipped knives found so abundantly associated with them. In whatever way some of the ruder and larger specimens may have been used, it is evident that fig. 70 would make a better harpoon-point than knife.

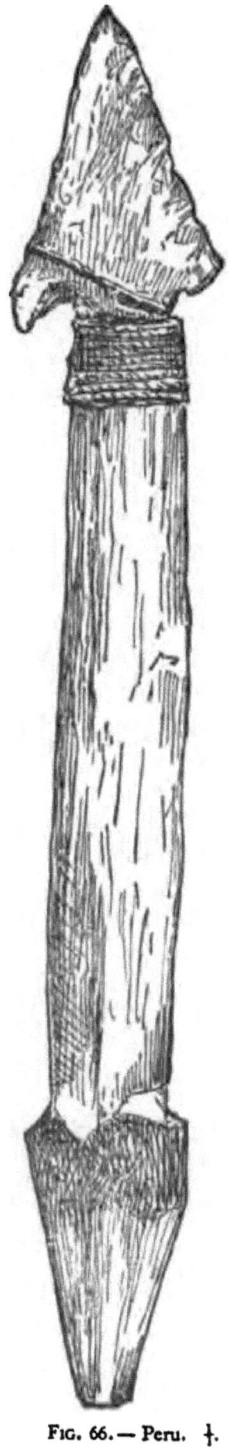

FIG. 66.—Peru. ⅓.

---

[23] American Naturalist, vol. v, p. 16, fig. 5.

[24] *Loc. cit.*, vol. xii, p. 746.

Ground or polished slate arrowheads and knives of this pattern are not confined to Vermont. In western New York, several specimens have been found, which so far vary in size, as to suggest that the smaller were used as arrowheads and the larger as cutting implements.

A single specimen, identical with fig. 70, was recently found in Morris Co., New Jersey, on the shore of Lake Hopatcong. It was made of the roofing slate that occurs at the Delaware Water Gap, and vicinity. The locality where this knife was found is one that suggests that an implement so well adapted for spearing fish, as is this pattern of so-called knives, should have been used in this manner, and not as

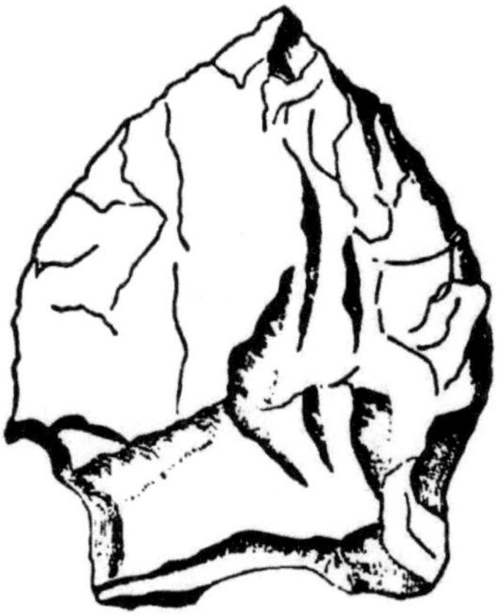

FIG. 67.—New Jersey. ½.

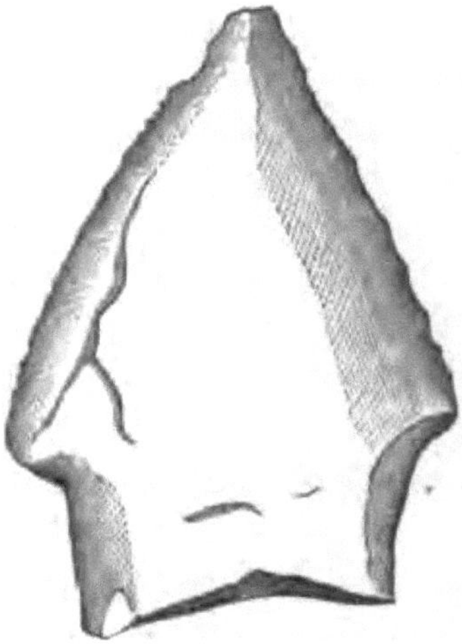

FIG. 68.—New Jersey. ½.

a cutting implement of any kind; especially as an abundance of delicately chipped jasper knives have been gathered in this neighborhood.

Another class, if they may be so called, of stemmed knives, is represented by figs. 71 and 72. Knives of this pattern, which are characterized principally by the very acute point in the middle of the blade, are not frequently met with along the Atlantic seaboard, and no interior locality is known where they are a characteristic form. Such as have been examined were all from New Jersey except two—one from Massachusetts, and the other from Texas. Of a series of eleven from New Jersey, seven were facsimiles of fig. 71, and the others more like fig. 72. These knives are all made of jasper, and show an unusual amount of care in the chipping, not always in the small size of the flakes detached, but in the care that appears to have been taken to bring out and preserve the characteristic point in the blade.

For what particular purpose, if any, these curiously shaped knives were designed is a matter of conjecture. That they were used as cutting implements is almost a certainty.

Circular and square knives also, carefully chipped from jasper and chert pebbles, are often found. They are usually of small size and have such evenly chipped edges, that there is no evidence that any portion has been inserted in a handle of any kind. Similar square

FIG. 69.—New Jersey. 1/1.

FIG. 70.—Vermont. ½.

FIG. 71.—New Jersey. 1/1.

knives, made of chert, have been found in the ancient graves in the islands off the California coast, that measured less than an inch in length or breadth.

Certain irregularly chipped, flint objects are of quite common occurrence, wherever the better known objects are found, and it is a matter of much uncertainty whether their purposes, in all cases, can ever be determined. Some indeed, notwithstanding the care with which they have been chipped, are wholly unintelligible; but others, though seemingly of little value as knives, have sufficient margin in a continuous line to give them a cutting edge; and so with some feeling of doubt, they are classed among the cutting implements. Fig. 73

represents one of these ill-defined specimens, which bears some resemblance to certain of the fossil sharks' teeth found in the New Jersey greensand marls, and which were frequently used as ornaments, and occasionally as weapons. These shark-tooth shaped implements have been supposed to be knives, rather than arrowheads, in consideration of the great objection the curved point would offer to their use in this latter capacity.

FIG. 72. — New Jersey. 1/1.

Fig. 74 represents another example of these apparently meaningless implements, yet one certainly not of accidental occurrence, as every portion is most admirably chipped. Whether knife or drill, or both, it is difficult to determine; but it would serve the purpose of a knife as well as it would that of any other form of chipped implement, particularly, if securely fastened to a wooden or bone handle.

Knives of shell and bone were also in use by the Indians, but none

FIG. 73. — New Jersey. 1/1.

FIG. 74. — New Jersey. 1/1.

have been preserved, so far as I have been able to ascertain. Kalm remarks of the New Jersey Indians, that besides knives of flint, they "were satisfied with a sharp shell, or with a piece of a bone which they had sharpened" (Travels in North America, vol. ii, p. 39, London, 1771), and refers to the use of "sharp shells," in their canoe-making. It is probable that the stouter Unio or mussel shells are here referred to, and if so they would necessarily be beyond recognition, if indeed, they were not destroyed by use. Fragments of marine shells, with ground cutting edges, as before stated, have not been found on the Atlantic seaboard, or, if so, but very rarely.

CHAPTER VII.

DRILLS, AWLS OR PERFORATORS.

WHILE single objects are constantly being discovered, the purpose of which it is difficult, if not impossible, to determine, it has not happened that any numerous class of objects has been gathered, the use of which could not be shown with some degree of probability. Possibly the smaller polished celts may come under the head of objects of unknown uses; for, as yet, it is not demonstrable how many of the very diminutive examples were used, although, as in all such cases, it is easy to conjecture. In another publication[25] they have been called "skinning knives," not because there was actual evidence that an Indian had ever used one as a "skinner," but simply because it was found possible to skin a small mammal with one, with great ease. It is not desirable, however, to be guided by one's own experience in this matter, and when the use of any pattern of stone implement is not apparent from its size and shape; then we must admit our ignorance and confine ourselves to such comprehensive terms as "celt" or "implement;" unless indeed, we find a similar implement in use among existing races. In such a case, we have usually a safe guide.

The series of objects now under consideration, it is probably safe to classify as drills or perforators, because their principal feature suggests no other use whatever; and our faith in this classification is confirmed by experiment, for it is found that they can be used as such, even when the material perforated is of an equally compact or even harder stone than that of which the implement itself is made. Certainly, for drilling thin plates of stone, of which such large num-

---

[25] Smithson. Ann. Rep., 1875, p. 304–6, figs. 115 and 120.

bers are everywhere found, they are admirably adapted, though perhaps not more so, than are many of the chance flakes found in the refuse heaps where arrowheads have been made.

Admitting this use for the more common sizes, which have, at the same time, the strength and durability required for this purpose; what are we to consider the use of the more delicately formed examples of this same implement? In delicacy of chipping, many of these latter approach very closely, if they do not equal, the finest Danish arrowheads, though none, however, show any polish on the point, as occurs on many of the larger examples. Their appearance suggests that they were used to pierce, and not to perforate gradually, by a rotary motion; in a word, that they were awls used in sewing skins. This seems the more probable, inasmuch as the bone awls or needles, common to some localities, occur but very rarely in New Jersey, whilst in New England, they are common, especially in the shell heaps; and stone drills, whether large or small, are less frequent. This is not conclusive evidence, however, for in Ohio, both the bone awls and stone perforators of unusually delicate workmanship, are more abundant than in any known locality on the northern Atlantic seaboard. Why the two forms should be found associated, when the bone examples are not only better, but certainly more easily made, is something of a mystery, if it be true that they had identical uses.

It will be noticed, on examination of the several illustrations of these drills or perforators, that many of the small and delicate examples have very elaborately wrought bases, the majority of which, judging from the character of the design, were not fashioned for any useful purpose. The length of the drill, also, varies indefinitely, many having but a sharp slender projection of less than one-fourth of an inch, from a large square base, which is often beautifully chipped; others, again, are three and even four inches in length, exclusive of the base.

While the purpose of the larger drills, as figs. 75 and 76 may have been to drill thin plates of stone, such as the pendants and other ornamental objects described in Chapter XXVII, and the others may have been used as awls or perforators of leather, there is reason to believe that the manufacture of shells and beads explains the occurrence of so very many of the smaller sizes. Beads, it is well known, have ever been a favorite ornament with savage races, and whether made of shell or bone, or of clay or stone, they are found in greater or less abundance, wherever the ordinary forms of stone implements are found. Considering that thousands of small shell beads and strings of wampum are not uncommon, it is natural to conclude that the implements, wherewith they were perforated, should be correspondingly abundant, and that the intermediate sizes at least could have been so used. Probably even the more delicate would not break under the rotary motion if used only for perforating shells and bone. The former is the material of which wampum is made, and is quite as wearing to flint, as any mineral that we find, that has been perforated by a stone-drill. Still, this latter use of the small drills is altogether conjectural; and the suggestion that they were solely for piercing leather, and only used by the women, is certainly the most plausible explanation of their purpose.

FIG. 75.—New Jersey. ½.

Fig. 75 represents what may be considered a typical example of a flint or jasper drill, such as is found in abundance in New Jersey, and in fewer numbers in New England. This specimen, which is of about the maximum length of these implements, is carefully chipped from chalcedonic quartz, and is as symmetrical as are the best finished daggers of flint. There is no trace of wear upon the point, and for whatever purpose it may have been designed, it is evident that it has been but little used. Of the hundreds of such drills as fig. 75 that we now find in our fields, the great majority are broken near the middle of the shaft, but whether this fracture occurred while the implement was still in use, or after it was discarded, and in very recent times, cannot be easily determined. That there are found, however, many more of the bases of drills, than of the pointed ends, is certain; and from this fact, it is safe to infer that a considerable portion of those that have been broken, at or near the middle of the shaft, were fractured while in use.

Fig. 76 represents a second example of this form of drill, of a more common size and character of finish. There is but little evidence of wear at the tip. Unlike slender, tapering spear-points, especially such as have elsewhere been called fish spears, this and the preceding specimen are not in section, a flattened oval, but are distinctly quadrangular. This feature, which is characteristic of the great majority of these drills, is to some extent indicative of their use. The four sides, while giving four cutting edges, and thus facilitating the perforating power, if rotated, would largely decrease the penetrating power if used as an arrow or spearpoint. Such a use, however, would scarcely be suggested for this implement, even if unquestionable arrowheads were not known.

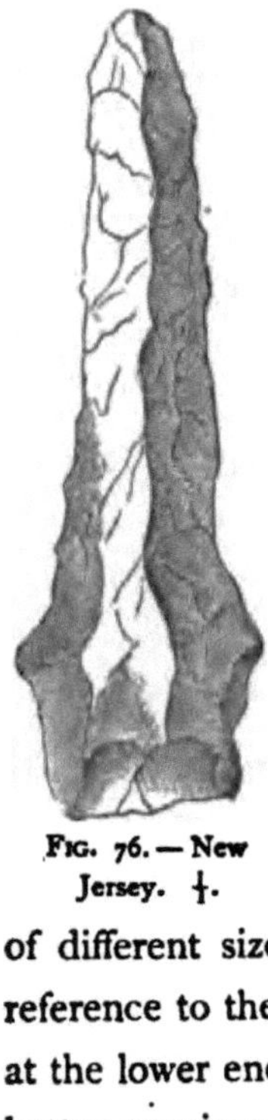

FIG. 76. — New Jersey. 1/1.

The mineral, of which fig. 76 is made, is a dull yellowish-brown jasper, such as occurs in pebbles of different sizes in the river gravels. The base, as will be seen by reference to the cut, while simple in design and convenient in size, has at the lower end a well chipped cutting edge, such as characterizes the better specimens of jasper knives. Whether this edge was designed to be used as a knife, or whether the implement was inserted in a handle is not determinable; but on examination of a large series, it seems probable that the broad bases, such as figs. 77 to 81, inclusive, possess, bear some relation to the uses to which the drill proper was put.

Figs. 77 to 81, inclusive, illustrate very nearly to what extent the shape and size of these broad based drills vary, although the examples here figured do not fairly exhibit the great beauty of finish along the edges of the base shown by a few specimens more recently collected. Did there seem to be any limit to the numbers of these stone drills still to be gathered in many localities, it might be supposed that the makers of these implements were economically disposed; and, foreseeing the early destruction of the points or drilling parts, had ready the implement of such shape that, with little or no additional labor, it could be

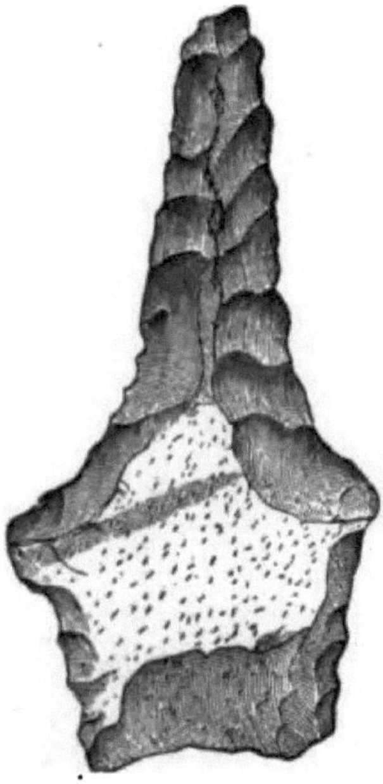

FIG. 77.—New Jersey. $\frac{1}{1}$.

FIG. 78.—New Jersey. $\frac{1}{1}$.

converted into a knife. That the broad based examples may be considered a combined knife and drill is no tax upon one's credulity, for in no other way can it be explained why such a "drill," if such it only was, should have so elaborate a base, as in fig. 79 for example. When we come to consider the smaller drills, or "awls," we shall find even more notable examples of these knife-like bases. These bases, it should be further stated, do not only occur on such drills as are chipped from jasper pebbles. Of the five examples figured, figs. 77 and 78 are of slate and comparatively soft, while 79, 80 and 81 are ot jasper. The relative merits of these minerals for producing the best results in the "flint chipping" art have been quite fully referred to, when considering the various patterns of so-called knives; but it is not out of place to mention at this time, the fact, that slate drills, when newly chipped, were much more unyielding and sharp about the point and edges than they are now after centuries of exposure to the weather. For an inconsiderable depth, a process of decomposition has taken

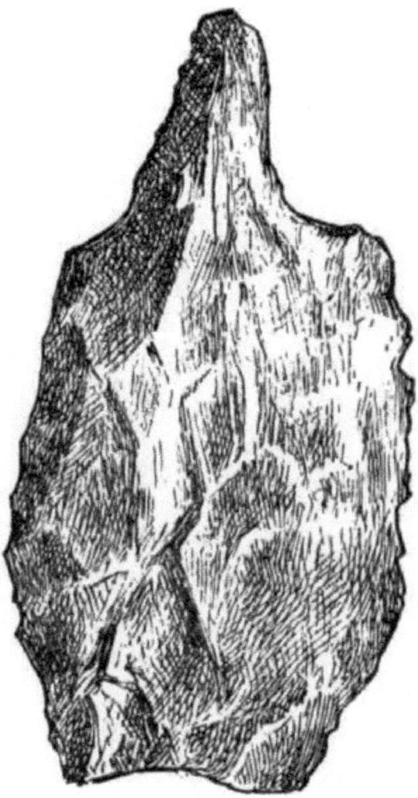

FIG. 79.—New Jersey. ½.

FIG. 80.—New Jersey. ½.

place, which gives to many objects, made of slate, a very chalky appearance; but this is so superficial, that the mere handling will in some cases remove the decayed portion, and exhibit beneath, a dense, sharp surface that readily scratches hard mineral substances, and proves the material available for making "flint implements."

In conclusion, as will be seen by reference to the illustrations, the points of the drills are in most cases wanting. This, as has already been mentioned, arises from the fact that they were broken when in use, rather than by any subsequent exposure to ploughs and harrows, or to the hoofs of horses.

In figs. 82 and 83, we have examples of a common pattern of drill, that differs somewhat from the preceding. Both are of jasper, similarly shaped, and vary but little in size. In these examples of stone drills, we have a peculiarity which brings them nearer to the arrowheads. in

FIG. 81.—New Jersey. 1/1.

FIG. 82.—New Jersey. 1/1.

FIG. 83.—New Jersey. 1/1.

that, instead of being distinctly quadrangular in section, they are of a flattened oval shape in section, and have the thin, chipped cutting edges of knives or arrowheads. Elsewhere[26] the suggestion has been made that these bases, when distinctly oval, may have been used in drilling the bowls of steatite smoking pipes, but subsequent examination of many pipe bowls now convinces me that this is an error; as the tool marks or *striæ* in the bowls of such pipes are so regular, that a hollow

[26] Smithson. Ann. Rep., 1875, p. 323. See Rau, in Smith Ann. Rep., 1868, p. 392.

drill was in all probability used, except in the largest pipes, which show that the excavation was first pecked or gouged out, and then smoothed in some undetermined manner. The thinner edges of these flattened drills do not show any indication of wear or polish. Probably they were chipped off as the implement was rotated, and these minute flakes acted similarly to the fine grains of sharp sand, which it is well known was used in connection with hollow drills.

Fig. 84 represents a roughly chipped implement much like an arrowhead, but of such an irregularly shaped base, that it could scarcely be so secured to a shaft, as to have been made available for that purpose. As in the preceding, it is thinner and flatter than the majority of drills, but shows some traces of wearing along the edges and at the tip, and was probably used as we have here suggested. It differs from drills of both the quadrangular and flattened forms, in being quite flat upon one side, and with a moderately well defined ridge upon the other.

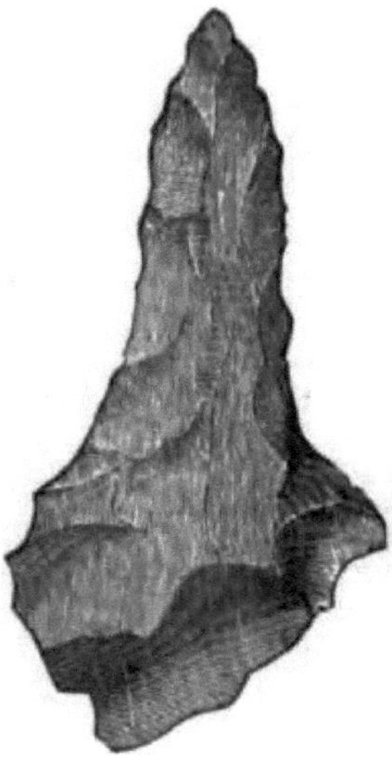

FIG. 84.

It will be found, however, that in all large collections of stone implements of every class, other than ornaments, many examples will be of such indefinite shapes, as to make it exceedingly difficult to classify them,—a labor not without its difficulties even when the objects are seemingly so distinct as are the axes, spearpoints, or polished celts. Until our knowledge of the habits and customs of prehistoric man is largely increased, all classification must to some extent be conjectural, and considered more as a help to archæological studies, rather than of scientific importance.

In fig. 85, we have a second example of a chipped implement, which is, judging from our own experience, rather a drill, than either an arrowhead or a knife. It is too long and narrow for the latter, and without such a base as is common to the spearpoints of that size. It is of slate, somewhat weathered, and exhibits no trace of polish on either the tip or sides.

Fig. 86 is, in most respects, similar to the large based drills already described, but it is as thin and sharp upon the edges as an ordinary arrowhead. The long, narrow point and stem suggest the drill rather than the spearpoint, and the well defined cutting edge of the base is

FIG. 85.

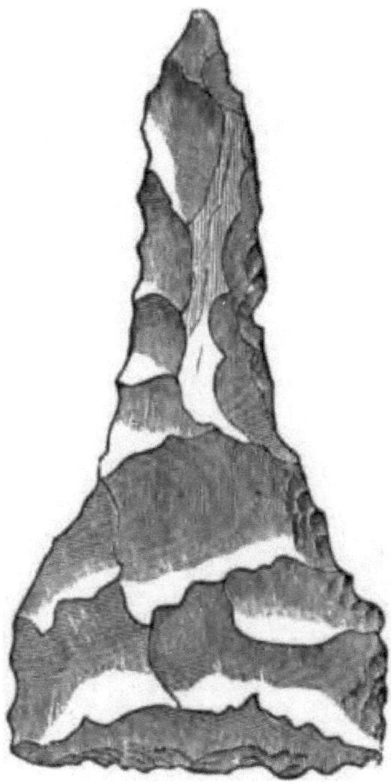

FIG. 86.

too carefully worked not to have been utilized subsequently; and this could not have been the case, if the specimen had been used as tip to an arrow or point to a spear.

Fig. 87, on the other hand, appears to have been attached to a handle of some kind; the double notching at the base—a feature of rare occurrence—indicating that it was desirable to make it very secure. Had the specimen a well defined, sharp point, it would be very doubtful if it were intended for a drill; but while the crooked point is a drawback to its use in this capacity, it is probably less of an objection than it would be if used in any other way.

A further peculiarity of this specimen is that the under surface is almost perfectly flat, while the side shown in the illustration is ridged and slopes from a central line as in the ordinary spear and arrowpoints. The flat side is not a single plane of cleavage, but has been carefully chipped, and presents fully as many distinct planes as the other side.

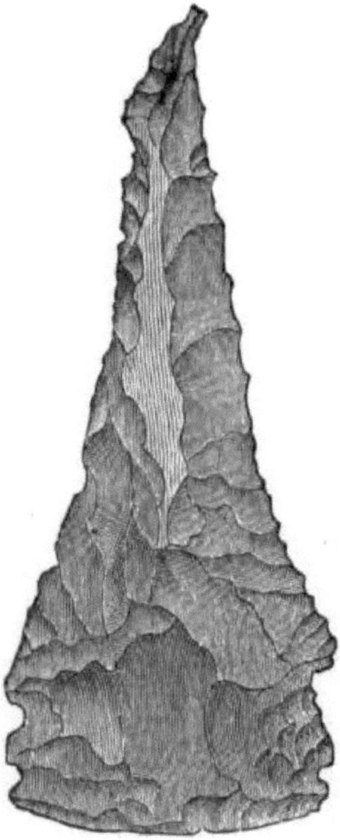

FIG. 87.

Fig. 88 represents a connecting link, if we may so call it, between the ordinary arrowpoints and those smaller drills, which, as "awls," we shall consider in a subsequent section of this chapter. While the size, shape and material of this specimen are such as are common to many arrowpoints, the once highly polished point shows conclusively that whatever may have been the purpose of the maker, it was undoubtedly used as a drill, and not always upon such yielding material as leather. The marked indications of its use, so seldom seen on the ordinary drills, make this specimen unusually interesting. Occasionally it happens, especially when collecting on what was once a village site, that several drills with distinctly worn or polished points are found, and hence it is sometimes supposed that a majority of drills and awls show similar indication of use. This, however,

FIG. 88.

is undoubtedly an error. Indeed, apparently unused specimens really do occur in such excess, that the question may well be asked, whether, in perforating stone disks, and drilling ceremonial stones, implements of wood were not always used.

Fig. 89 represents a perfect example of a form of implement of which but few specimens have been found. It is, without hesitation, classed as a drill, and is supposed to be of the maximum size of this pattern. The specimen is of compact slate, well chipped, and measures four and seven-eighths inches in length. Other examples, but of jasper, are each nearly an inch shorter.

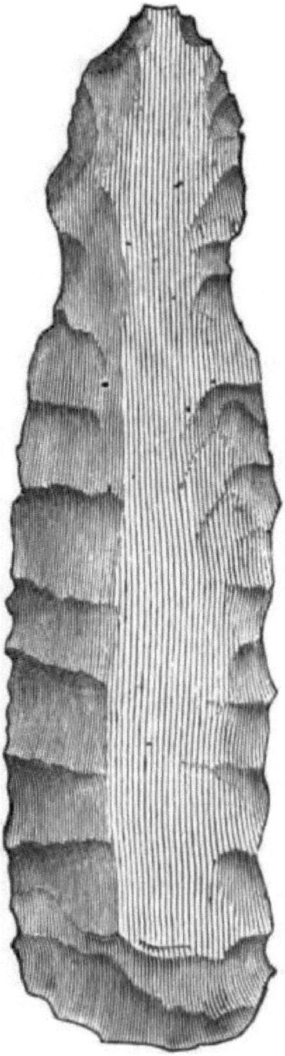

FIG. 89.—New Jersey. 1/1.

It will be noticed that the head or pointed end of fig. 89 is narrower than the main portion of the implement. This feature is common to all those collected in New Jersey. In the following illustration of a broken specimen, this is even more marked than in the present instance. As in the case of those smaller drills with large bases, this specimen, also, has the margins of the handle or lower portion brought to a well defined cutting edge by chipping.

It is, indeed, not altogether improbable, that it is a misconception to consider fig. 89 a drill. If we reverse the implement, and assume that the pointed end is so shaped for the better attachment of a handle, we have then, thus hafted, a most admirably designed flint knife, and one that on a smaller scale occurs along the Atlantic coast.

Whether the preceding specimen is a drill, a knife, or whether it was intended for some unknown use, it is quite certain that the fragment, fig. 90, is a portion of a similar implement. In this instance, the material is quartz, and it is worked with more care. There seems to be, also, too much work upon the pointed head—assuming the specimen to be identical with the preceding—to have been intended merely for a stem to insert in a wooden or bone handle. Its finished condition suggests that it was the principal feature of the implement, and that which decided its purpose in the mind of the ancient flint worker who fashioned it.

Fig. 91 represents a very roughly chipped implement of slate which very strongly resembles the preceding specimens, although there is sufficient difference to justify the supposition that they may have had different uses. In the specimen here figured, however, we find that the pointed end is quite smoothly polished, and this at once suggests its use as a drill. The narrowed portion or waist is nearly in the middle, and this gives the specimen a much larger head than in fig. 89; though, like it, the sides are brought to a moderately well defined cutting edge. Rude as it is, there is much evidence that it was used as a drill, although among the hundreds of perforated objects that have been collected, there are none that suggest the use of a drill of this size, rather than that of a smaller form, such as fig. 76. This, of course, is on the supposition that the ceremonial objects and other similar specimens were perforated with a hollow reed, in connection with sand and water. This, however, was not always the case, as unfinished objects of this character are found which show that a solid, pointed drill had been used. But of the many examined, none have the perforation equal in diameter to the width of the specimen, fig. 89.

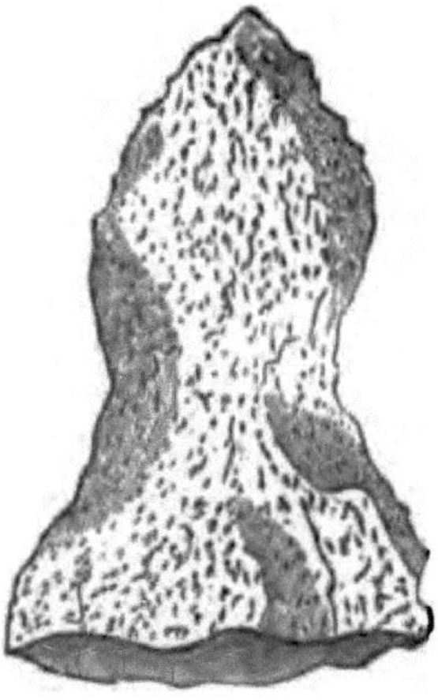

FIG. 90.—New Jersey. ½.

In figs. 92 and 93, we have, in the former, a half of a stone gorget, in the upper left hand corner of which is a countersunk hole, such as characterizes these objects. In order to test the efficiency of even the softer drills, the writer made the series of nine holes with the drill, fig. 93. A glance will show that the one perforation, made by the Indian, was in all probability made with a similar tool. The drill itself became highly polished by so much use, but suffered less than a similar drill of jasper, which was used but a short time, and which from constant splintering was soon too blunt and worn to be efficacious. This was no doubt in part due to very rapid rotation, and the fact that the first experiment was made without having the stone wet. The other eight holes were drilled with the slab under water.

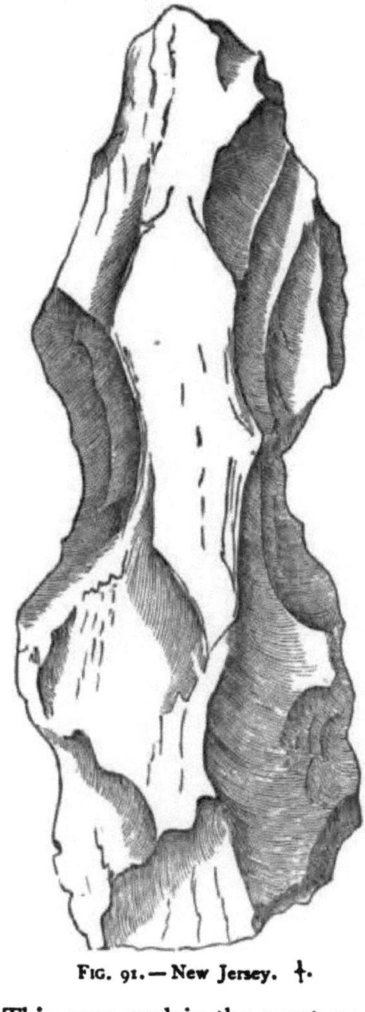

FIG. 91.—New Jersey. ½.

Is it not probable that the Indians did not use water in connection with these stone drills? If they did not, we can readily see why so many of the jasper specimens should be broken at the tips, and free from polish. The friction of *dry*, rapid rotation causes a steady splintering, and keeps the drill in a newly chipped condition. When the same drilling is done under water, jasper soon acquires a magnificent polish. This may explain the great excess of unpolished jasper drills, over such as are worn and smooth.

Fig. 94 represents a beautiful jasper specimen of the smaller drills which will be here referred to as awls, under the impression that their purpose was more for perforating leather, and such yielding substances, than for boring through stone or bone. Why such an exaggerated base should be given to an implement of this character it is difficult to conjecture, unless the base was used as a knife, which in this instance is not improbable, for it is seldom that chipped jasper knives, with more accurately finished cutting edges, occur. The extreme tip of the awl in fig. 94 is slightly worn, and it is reasonable to suppose that,

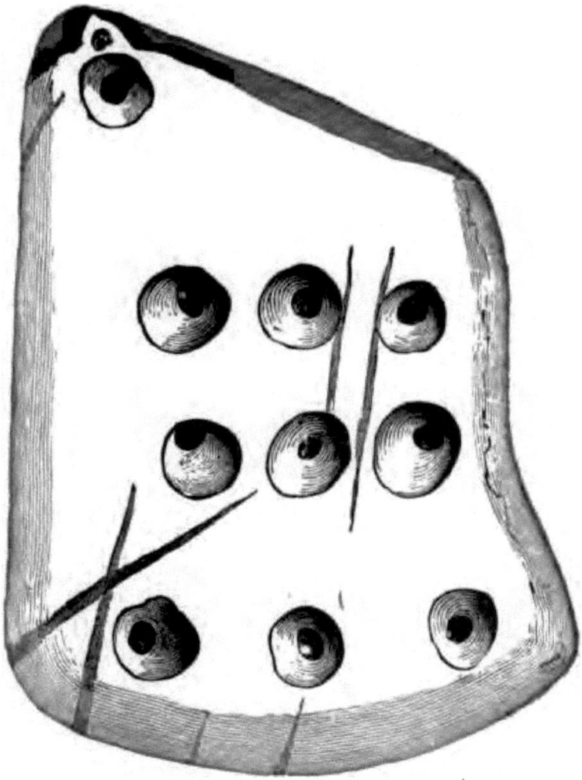

FIG. 92.—New Jersey. ½.

when first chipped, this portion of the implement was considerably longer than at present. Even when used solely as an awl for piercing leather, these points would in time become dulled and wear away, or if re-chipping became necessary, they would be materially shortened. On the other hand, the knife would not be noticeably dulled by use, but would receive in time a polish sufficient to glass the surface of every facet, without obliterating the delicate ridges that define the outline of the flakes detached in chipping the implement. This slight polish from use is noticeable on this specimen, and, to a less degree, on many similar ones, from the same locality. In the New England states, exclusive of the valley of the Connecticut, these drills are scarcely known. In the latter locality, they are sometimes found.

Fig. 95 represents a small flake of jasper, unaltered anywhere, ex-

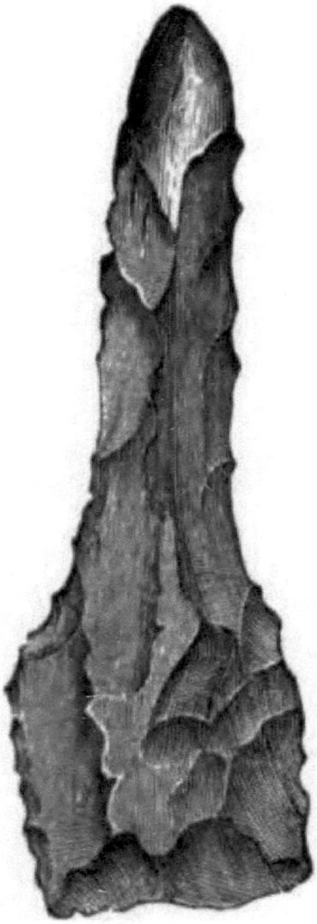

FIG. 93.—New Jersey. 1/1.

FIG. 94.—New Jersey. 1/1.

cept in its more slender portion, which has been carefully chipped until an awl or drill of great beauty has been produced. There is, at the extreme point, no polish or other trace of use. Drills or awls, made from flakes of about this size, are common in New Jersey and New York, and are found in considerable numbers in the valley of the Connecticut river; but they are rare in Massachusetts and northward. This assertion is based upon collections in various institutions, especially those of the Museum of Archæology at Cambridge, and of the Academy of Science at Salem, Mass. Inasmuch as the collections here mentioned, as well as those in other institutions, are so largely made up of contributions from innumerable sources, it is possible that the collectors of such material, as has been preserved, have overlooked so small and unattractive an object as the flake drill in fig. 95. To some extent certain objects may be thought to be rare or unknown in given localities, simply because they are not readily found; when, in fact, they are really fairly abundant, but require careful search to discover them. This is not applicable, however, to the region about Salem, Mass., which has been most carefully hunted over by those who, of all others, have had that experience and preliminary training requisite for this important element of archæological research.

FIG. 95.—New Jersey. 1/1.

Figs. 96 and 97 represent small stone awls, such as the preceding, except that in this instance (fig. 96) there is a finished base, which is not, as in fig. 94, at all knife-like in character. There does not appear to be any special object in giving such a finish to the base, and it can only be looked upon as the whim of the maker. This specimen is made of light blue-gray jasper, and is chipped with great care, the workman having preserved very nearly a uniform width throughout the entire length of the stem. Larger drills of this same pattern of base are moderately common. Fig. 97 represents what appears to be the stem of a drill that has been broken near the base. Such fragmentary specimens are not very abundant. This specimen has some trace of wear on the extreme point, though it is not usual to find such indications of use as has been elsewhere stated when referring to this same specimen.[27] Atten-

FIG. 96.—New Jersey. 1/1.

---

[27] Annual Rept. Smithson. Inst. 1875, p. 323; fig. 154.

tion may be called to these apparently broken awls, as being possibly intentionally unfinished or broken at one end, for the purpose of inserting this blunt termination into a bone or wooden handle. Many flint knives are in this same manner broken off abruptly, apparently by accident; though there is reason to believe that they were really left in this condition by the workman. These have already been noticed.

Fig. 98 represents a third example of these small and slender drills, with its extreme point lost by use or accident. Like the preceding, it is chipped from jasper, probably a pebble of about its present size. Such jasper pebbles, partially chipped, have occasionally been found,

Fig. 97.—New Jersey. 1/1.

Fig. 98.—New Jersey. 1/1.

Fig. 99.—New Jersey. 1/1.

and the entire series presents one most interesting peculiarity, viz.: that the stem is invariably first chipped, not roughly, but delicately, and then the base is worked into such shape as is desired, subject of course to the general shape and size of the unworked portion of the pebble. This clearly shows that in all cases the stem and point are the principal features of the implement, and the shape of the base, whether it be fashioned for use or ornament, is altogether a matter of minor importance.

Fig. 99 represents a small drill with a square base, such as is occasionally found in New Jersey, but is not abundant in any one locality. In the Connecticut valley, a few have been found, but usually the drills from that section have bases more like that in fig. 101, or what is known as the arrowhead base. Fig. 99 is made of slate, is neatly chipped, and has the point more than usually well polished. The majority of the awls of this size are made of jasper or quartz.

Fig. 100, represents a curiously designed specimen of a jasper awl or drill, such as is found occasionally associated with the regular patterns of this implement. Being irregular forms, there is little to be said about them. In the present instance, it was at first supposed to be an arrowhead, with a single exaggerated barb; but such can hardly have been the case, as the base is not of such shape as to render the attaching of the arrow-shaft practicable. Further examination of the specimen shows, also, that the end of the awl, as it is now believed to be, is quite smooth; and, from this fact, it is safe to infer that the implement is an awl or perforator, notwithstanding the peculiarly shaped base. The point of the base is quite sharp, and may also have been used for piercing leather. In such a case, the implement becomes a double awl, and there are others of this character, of even more pronounced shape than fig. 100. The double awls from New Jersey, now in the Museum at Cambridge, have usually a square or oval base, from which, at opposite ends, extend short but narrow and neatly chipped projections. These double awls are rare, in comparison with those of the ordinary patterns. Of those examined, one-half were made of argillite and slate, the others of jasper.

FIG. 100.—New Jersey. ½.

Fig. 101 represents a small awl with a base of the highest degree of finish. While the general appearance of this specimen is that of an arrowhead, it is evident from the polish of the point, and the heavy, exaggerated base, that its purpose was as stated, and not that so undesirable a chipped flint should have been used to tip the shaft of an arrow.

In Massachusetts, there are found but few of these small awls or drills, and nowhere are they so abundant as in Ohio and Indiana. In New Jersey, they constitute about one per cent. of the whole number.

In the valley of the Connecticut, they are well represented; although none that I have seen from there have been quite so short as the one figured above. The bases generally, of the Connecticut specimens are either stemmed or notched, and are in every respect identical with the stemmed and notched bases of the arrowheads from the same locality.

The same may be said of the larger drills, such as fig. 75, which is here taken as the type of the drills proper. Those from the Connecticut valley are of identical patterns, and usually of the same material. Indeed, so far as ordinary Indian stone implements are concerned, the specimens from the valleys of the Delaware and Connecticut are so similar, that it would be impossible to determine from which of the two localities any given specimen had been brought.

FIG. 101.—New Jersey. 1/1.

In the valley of the Susquehanna river, drills of all patterns and sizes are apparently less abundant than in the valleys of the two rivers mentioned. In the collections of the late Professor Haldeman, there are but few specimens of drills, and none are of so elaborate a finish as the finer New Jersey examples here figured.

In the large series of stone implements gathered from the Chickies Rock retreat, described by Prof. Haldeman in the Compte Rendu of the Congress of Americanistes, at the Luxembourg session, in 1878, there are a number of pointed flakes and fragments of quartz and jasper which were considered by Prof. Haldeman to have been "borers" and are so labelled in the collection, but none of these exhibit any indication of use. They may have been reserved for use as drills, but they are simply accidentally pointed flakes which possibly were gathered from the refuse chips of an arrowmaker's workshop, and intended for use as drills or borers.

The large majority of the drilled ceremonial objects, found in this Rock retreat and in the neighborhood, show by the character of the perforation, that a reed or hollow drill of some character was used.

Fig. 102 represents a peculiar implement which seems more likely to have been used as a drill, than as a knife or any other kind of stone implement. There is in all that have been examined, not only a well defined point, sufficiently acute to pierce leather readily, but this point has, in the example here figured, a degree of polish, from use, which clearly shows that this implement, at least, has been used as a drill or awl. The flattened bottom, and evenly curved back are so fashioned as to make it very easy to hold the specimen, and indeed, a much better grasp can be secured when such an implement is used, than with any of the long slender drills, such as have been described.

All that have been gathered of this pattern are made of argillite; and, as will be seen by comparing the illustration below with some

FIG. 102. — New Jersey.

of those palæolithic implements from the gravel beds, there is a marked resemblance between them. This similarity, however, indicates no relationship between the two forms, though it has misled superficial observers into numerous errors as to the significance of those found in the gravel. Nothing from the latter beds evidences any such use as that of drilling stone with stone, or of utilizing sharpened flakes of argillite as awls or perforators of leather; yet that palæolithic man used the skins of animals for clothing can hardly be doubted.

Fig. 103 represents a very interesting specimen of stone awl, which presents a feature that was wanting in all those previously described. This specimen is polished over its entire surface. It is made of white quartz, and has been ground down until every inequality has disappeared. The point is very sharp, and it is questionable whether for piercing skins of animals, or equally yielding substances, any implement of metal would be preferable. Implements of this pattern are very rare. Of five hundred drills and awls of all patterns from New Jersey, in the Museum at Cambridge, there is no other example of this pattern; nor is there anything approaching it from the New England states. As has been suggested, it is probable that in New England, bone needles generally replaced the stone awls.

FIG. 103.—New Jersey.

Before concluding this subject of the stone implements that are supposed or known to have been used for drilling through stone, and perforating leather, in lieu of the modern steel needles, it is proper to call attention to other forms of polished stone implements, which, like the preceding (of the purpose of which, there can be no doubt), may be properly classed as drills of some particular pattern, and as such designed for peculiar purposes. Whatever may be the character of these "uncertain forms," they are not unique, and hence it is evident that they are *not* merely the result of a whim on the part of the maker.

Fig. 104 represents a polished stone implement, of peculiar pattern, from Concord, Mass. As the illustration plainly shows, there is a well defined, piercing point at each end of the implement, and these were evidently intended to be used. Of the large series of stone implements from Massachusetts in the Museum at

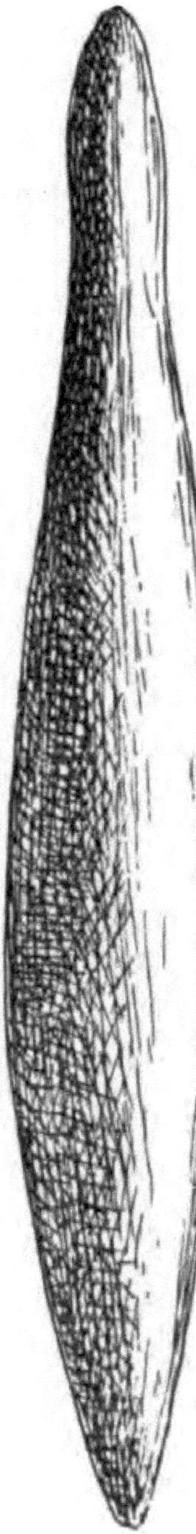

FIG. 104.—Massachusetts.

Cambridge, there is no other specimen that is similar to this. From the shellheaps, however, have been procured many large bone implements, pointed at one end, which are practically the same as the above. A broken specimen of an implement of this pattern has been found in New Jersey, and another, not broken, but smaller, near Columbia, Penn., on the bank of the Susquehanna river. Fig. 104 is circular in section at all points. Except in this, the larger chipped drills, such as fig. 89, would, if ground down until every trace of chipping disappeared, be similar in form. Fig. 104 has evidently not been first chipped and then ground or rubbed down. The surface of the widest portion is apparently the unaltered surface of the natural cylindrical pebble, selected for the purpose of making a long, double pointed drill, if such the implement really is.

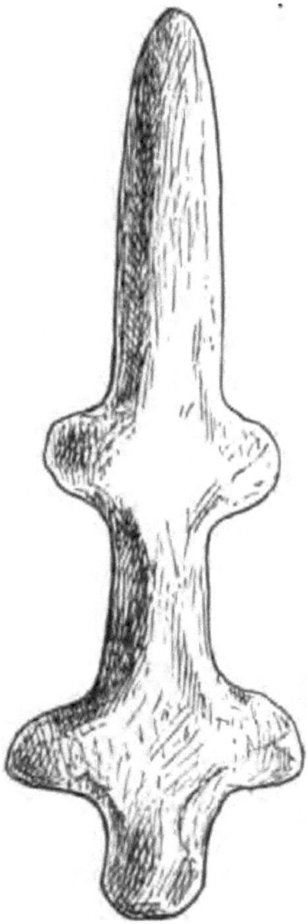

FIG. 105.—New Jersey. ½.

Fig. 105 represents a supposed drill, which, in its design, bears much the same relation to those of the largest size already figured, that the diminutive awls bear to these arrowheads, the bases of which are reproduced. Were there only the lower side projections, this implement would be of peculiar interest, in that it has been worn to a positive polish, although originally chipped into shape. The upper, smaller projections lessen very much the value of the implement, if looked upon as a drill, and do not of themselves offer any suggestion as to the purposes for which the implement may have been used. The extremely dull point shows conclusively that whatever may have been its condition originally, it now has no penetrative power.

Objects like fig. 105, although few in number, have been met with in widely separated localities. Specimens, differing in no important feature, not even in size, and all with a polish produced apparently by long usage, are known from Ohio, Wisconsin, western New York, central Pennsylvania, and New Jersey.

Notwithstanding some slight resemblance thereto, it seems wholly improbable that these polished implements should have been weapons of any kind. On comparing the two forms, fig. 105 will be found to vary very much from the flint daggers found in Tennessee and Arkansas.

Unquestionable stone drills of this pattern, but smaller, have been found in Illinois, which have the barb-like projections distinctly curved, and much longer than those of fig. 105. Others, again, have these lateral projections of greater width than length, thus giving the implements the appearance of birds with outspread wings. No examples of these have been found in New Jersey, except a fragmentary specimen of the latter, made of argillite, which is now in the Museum of Archæology at Cambridge, Mass.

CHAPTER VIII.

SCRAPERS.

To an easily recognized class of chipped stone implements, found more or less abundantly throughout Europe and North America, has been applied the name "scraper," a term that at once gives the reader a full insight into the object of the implement. These so-called scrapers have been described[28] as "oblong stones, one end of which is rounded and brought to a bevelled edge by a series of small blows. One side is flat, the other or outer one is more or less convex; sometimes they have a short handle, which gives them very much the appearance of a spoon. They have been found in England, France, Denmark, Ireland, Switzerland, and other countries. They vary from one to four inches in length, and from half an inch to two inches in breadth. * * * * Modern specimens (Esquimaux) are in form identical with the old ones."

Along our northern Atlantic seaboard, the abundance of these scrapers varies exceedingly. In New Jersey, they are very common wherever relics of any kind occur; and the various forms are as well represented as in Ohio, whence come the finest examples of this, as well as other forms of chipped implements. In the Connecticut valley, they are less abundant, though in nowise scarce; while in eastern Massachusetts they are very seldom found. In the various river valleys of New York, they occur frequently; but judging from the collections from that section, they are not as abundant as in the Delaware and Connecticut river valleys; although I learn from Rev. W. M. Beauchamp of Baldwinsville, New York, that in Onondaga

[28] Lubbock, Prehistoric Times, 2d ed., p. 92.

county, scrapers are abundant, and of many forms, and often combined with knives.

It is not impossible that one reason for the absence of this useful implement in some localities is that other material than stone was used in their manufacture, and they have been destroyed. Nature provides in the shells of certain bivalves, excellent scrapers which would effect all that the Indian ever accomplished with those made of flint. When resident throughout the year, in the immediate vicinity of the sea, it is well known that many implements of shell were daily in use; and hence, to-day, as we wander along the shellheaps and coastwise haunts of the Indians, we find fewer stone implements and a smaller variety of them than occur in the fields that border on our inland streams.

While the term "scraper" suggests at once, that something was scraped, it does not imply any particular object, although the fact that the skins of mammals needed careful working, and scraping particularly, to be available for clothing, is so well known, that skinscraping is the supposition on the part of every one who meets with the term "scraper," so far as it describes a class of stone implements. Many of these scrapers, however, are of such small size, that it is doubtful if they could have been used advantageously for any such purpose; yet their abundance and the care with which they are finished show conclusively that they played an important part in the production of some object in constant use. It is more probable that they were used in making bone beads and similar trinkets, than in any other way.

Fig. 106 represents a jasper scraper, of the simplest form and of the maximum size. Of a series of over five hundred examples of this pattern of stone implement, none are larger, and but three are more than half its size. If classified according to size, this would be taken as a typical specimen; but it cannot be so considered among the series of scrapers as represented in the area of the northern Atlantic seaboard. Certainly, scrapers of this size are of very exceptional occurrence.

As is the case in ninety per cent. of the scrapers from New Jersey, and in all such as have been examined from New England localities, the working edge of this implement shows no indication of wear as might be expected, even if a substance no harder than deerskin was rubbed with it. The grit that would be certain to be upon such skins would

FIG. 106. — New Jersey.

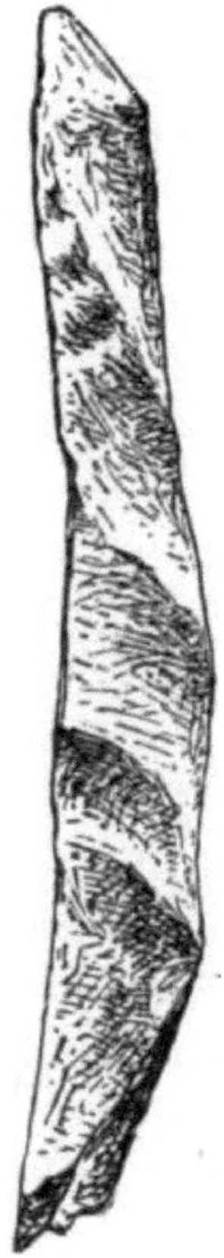

FIG. 106a. — Side view.

be sufficient to produce a few striæ ; and yet on this, as upon hundreds of others, there is no indication that the specimen was ever in use. An implement as large as fig. 106 was probably used without a handle of any kind ; at least it needs none, if used in the manner suggested. This pattern, but of smaller size, when used by the Eskimo, was mounted in a small wooden or bone handle.

Fig. 107 represents a flake of green jasper, that has been utilized as a scraper, by giving to it a distinctly bevelled edge. The under side presents but a single plane, and but little alteration was needed to convert the flake into the excellent scraper it now is. Flakes, thus used, constitute about one-fifth of the whole number found. This, at least, is true of these implements as found in New Jersey, where of a series of five hundred and five, ninety-seven were flakes similar to the preceding, and the one here figured. There is such an abundance of available flakes, like fig. 107, among the chips made by the ancient arrowmakers, that it remains a wonder why such elaborately wrought scrapers should ever have been made; and yet, in fact, they outnumber the no less desirable flake-scrapers, five to one. Flakes converted into scrapers, like fig. 107, are not of such irregular and indefinite shapes as the term "flake" implies. Fully three-fourths of them are symmetrically triangular, or vary therefrom towards a four-sided implement. The majority are not as irregular even as fig. 107 in outline.

FIG. 107.—New Jersey.

Occasionally, flake-scrapers like the above, are found with a distinctly knife-like, or cutting edge on the end opposite that which is bevelled, showing that a combination of the two forms of knife and scraper was desired. It may not be mere fancy to suppose that such implements were used in scaling and cleaning fish; especially as some of the larger forms have been found in shellheaps with a large quantity of fish bones. It may be well here to remark that fish constituted a very important factor in the food supplies of the Indians of the Atlantic coast, and our principal river valleys; and it can scarcely be questioned that a large proportion of the stone implements now found along these river valleys, and in the shellheaps of the seacoast, were designed for the capture and subsequent conversion into food of the hundreds of edible fishes found in those waters.[29]

[29] In a notice of various shellheaps along the Massachusetts coast, Professor Putnam refers to the great quantities of the bones of the Monkfish, or Wolf-fish (*Anarrhicas lupus*), now regarded as unfit for food. More than two-thirds of the fish bones of some heaps were those of this species. Bulletin, Essex Institute, Salem, Mass., Vol. iv, p. 123.

Fig. 108 represents a large slate scraper of a pattern quite common in Europe, but less frequently met with in eastern North America. The bevelled edge is not as distinctly chipped as in jasper specimens; or, if it were originally well wrought, it has by use become very jagged and rough. The material, slate, however, does not flake readily, and the objects made of it are usually quite rough, and compare very unfavorably with the jasper specimens of similar patterns. Where slate

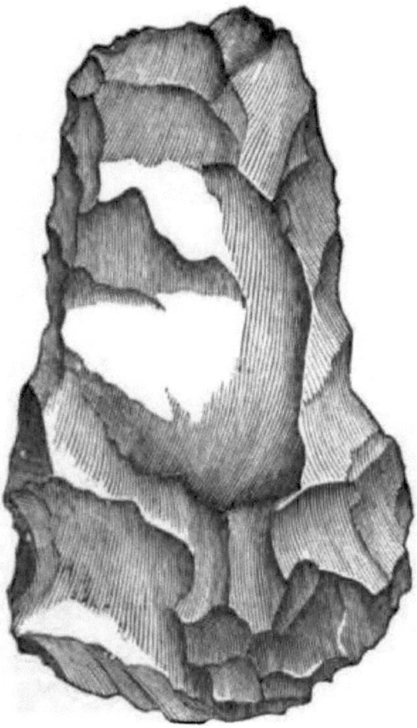

Fig. 108.—New Jersey. $\frac{1}{1}$.

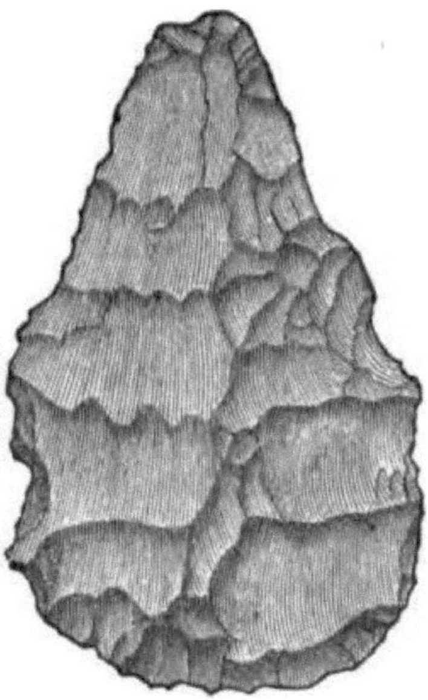

Fig. 109.—New Jersey. $\frac{1}{1}$.

is in place, it was used to a large extent, even when other and more desirable stone could be readily obtained.

Fig. 109 represents a symmetrical specimen of a scraper. The material of which it is made is jasper, and the care shown in the chipping is very marked. The front or working edge is beautifully bevelled, and shows (as is so rarely the case) a faint trace of polish and striæ, indicative of the presence of grit, or of its use in some other capacity than scraping skins. The handle, or narrower portion of the implement is evenly chipped and brought to a sharp edge, both along the sides and at the end. It is probable that this trimming of the handle was for the ready insertion of the implement into a bone socket, rather than that the delicately chipped margins were intended for cutting. Of the various patterns of scrapers that are found in New Jersey, none are more suggestive of use in connection with a handle than such as this specimen. Fig. 109 is almost identical with the modern Eskimo scraper, figured in Prehistoric Times, p. 93, figs. 106–108, and varies but little from an ancient specimen from the south of France, figured on p. 92, of the same work. It would appear then, that these scrapers occur, as do arrowheads, over a large portion of the globe, and are of essentially the same shapes and sizes.

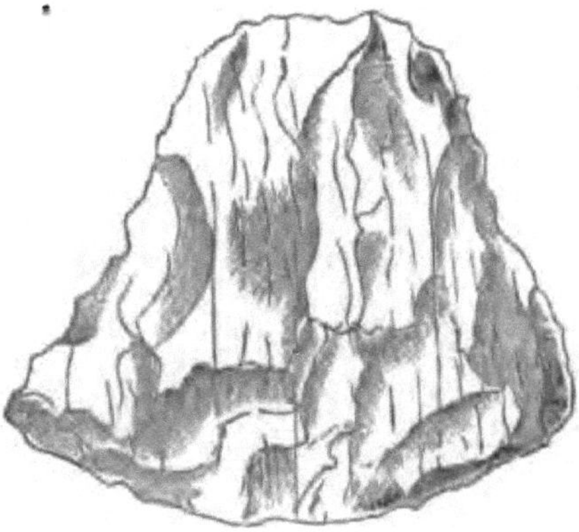

Fig. 110.—New Jersey. ½.

Figs. 110 and 111 represent two smaller scrapers that differ from such as have been described, in having an edge not bevelled from one side only, but from both, thus giving it not an ordinary straight chipped cutting edge, but an oval one. Even those that are most like the larger knives, do not seem to have had cutting edges, which have become dulled by use. This rounded rather than bevelled edge does not make of these implements a desirable form of scraper, as we understand their use; but it seems more rational to class them as such rather than to consider them as cutting implements, as has been done.

While implements with an edge similar to that of the "round-nosed turning chisel" may be considered as typical scrapers, those here described must be classed as a modification of the former, until it is shown that they were intended for some purpose of a wholly different character. In fig. 111, we have an example of this pattern with the edge slightly bevelled, or rather somewhat approaching the characteristic bevelled edge,

As a class, these specimens constitute but a small percentage of the whole number of typical scrapers, found in New Jersey.

In figs. 112 and 113 are represented examples of the smallest-sized scrapers, found in any locality along the northern Atlantic seaboard. These small scrapers are invariably made of quartz or jasper, and have all the symmetry and care in finish that characterize the specimens of larger size. In many of them, the working edge is even more carefully chipped, and when the specimen is a flake, or chipped only on one side, the characteristic features of the scraper are better shown on these smallest specimens, than on the majority of those of the medium size. In fig. 112 we have a specimen that has been chipped upon both sides, but the under side is almost as smooth and even as if it were a single plane. Fig. 113 is a flake, and has the under side perfectly smooth, but is slightly curved in the direction of the bevelling of the working edge.

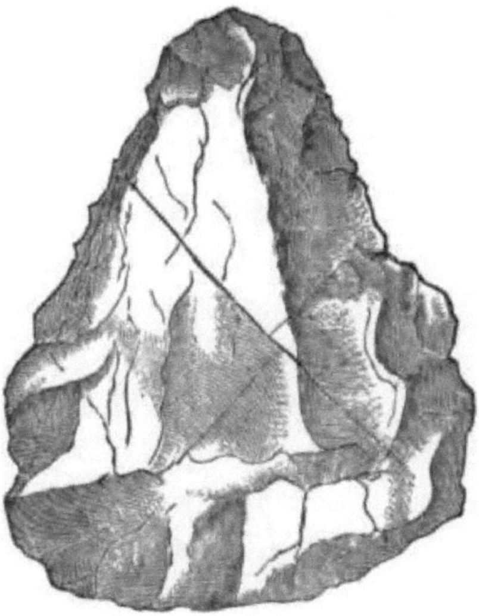

FIG. 111.—New Jersey. ½.

In New Jersey these small scrapers are comparatively abundant. Of a series of five hundred, eighty-six are less than an inch in length. and seventy-five per cent. of these are triangular flakes, chipped only upon one side.

So large a number of these diminutive objects have been found, that it is evident they were in common use for some purpose; and the fact that they are almost always found on known village sites would indicate that they were used in the manufacture of some other article, and not, like a knife or spearpoint, carried about the person.

Fig. 114 is supposed to be a jasper scraper, although the working edge is not bevelled, but has been produced by detaching a single flake along the entire width of the specimen. This gives the implement a curved edge which is exceedingly well adapted to scraping, whether it was ever so used or not. The end supposed to be inserted in a wooden or bone handle is pointed, and, were it not so short, would seem of better shape for secure hafting, than when broader. Scrapers of this pattern occur in Europe, and in some instances they have the pointed end more tapering and drill-like, than in the example here figured.

FIG. 112.—New Jersey. ½.

Fig. 115 represents a scraper made of slate, and is a modification of the triangular pattern. It is even more distinctly stemmed than in the preceding instance. The working edge is not distinctly bevelled, but it has been apparently, and is remarkable as being unusually curved. This scraper is greatly weathered, and in places is so smooth that all traces of the originally chipped surfaces have disappeared. This material, slate, was used for scrapers much less than argillite, and all the specimens of this mineral that have been gathered have been of the rudest character, both as to shape and finish. Possibly they were made for some emergency and then cast aside.

FIG. 113.—New Jersey. ½.

FIG. 114.—New Jersey. ½.

Fig. 116 represents a handsomely shaped and carefully chipped scraper, of brown jasper. It has some points in common with the preceding, but differs materially in others. The working edge is unusually narrow for the length, but the bevelled edges are continued along the sides, nearly to the commencement of the narrowed portion or handle of the implement.

Of the few examples of true scrapers from New England, none appear to be of this pattern, which is quite common in New Jersey. Even such as are found in the Connecticut valley seldom have a stem or handle, which portion is often of better finish than represented in the figure.

The ordinary scrapers found in New Jersey and northward do not

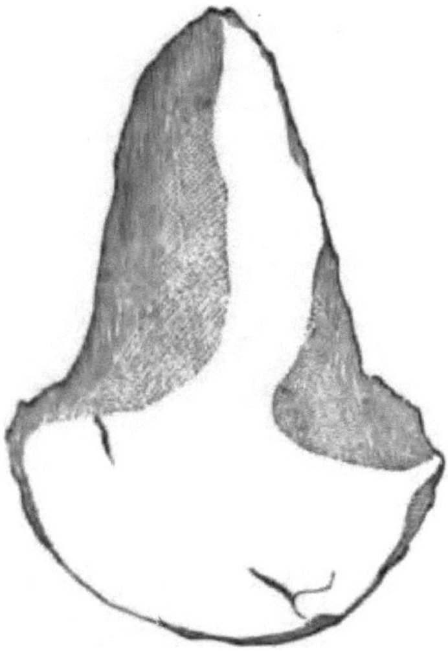

FIG. 115.—New Jersey. 1/1.

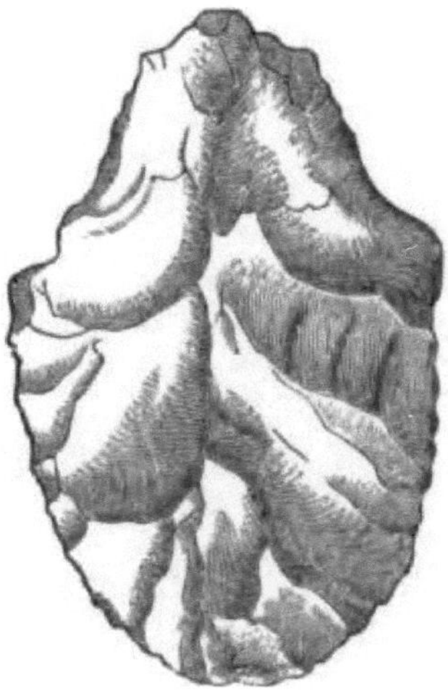

FIG. 116.—New Jersey. 1/1.

appear to be of common occurrence in the more southern states, although in Ohio and Indiana, they are even more abundant. In the series of stone implements described by Col. C. C. Jones, jr., in "Antiquities of the Southern Indians," there are no examples of stemmed, or oval scrapers, nor of the other more finished patterns that occur in New Jersey; except such as are supposed to be made from the bases of broken arrow and spearpoints. On the other hand there have been no examples found here of the large polished stone scrapers, with a celt-like edge, and a square stem-like handle. These so-called scrapers, which are sometimes perforated at the junction of the blade and handle, are peculiarly a southern form.

Fig. 117 represents a large stemmed scraper, of about the largest size. This specimen is made of slate, rudely chipped upon both sides,

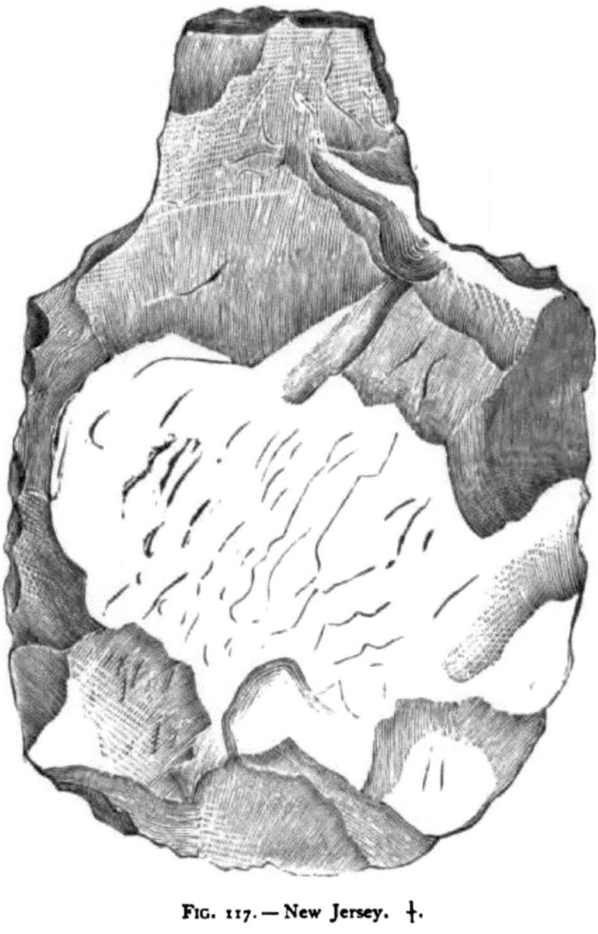

FIG. 117. — New Jersey. ½.

and nas not a distinctly bevelled edge. The surface is now much weathered and of a yellowish hue; but the slate of which the specimen is made is of a deep bluish-black color. There are some slight traces of wear upon the working edge, and there may have been many more, before the surface became so decomposed from long exposure.

When referring, in a preceding paragraph to the absence of any examples of polished stone scrapers, such as those described from Georgia, it was more with reference to their being a polished instead of a chipped implement; for in size and outline, fig. 117 is practically the same implement, and, if polished, would be indistinguishable from those found in Georgia. Not all of the sou hern polished scrapers, however, are of as hard a stone as diorite. There is a specimen from Arkansas, in the Museum at Cambridge, made of cannel coal, and another from Tennessee, made of a comparatively soft, slaty rock. Scrapers of this pattern of fig. 117 are very rarely met with in New Jersey, and are all the more interesting, since in shape and size they bear so great a resemblance to the polished specimens from the south. In the three following illustrations, figs. 118, 119, and 120, we have excellent

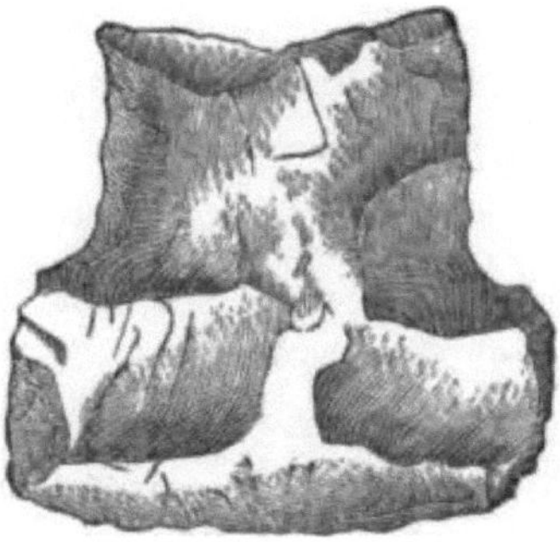

FIG. 118.—New Jersey. 1/1.

FIG. 119.—New Jersey. 1/1.

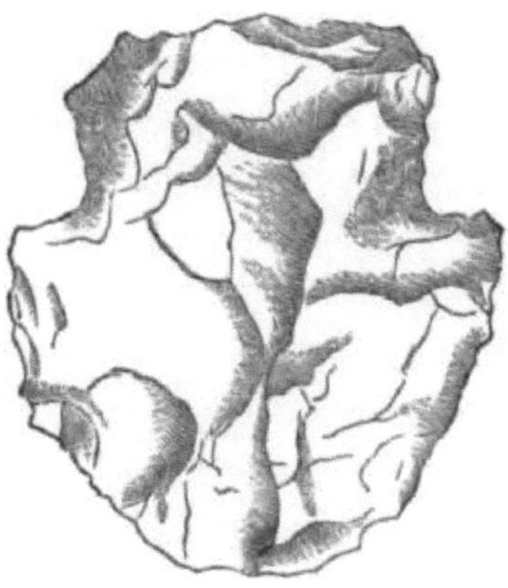

FIG. 120.—New Jersey. 1/1.

average examples of the ordinary jasper stemmed scrapers, such as occur in wonderful abundance, throughout much of the area of New Jersey, and less abundantly in New York and New England. These scrapers are quite uniform in size and vary but little in design. The great majority have a distinctly bevelled edge, which in some cases is of remarkable finish, from the small size and uniform direction of the facets. In some, the chipping of the edge has been from both sides. Whether such are really scrapers, or stemmed knives, is a matter of doubt. As we find, however, a well-marked form of stemmed knives which are always much thinner, and differ materially from these in other respects, it is probable that although not bevelled, the edges of such as are chipped upon both sides were used as scrapers, and not as cutting implements.

From an undoubted likeness to the bases of spearheads, it has been very generally supposed that scrapers of this pattern were usually, if not always, made by utilizing the bases of such spearpoints as happened to get broken. To some extent this may have been true, but that it was generally so is evidently a mistake. A careful examination of a large series of these stemmed scrapers shows, in very many, a gentle curve of the whole implement in the bevelling of the edge, which, if continued throughout the entire length of a spearpoint of a size proportionate to the supposed base, would make the implement too crooked to be of any value as a weapon. Again, we see in every large series, a gradation from the triangular or quadrangular flake to such as are distinctly stemmed; and more important than all, numerous specimens have been found in refuse heaps of flint chips, which have had the bevelled edge complete, but in consequence of some unseen flaw in the mineral, the stem had not been made, and the specimen in this unfinished condition had been rejected. It should be remembered too, that spearpoints would most likely be broken in hunting or in warfare, and in either case, the chances of finding the broken weapons would be quite small, as they did not then fall upon cultivated fields or stretches of grassy meadows, but in dense forest growths or tangled thickets, where they would remain unnoticed even by the keen-eyed natives.

That stemmed scrapers were as much an independently designed implement, as spears or arrowheads, is proved by the occurrence of such flakes as fig. 121 which is carefully chipped along the sides for the express purpose of producing a well-marked, stem-like projection. Flakes, thus worked, are not chance occurrences, but are quite frequently found, and so have a direct bearing upon the question.

FIG. 121. — New Jersey. ½.

Figs. 122 and 123 represent two examples of another form of stemmed scrapers which are much less common in New Jersey, and are very rarely seen in New England. The same objections to classing them as rechipped arrowpoints, that have been urged with reference to the preceding specimens, hold good in these cases. In fig. 122, we have, it is true, a specimen equally chipped upon both sides, and possessing no feature which may not have been equally common to an arrowhead, but in fig. 123 we have an instance of the same pattern of scraper chipped directly from a flake of quartzite, which shows that such scrapers were not always made from the broken fragments of other kinds of implements. It is very seldom that we meet with scrapers of this pattern as small as this, and with the bases so very deeply notched. The object of this is difficult to determine. Indeed, were these scrapers, in all cases, the utilized bases of spear and arrowpoints, they would be much more intelligible; for certainly as a simple instrument for scraping either skin or bone, those of plain triangular outline or with short straight stems have every advantage possessed by figs. 122 and 123.

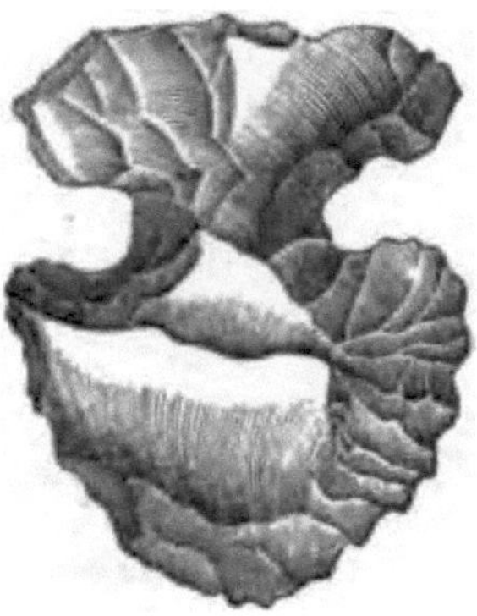

FIG. 122. — Indiana.

FIG. 123. — New Jersey.

Figs. 124 and 125 represent two examples of a form of supposed scraper, of which a number of specimens have been found in one limited locality. The bases are chipped so as to give them distinctly bevelled edges, and the pointed ends are so shaped as strongly to suggest the idea that these implements are combinations of the scraper and the drill. If so, they were probably used in the manufacture of small articles from steatite and other soft minerals, and not as scrapers of the common patterns were used. The pointed end of fig. 125 clearly shows evidences of wear, such as would be produced by constantly boring or drilling other stones; and there seems no reason why scrapers like these should not have been used to rub down, to a uniform surface, such slabs of slate and sandstone as were used for pendants, gorgets and ornamental objects of that character. Until we find the workshop-site and refuse of the worker in stone, who made ornaments instead of arrowheads, it will be impossible to say just what

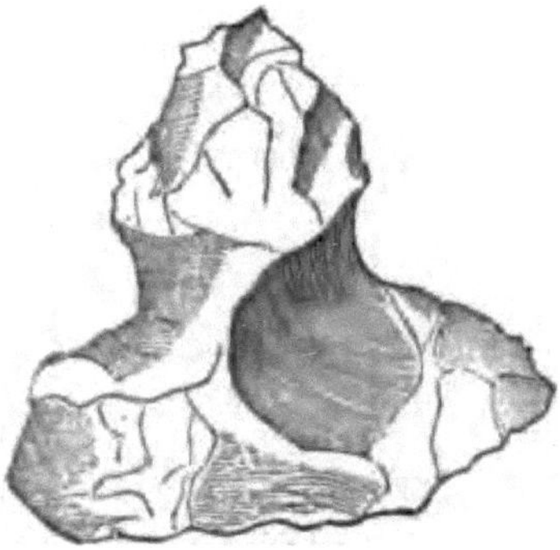

FIG. 124.—New Jersey.

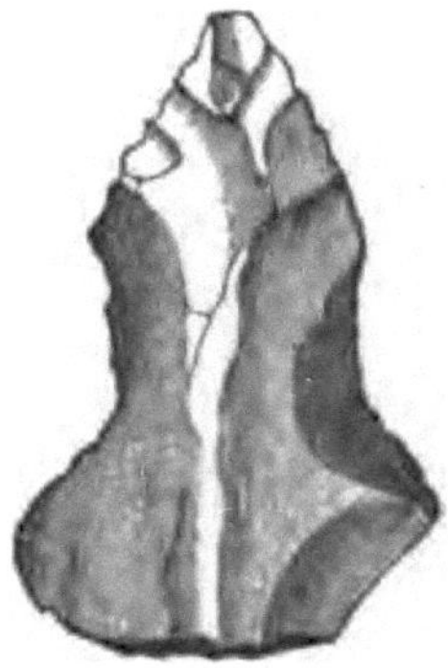

FIG. 125.—New Jersey. ½.

methods were pursued in making some of the commonest objects of this character; but that the combined drills and scrapers here figured were put to some such use seems far more probable, than that they were simply skin-dressers.

Fig. 126 represents a chipped implement that has much the appearance of a knife, and also closely resembles the chipped flint "sling-stones" that occur in Europe. In this instance, it is believed to be a scraper. Scrapers with battered edges, and of a quadrangular outline, very similar in general appearance to the modern "strike-a-lights" are rare. Fig. 126 represents such a "scraper." It is of yellow jasper, an inch and a half long, an inch wide, and half an inch thick near the middle of the specimen. The front edge is much battered and has every appearance of having been struck against a mineral as hard as pyrites.

Pyrites, in masses of various sizes, is very abundant about Trenton, N. J., where these short, thick scrapers are found. It occurs in the beds of Jurasso-cretaceous clay which crop out of the hillsides along the New Jersey shore of the Delaware, near Trenton, New Jersey, being there attached in large masses to the fossil trees embedded in these strata.

The conclusive evidence brought forward by Mr. Evans,[30] that many of these short, thick "scrapers" were used in connection with pyrites for producing fire, has rendered it probable that in this country also, such may have been the use of these thick, quadrangular masses of jasper, which, while resembling scrapers, have battered edges, and in all respects are just such forms as might be supposed to have been used for this purpose; nor do we lack historical evidence that the Indians of the Atlantic coast, to some extent at least, were accustomed to procure fire in this manner. John Brereton, in his Brief and True Relation of the Discovery of the North Part of Virginia (London, 1602) says of the Massachusetts Indians, "they strike fire in this manner; every one carrieth about him in a purse of tewed leather, a mineral stone (which I take to be their copper), and with a flat emery stone (wherewith glaciers cut glass, and cuttlers glaze blades), tied fast to the end of a little stick, gently he striketh upon the mineral stone, and within a stroke or two, a spark falleth upon a piece of touch wood (much like our sponge in England), and with the least spark he maketh a fire presently."

FIG. 126.—New Jersey. ½.

While it is possible that knowledge of this method of producing fire was derived from the Europeans, who, for more than a century

[30] Ancient Stone Implements of Great Britain, p. 280.

previous to the visit of Brereton, had occasionally visited the New England coast, it is not clear how the Indians learned to use pyrites, which Brereton evidently mistook for their copper, as a substitute for iron or steel. It is highly improbable that the knowledge of the fact, that iron pyrites would answer the same purpose as steel in producing fire, was likewise derived from European visitors to our coast; and, if not, the statement of Brereton gives us evidence of a custom, which subsequently fell into disuse, as it is not known that any tribes of Indians, either south or west of Massachusetts, ever procured fire in this manner, unless we accept the abundance of these short, thick scrapers as evidence of a custom, which is so clearly described as, at one time, common among the New England tribes.

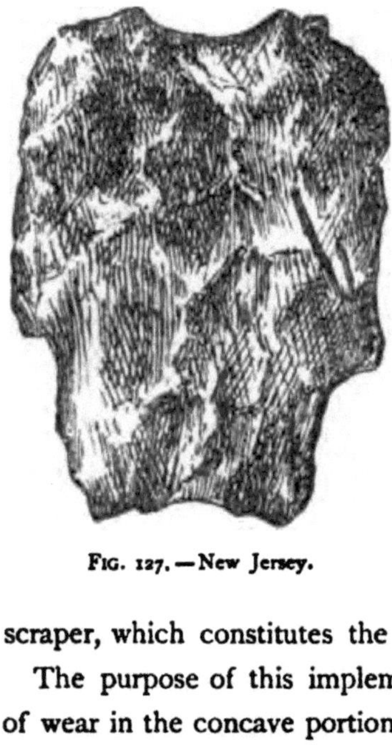

FIG. 127.—New Jersey.

Fig. 127 represents a form of what may properly be called a "scraper," although essentially different from the ordinary form of stone implement so called. The material is slate, but of so compact a nature, that it is susceptible of a considerable polish, which is shown in the margin of the concave surface of this scraper, which constitutes the peculiar feature of the specimen.

The purpose of this implement is indicated by the evident traces of wear in the concave portion of the front or scraping margin; this is, that of scraping the shafts of arrows, cylindrical bones, and objects of that shape.

Mr. John Evans has given the designation of "hollow scraper" to an implement of this character (Ancient Stone Implements of Great Britain, p. 287, fig. 226), and remarks of them "I have two specimens with the hollow as regular in its sweep as any of the scrapers of the ordinary form. Tools of this kind seem well adapted for scraping into regular shape the stems of arrows, or the shafts of spears, or for fashioning bone pins."

As yet, but a single specimen from New Jersey of these concave scrapers has been brought to the writer's notice, although, next to arrow-points and knives, ordinary scrapers are the most abundant of all the forms of chipped implements. From this, it is safe to conclude that this form was one not in general use.

In the Susquehanna valley scrapers are apparently less abundant than in the valley of the Delaware, but some interesting examples have been obtained. In the collection of the late Professor Haldeman are, among many forms, two jasper scrapers possessing the peculiar feature of fig. 127. One of them is a flake one inch in width and two and one-half inches in length. One side is nearly straight; the other, with a deep, accurately curved concavity. The bevelled edges are well worked, and the specimen clearly shows that it was intended for scraping convex surfaces.

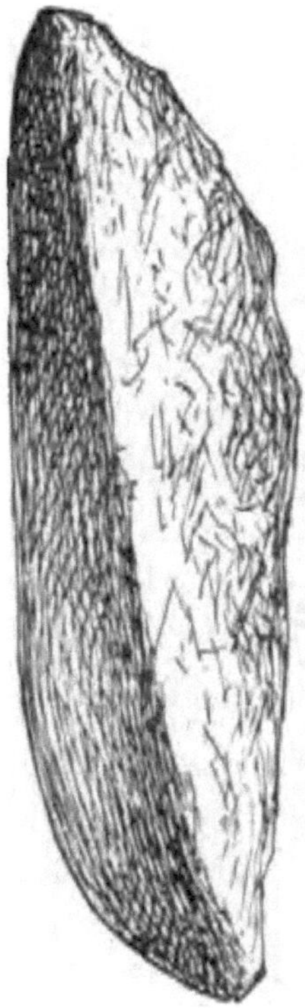

FIG. 128.—New Jersey.

Fig. 128 represents a simple form of scraper that is of exceeding interest. As the illustration shows, this implement is not chipped, but a simple flake-like fragment of an oval pebble.

Attention was first called to these implements, as found in New Jersey, by the publication by Prof. Jos. Leidy, of a notice of various rude flint implements, found near Fort Bridger, southern Wyoming. Of these, Dr. Leidy remarks,[31] "In this relation I may take the opportunity of speaking of a stone implement of the Shoshone Indians, one of so simple a character that had I not observed

[31] Geol. Survey: Montana, etc. Hayden's Annual Report, 1872, p. 653, pl. 12, fig. 13.

it in actual use and noticed it among the materials of the buttes, I should have viewed it as an accidental spawl. It consists of a thin segment of a quartzite bowlder, made by striking the stone with a smart blow. The implement is circular or oval, with a sharp edge, convex on one side and flat on the other. It is called a 'teshoa' and is employed as a scraper in dressing buffalo skins. By accident, I learned that the implement is not only modern, as I obtained one of the same character, together with some perforated tusks of the elk, from an old Indian grave, which had been made on the upper side of a butte, and had become exposed by the gradual wearing away of the latter." It is not improbable that these primitive scrapers have a wide distribution along the northern Atlantic seaboard, but that thus far they have been generally overlooked. Attention being called to these western specimens, search was made for the same form in central New Jersey, and on several village sites numbers were found, some of them showing more traces of wear, from long continued use, than is shown on any of the flint scrapers previously figured and described. These "teshoas" from New Jersey differ from the same implement as described by Dr. Leidy in that the detached surface is not generally flat, but is as convex as the corresponding, or natural surface of the pebble. From the smoothness and general appearance of this comparatively newer surface, it is evident that the implement has been detached by a single blow from the pebble, and that this convex surface was then produced and was not the result of subsequent secondary chipping. As yet no examples of pebbles, from which these implements have been taken, have been collected. One such would, by its concavity, at once determine whether these bi-convex "teshoas" were made by a single dexterous blow, or whether subsequent work was necessary to give them the outline they now have.

CHAPTER IX.

SLICKSTONES AND SINEW DRESSERS.

When we remember how important to the Indians were the well dressed skins of the elk, bear, deer, beaver, otter and muskrat, all of which mammals were once abundant throughout the northern and middle states, it is not strange that among the many curious forms of stone implements, that we now gather from the long deserted haunts of the country's primitive occupants, there should be numbers of such as were used in dressing and preserving the skins of these animals. As we have already seen, one kind of flint implement, carefully chipped, was used at one stage of the process of curing leather. The polished stone implements described in the present chapter are supposed to have been used subsequently to the scrapers, and also, after the cured skins, or leather, had been "made up" into clothing, while others, curiously grooved, were doubtlessly of use in reducing sinews to a more compact condition in which they served as thread.

We are informed by Holm, that the Delaware Indians could "tan and prepare the skins of animals, which they afterwards paint in their own way." This knowledge of curing leather was, of course, common to all our coast tribes, as it is to-day to the Indians of the far west, and hence it is no mere fanciful speculation to treat certain highly polished but otherwise unworked pebbles as the "slickstones" used by the savages in rubbing their leather to make it soft and pliable.

Burnishers or slickstones, as they are generally called, are of common occurrence in England, and their purpose has been carefully explained, and their history in later times given by Mr. Evans.[32] He remarks

[32] Evans. Ancient Stone Implements of Great Britain, p. 394. London, 1872.

that one "purpose to which stone implements seem to have been applied, in connection with * * * the preparation of leather, is that of burnishing or smoothing, somewhat in the same manner as is now effected by the flat-iron;" and also records the interesting fact, that "Mr. Greenwell has a *celt* from Yorkshire, which was used by a shoemaker for smoothing down the seams he made in leather," and adds "the old English name for the smooth stones used for such purposes is slickstone." Following the example of so safe a guide, these same stones, which are of all sizes and a great variety of patterns, but of comparatively uniform material, are treated as slickstones.

The form of this stone with the series of deep, narrow, parallel grooves, which is in many cases, an apparent combination of the slickstone and sinew-dresser, does not appear to occur in Europe, but it is of very general distribution on the North American continent.

Practically the same implement as those here found in great abundance, Professor Nilsson[33] has figured and described as a "stretching implement." He says of the illustration which he gives, "The widened part, representing the edge, has been rounded off by *constant wear*, probably from being rubbed against leather or something of that kind. A person, who has lived many years as a mechanic in Greenland, thinks he has discovered a great resemblance between this stone implement and the bone implement, provided with a handle, which is there used for stretching skins in order to give them the requisite softness. A somewhat similar stretching implement of iron is still used in those parts of Scania where the winter dress of the peasantry consists of sheep-skin coats."

Fig. 129 represents an example of polished pebble, that has been altered little, if any, in shape. A noticeable feature is in its being perforated by five small holes, which are natural, however, being thread-like veins of softer mineral which have been drilled out. One of these perforations occurring near the margin of the stone, the stone itself has been worn off at that point until much thinner than elsewhere,

---

[33] Nilsson. Stone Age in Scandinavia, p. 77, and pl. ix, fig. 185. London, 1868.

and the hole then enlarged. A cord was probably passed through this hole to suspend the implement.

Fig. 130 represents a second example of a pebble, which is not only highly polished, but the curved margin has been worn away, until it

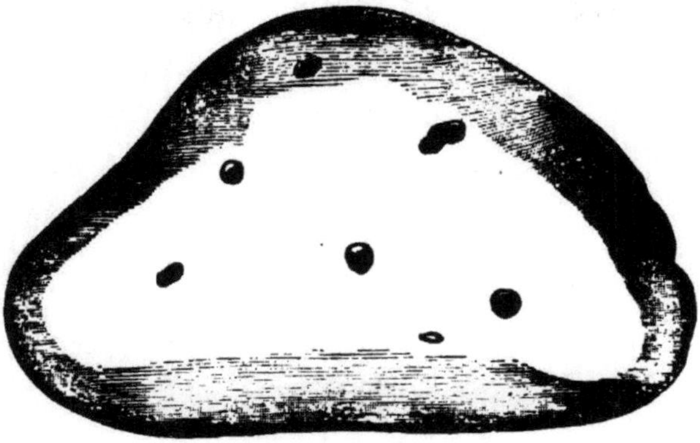

FIG. 129.—New Jersey. 1/1.

closely approaches an ordinary small celt. This curved margin has been worn more upon one side than the other, and were it not for the faint striæ that dull the surface slightly, it would possess an extraordinary polish. Knowing the purpose of these polished pebbles, one can readily imagine that the exceeding smoothness is still due to their greasy condition when in constant use.

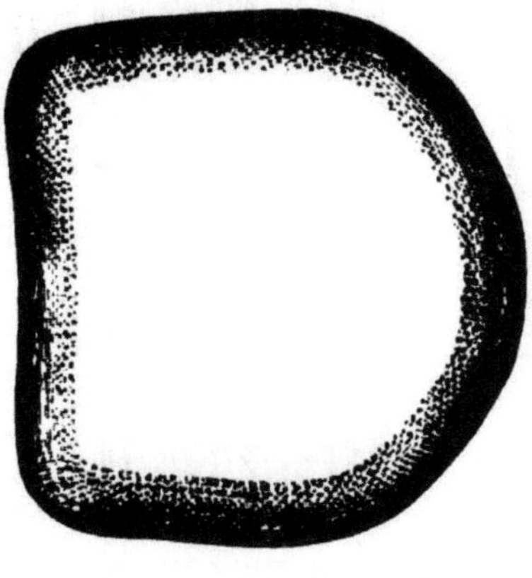

FIG. 130.—New Jersey. 1/1.

While the great majority of slickstones are natural pebbles, such as the preceding specimens, others are wholly artificial in shape, and possess in most cases, a polish only equalled by the finer celts. Fig. 131 represents a polished porphyry pebble, perfectly symmetrical, and very tastefully designed. This specimen may be considered as an example of the highest type of these implements.

Slickstones of this pattern are of frequent occurrence, although, greatly outnumbered by the natural pebbles that have been used for the same purpose. Of a series of eleven, all but one are of this material, and all but two, well polished. One specimen was drilled in one corner, as are many of the similarly shaped "sinew-dressers." None of them appeared to have been worn upon any one surface more than another, and all showed traces of scratches, as fine as hair lines, which were possibly due to the grit that had doubtlessly adhered to the leather while in process of manufacture.

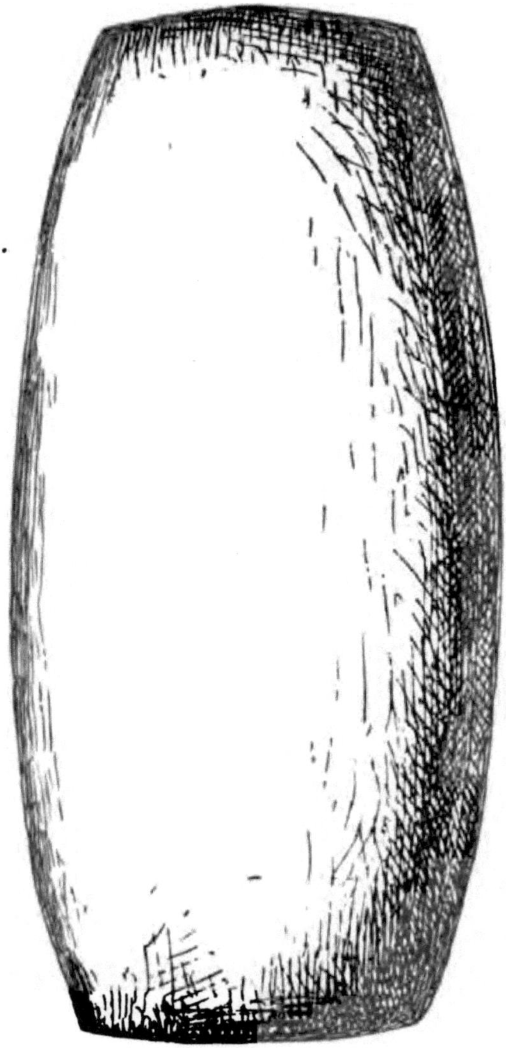

FIG. 131.—New Jersey. $\frac{1}{1}$.

Slickstones of this elaborate pattern are found over much of the area of the middle states. As they are of so uniform a size, when of this pattern, it has been questioned whether they had not some other use than that of rubbing-stones for dressing leather. The labor of grinding to their present shape, and subsequently polishing the objects, was very considerable, and yet, when finished, they possessed no advantage over the smaller, irregular pebbles used for the same purpose. This objection, however, applies to many other forms of stone implements, which also exhibit, what to us seems evidence of a vast deal of unnecessary labor. The remarkable uniformity in size and in material of these implements, on the other hand, deserves attention as a feature of great interest. All the examples that I have seen from New Jersey and New York, and those in the cabinet of the late Professor Haldeman, from Pennsylvania, were remarkably alike in size, and all made either of porphyry or hornstone. Does this uniformity of size, finish and material indicate some unknown use, not connected with skin-dressing?

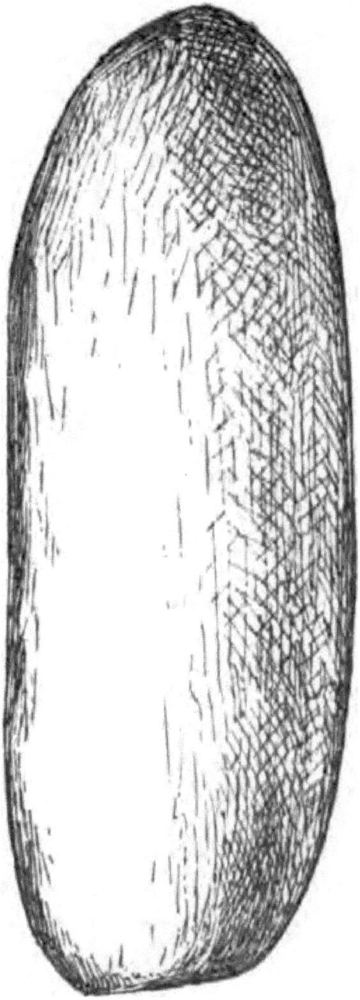

FIG. 132.—New Jersey. ½.

Fig. 132 represents a pestle-shaped pattern of these slickstones, made of a black hornstone pebble. The entire surface is very highly polished, and the lower end, as shown in the illustration, has been worn away until perfectly level. Of the various forms of natural pebbles chosen for slickstones, few have been found that are cylindrical, although, when of this size, they seem most admirably adapted for rubbing seams, and otherwise working leather in the course of its manufacture into clothing. Of a series of sixty-eight slickstones found on an extensive village site in Mercer Co., New Jersey, there are but three that are strictly cylindrical, and but eleven that are materially longer than broad. A few are less than two inches in length; but the majority are about five inches in length, by three to four in width, with one side more worn and polished than the other. Black stones are almost always chosen. In the entire series, but two are of white quartz, though this mineral is susceptible of being worn very smooth, and occurs as water-worn pebbles of proper size and shape.

Fig. 133 represents a white marble slickstone or smoothing implement of altogether different pattern from any previously described, but one that, while rare in New Jersey and New England, is of common occurrence in the south and west.

This specimen is made of a pure white marble found near Attleboro, Bucks Co., Pennsylvania. The entire surface is worked, and is quite smooth, though only the flat, under surface has any degree of polish.

FIG. 133. — New Jersey. ½.

As will be noticed in the illustration, this implement has a groove extending across the back and sides. Other examples have the groove extending lengthwise. The purpose of these grooves, considering the implement to be a slickstone, is not clear.

Col. C. C. Jones, jr.,[34] in his elaborate volume on the stone implements and other relics of the southern Indians, figures a specimen of this pattern of slickstone, but without the groove. He remarks, that the specimen he has figured "typifies a large class, examples of which abound in the relic-beds on the Savannah river. Their use is not well ascertained, but their flat surfaces are very smooth as though they had been constantly employed in rubbing."

---

[34] Jones. Antiquities of the Southern Indians, p. 292, pl. xvi, fig. 9. New York, 1873.

Associated as relics, that mark the former sites of Indian towns, and similar as they are in many particulars, it seems proper to place in the same chapter, although separately considered, those interesting stone implements that are identical with the preceding in every respect, save that of having a series of short and narrow, but deep marginal grooves, extending obliquely from a short distance within the margin to the edge of the implement. These grooves are always on both sides of the implement, and usually slant from right to left.

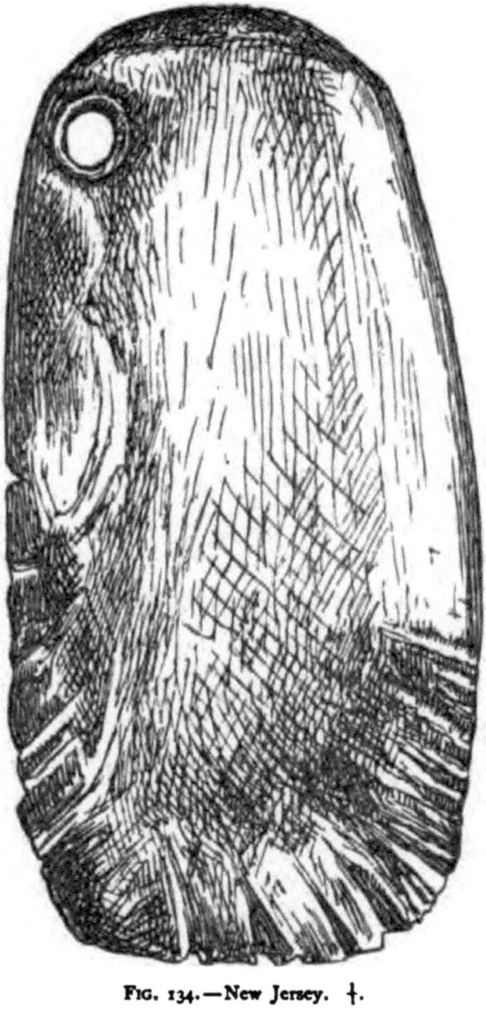

FIG. 134.—New Jersey. ½.

Fig. 134 represents a very well marked example of this form of implement, and one, too, that possesses an additional feature of interest, in the hole drilled in the upper left hand corner. The number of the deep, oblique grooves, characteristic of these implements, is unusually large on this specimen, generally the series at the side being wanting. While the perforation of one corner of these implements is quite common, it is seldom that the hole drilled is as large as in fig. 134, or as far from the margin of the stone. In many specimens the perforation is but one thirty-second of an inch in diameter, and so near the edge of the stone, that the rim of stone, that makes the perforation complete, is but little broader than the diameter of the opening.

When the implement was drilled, of course this narrow rim of stone must have been somewhat thicker, and it has subsequently been worn away by the friction of the cord passing through it; but, at the time, the perforation must have been so near the margin, that the successful accomplishment of the drilling is a source of wonder. Of the series of drills of all patterns that have been gathered from the Atlantic coast states, not one has yet been found that could have been used for the perforation of these implements in the manner described.

Fig. 135 represents a plainer example of these objects, but one that is of more common occurrence. This and the preceding one are of about the maximum size of this implement as found in the New England or middle states.

These implements have been long supposed to have been used in dressing sinews, by pulling them to and fro along these grooves. Whether this be true or not cannot be shown perhaps, but it seems far more probable than that they were used in any way as sharpening tools, for either the edges of celts or the points of bone needles.

In a recent monograph by Maj. W. H. Dall,[35] that author describes a "rough figure of some four footed animal, with a forked tail, perhaps used for smoothing down the asperities of their sinew thread by drawing it through the sharp furrow in the tail. The natives on the mainland use a similar instrument for this purpose."

It is known that sinews were used as bowstrings, and these might have been drawn over such deep furrows, as suggested by Mr. Dall; and possibly these same stones were utilized by the Delaware Indians in dressing thread made of other material. Holm records of them, that "the women spin thread and yarn out of nettles, hemp and some plants unknown to us."

---

[35] Dall. On the Remains of Later Pre-Historic Man in Alaska. Smithson. Contributions to Knowledge, No. 318. Washington, D. C., 1878.

Kalm (Travels in North America, vol. ii, p. 131) mentions the use of a common plant for thread-making, as follows: "APOCYNUM *Cannabinum* was by the Swedes called *Hemp of the Indians;* and grew plentifully in old corn-grounds, in woods, on hills, and in high glades. The Swedes have given it the name of *Indian hemp,* because the Indians formerly, and even now, (1749) apply it to the same purposes as the Europeans do hemp; for the stalls may be divided into filaments, and is easily prepared. When the Indians were settled among the Swedes, in Pennsylvania and New Jersey, they made ropes of this *Apocynum,* which the Swedes bought, and employed them as bridles, and for nets. These ropes were stronger, and kept longer in water, than such as were made of common hemp. * * * The Indians likewise make several other stuffs of their hemp. On my journey through the country of the *Iroquese,* I saw the women employed in manufacturing this hemp. They made use neither of spinning-wheels or dis-

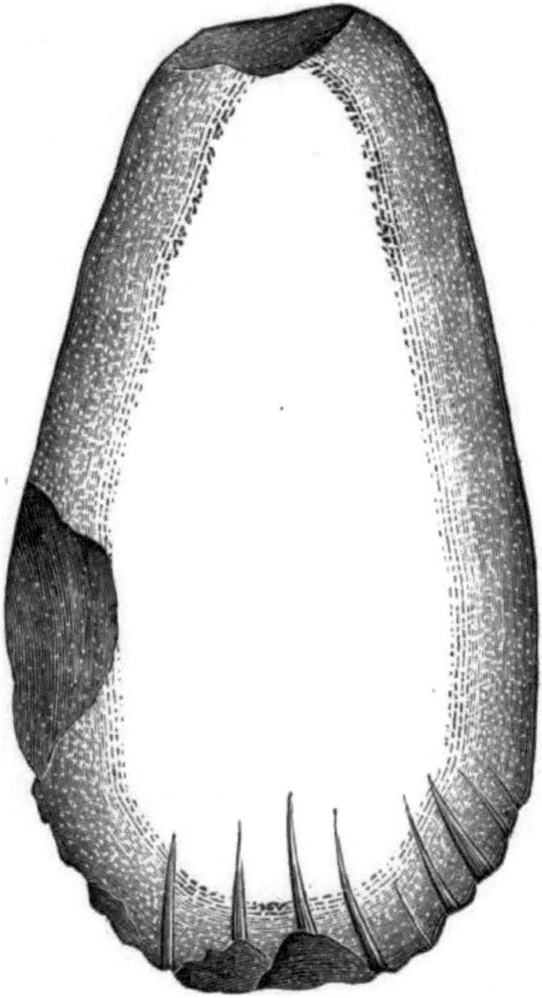

FIG. 135.—New Jersey. ½.

taffs, but rolled the filaments upon their bare thighs, and made thread and strings of them, which they dyed red, yellow, black, etc., and afterwards worked them into stuffs with a great deal of ingenuity. * * * Sometimes the fishing tackle of the Indians consists entirely of this hemp."

If such fibrous yarn was "dressed" with grease or, indeed, with wax, and then drawn over a groove in one of these stones, it would certainly be much strengthened.

Considering then, the facts; that sinews were utilized as bowstrings, that fishing with a line was a constant occupation, and that nets were woven, and we have evidence that "thread and yarn" were constantly in use, and there is much to lead to the inference that these grooved implements, if not simple "sinew dressers," were largely used for purposes so similar as to render the name not wholly inappropriate.

CHAPTER X.

MORTARS AND PESTLES.

One of the few objects that, by its mere presence, perhaps more vividly than all else, recalls the fact that these goodly lands were once tenanted by another and far different people, is the stone mortar, fig. 136, which, in its history, is so closely linked to that of the invaluable article of food, the maize or Indian corn. Whether the mortar is simply a slab of sandstone, with scarcely a depression upon it, a granitic bowlder with a deep cavity, or even a natural hollow in a stationary rock, its purpose, to the Indian, was the same. Within these hollows the corn was pounded into meal.

Holm,[36] speaking of the Delaware Indians, says that "that they make bread out of the maize or Indian corn, which they prepare in a manner peculiar to themselves; they crush the grain between two stones, or on a large piece of wood." Loskiel[37] states that they grind the maize "as fine as flour by means of a *wooden* pestle and mortar;" but it is not probable that many mortars of wood were used, considering the great number of those of stone that we find, especially in New Jersey.

Of the Delaware Indians the same author says: "They are fond of muscles and oysters, and those who live near an oyster-bed will subsist for weeks together upon them. They also eat the land-tortoise, which is about a span broad, and rather more in length; and even locusts are used for food. These come frequently in large swarms, covering and destroying even the bark of the trees."

As the oysters and mussels were also dried in large quantities for

---

[36] Holm. History of New Sweden, p. 121. Philadelphia, 1834.

[37] Mission to North American Indians, p. 67. London, 1794.

winter use, it is very probable that mortars were also used as a receptacle in which to reduce the dried shell-fish to a sort of powder or pulp. This, mixed with the corn-meal, made a dough, or batter, not unlike the modern "fritters." It was a common practice with the Indians to mix other articles of food with the meal made from maize. Loskiel mentions their mixing dried bilberries with corn-meal, and also smoked eels chopped fine.

"Hunter informs us that, in some of the Indian villages visited by him, there were one or two large stone mortars for pounding corn, which were public property. These were placed in a central part of the village, and were used in rotation by the different families."[38]

FIG. 136. — Massachusetts.

Mortars are of various forms. In the first place the Indians not unfrequently utilized for this purpose the natural hollows or depressions that are found in stationary rocks. Others of smaller size were made by pecking a deep cavity in a globular or cubical block of stone as in fig. 136, while in still other instances they simply made use of a flat slab of stone into which a shallow depression was worn, the result of long usage rather than of design.

The wooden mortars, having nearly all decayed and therefore known generally by the historical references to them, need only to be referred to as having once been used; but to what extent, in comparison with those of stone, is not known.

Fig. 137 represents an enormous glacial bowlder, portions of which still remain in view in Centre street, Trenton, N. J. The hollow or basin in this rock is of natural origin, being a "pot-hole," as such depressions are called by geologists. The pestle that is represented in

---

[38] Stevens. Flint-Chips, p. 546, quoting from Hunter's Manners and Customs of Indian Tribes, p. 269. London, 1870.

the illustration was found so near the supposed mortar, that it is quite probable it was used in connection with it; but the evidence that this "pot-hole" was used as a mortar rests upon the fact, that the rock occupies a position, once the very centre of a populous Indian village, and that in excavating cellars in the immediate vicinity of this rock, besides numbers of other relics, many broken and some short, unbroken pestles were found. It is probable, too, that the natural cavity has been somewhat deepened by use, but the rock being unusually hard and unyielding, it is difficult to determine to what extent, if any, this may have been done.

Bowlders with these natural basins, although too heavy to be moved

FIG. 137.—New Jersey.

by any one person, were occasionally transported to quite distant points, and there used in common by the people of the village, to whom they belonged. One such, to which reference is made in local history, formerly occupied a prominent position in what is supposed to have been the public square of a village, situated near Yardville, Mercer Co., New Jersey. This mortar, now in the Archæological Museum at Cambridge, Mass., is a large mass of Potsdam sandstone, weighing about one hundred pounds, with a deep, circular basin in it. It is known to have been used by the Indians, as recently as 1720.

Fig. 138 represents an average example of the small mortars. It is a flat, triangular piece of sandstone, somewhat less than nine inches along each side. The upper surface has been worn away until an oval depression has been formed about one inch deep at the centre. As very many of these mortars are even of less capacity than the speci-

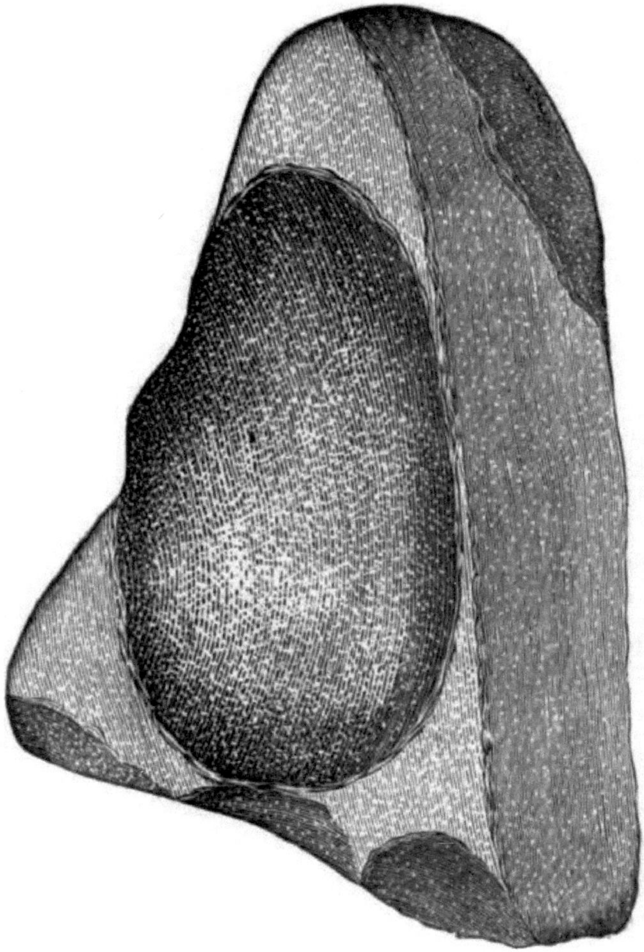

FIG. 138.—New Jersey. ½.

men here figured, it is evident that but little grain could be ground at a time, probably not more than was needed for immediate use. Why, as a rule, these mortars should be so much smaller than the Mexican metatès, which they closely resemble, is something of a mystery. The advancement of the Indians of the Atlantic coast, in the culinary arts, was as great as that of their more southern neighbors.

From the character of the wearing and scratches in the basin of fig. 138, it is evident that the "upper mill-stones" used with these shallow mortars, were the flat, oval pebbles, with one surface artificially worn, which are even more abundant than the mortars themselves. Fig. 139 represents one of these "upper millstones." It is an ordinary water worn pebble, with the under side worn very smooth by long continued friction against the sides and bottom of the basin of the mortar. Fig. 139 was found associated with the mortar, fig. 138, and as it fits the basin quite accurately, they are supposed to have been used together. This use necessarily consisted in a very limited rubbing motion, in the direction of the long axis of the diameter of the basin. With the supposed upper stone, fig. 139, there could not have been play enough to give the combined stones much grinding or crushing power.

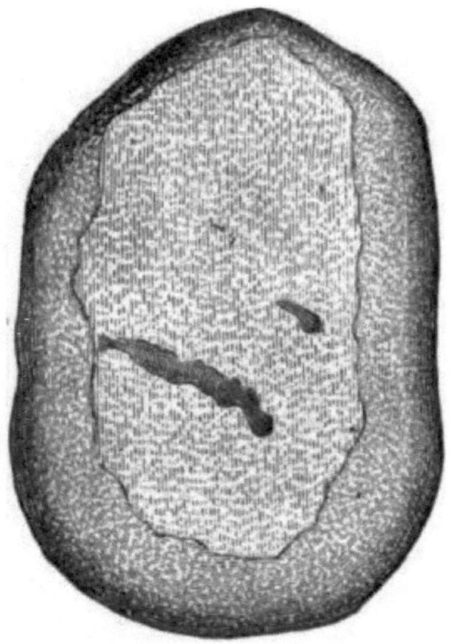

FIG. 139.—New Jersey. ½.

Fig. 140 represents a second example of these shallow mortars, which possesses many points of interest. This specimen measures fourteen inches in length by ten in width, and is of a uniform thickness of about three inches. The under surface is unchanged in any way from its natural condition, and the margins are but slightly smoothed or worn. The upper surface, on the contrary, is worn or intentionally ground down to a perfect level, and in the middle there is a shallow depression, perfectly circular, and marked by a series of well defined circular striæ. Mortars of this character are quite rare, and whether used in a different manner from the more common ones, such as fig. 138, is not known; but it is evident that the method of grinding on this specimen was very different from that employed in the preceding example. There it was evidently a rubbing motion, here a revolving one, and even the steady revolution of the end of a cylindrical pestle would not, unless guided with great accuracy, make such a regularly circular depression. The artificially smoothed surface about the basin in this specimen is a feature of much interest. Such mortars are not usually altered in any may. Among nearly a thousand of these objects found in the southern counties of New Jersey, but three were at all changed from their natural condition of subangular bowlders, except in the wearing away that had resulted in a basin-like depression.

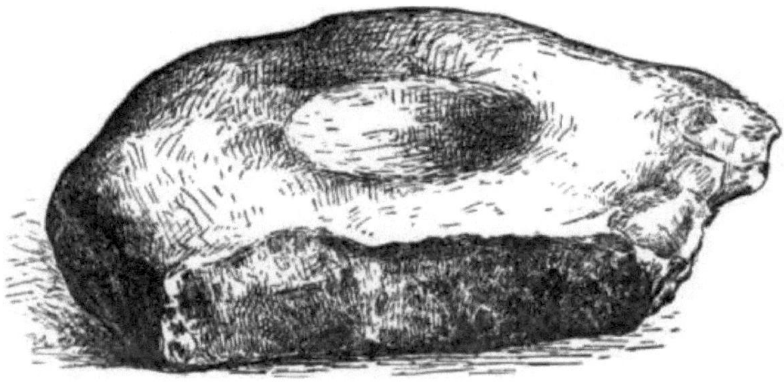

FIG. 140. — New Jersey. ½.

Shallow stone mortars, like the above, are common in the southern states. Col. C. C. Jones, jr.,[39] mentions, that "from a single relic-bed on the right bank of the Savannah river, a few miles above Augusta, I obtained, at one time, thirteen stone mortars made of flat bowlders taken from the bed of the stream and hollowed out on both sides to the depth of two or three inches. The average diameter of these shallow basin-like excavations was rather more than nine inches. No labor had been expended in shaping these stones. The natives took them as they found them, and simply formed the cavities. Placed upon the ground or held in the lap, with the assistance of the ordinary

[39] Jones. Antiquities of the Southern Indians, p. 312. New York, 1873.

dish-shaped crushing stones—large numbers of which were seen in the vicinity—the green corn could have been mashed, the parched corn pounded, or the husks beaten from the ripe grains."

Stone mortars, with a basin of such depth as required a long, cylindrical pestle, are of rare occurrence along the northern Atlantic seaboard. A single example has occasionally been found and referred to in local journals, but few appear to have been secured by the larger museums. Certainly, none of the character of the larger mortars, found in southern California, are ever met with east of the Mississippi river. In the southern states stone mortars of excellent workmanship have occasionally been found. Col. C. C. Jones, jr.,[40] describes one from Liberty Co., Georgia, as "made of yellow, ferruginous quartz with a flat bottom and circular walls gradually expanding as they rose; its general shape was that of an inverted, truncated cone. Entirely artificial, the exterior was well polished. About ten inches high, eight inches in diameter at the top and seven inches at the bottom, the interior had been excavated to the depth of nearly eight inches. At the top the walls were about three-quarters of an inch thick and increased in thickness as they descended." Mortars of this size are not known to occur along the northern Atlantic coast, though there have been found smaller examples of softer material, moderately accurate in outline and of careful finish, thus showing that the absence of the larger kinds was not due to any lack of capability on the part of the northern Indians.

Pestles, or the long cylindrical stone implements used in connection with the deeper mortars, particularly with the large wooden ones, are of very common occurrence throughout the entire area of the New England and middle states. Kalm (Travels in North America) speaking of the Delaware Indians, says that "they had *stone pestles*, about a foot long, and as thick as a man's arm. They consist chiefly of a black sort of a stone, and were formerly employed, by the Indians, for pounding maize, which has, since time immemorial, been their

---

[40] Jones, *l. c.*, p. 312.

chief and almost their only corn. * * * * They formerly pounded all their corn or maize in hollow trees, with the above mentioned pestles, made of stone. Many Indians had only wooden pestles." As a class of implements, while very distinct from all others, they vary much within themselves. Some are as carefully finished and as highly polished as the celts, while others are merely water-worn pebbles and can only be recognized as relics of the Indians by the battered ends, or other scarcely discernible marks of use. The majority of pestles were not made, but were chosen from the pebbles of the river. Such as were naturally cylindrical in shape were taken, and occasionally some slight modifications were made in the shape. The longest were, of course, made from long slabs of stone, and have their entire surfaces worked by pecking. This process is best exhibited in the grooved stone axes. What may be considered as the maximum length of pestles is difficult to determine, but those of a greater length than fifteen inches are not common.

Of a series of ninety-two pestles found in New Jersey, twenty-eight are plain cylinders, wholly shaped by pecking. Fifteen have a portion of their surface pecked, to bring them to a perfectly cylindrical shape. The others are natural pebbles with different amounts of alteration of the natural surfaces. The character of the ends of these pestles varies considerably. About one-third are polished, as if by use, while the others are rough and in some instances slightly battered. When we come to the smallest of these cylindrical pebbles, and especially such as have distinctly battered ends, it is necessary to exercise care that we do not confound pestles and possible hammers. Still, if used as pestles, these smaller examples would be used with shallow stone, and not deep wooden mortars, and so would be more or less liable to become broken or splintered at the ends.

Figs. 141 and 142 represent excellent average examples of the long, plain, cylindrical pestles, common everywhere. Fig. 141 measures seventeen and one-half inches in length and nearly eight inches in circumference. It is somewhat polished and bevelled at either end, which would indicate that it was used in the deep mortars for pounding, and not as a rolling pin upon a flat slab of stone; although this latter use appears to have been true of some, judging simply by the polish. The marks of the stone-hammer, in pecking the implement to its present shape, are plainly visible even to the very ends of the specimen. This pestle weighs seven pounds. Fig. 142 measures seventeen inches in length, and, in section, is flattened and quadrangular. Although smoother than the preceding, it shows the marks of the hammer-stone very plainly except at the ends, which are smoothed, but not polished, and nearly flat and square. Pestles of this pattern are not common. Schoolcraft[41] figures one that is similar.

Fig. 143 represents a third example of the ordinary plain pestles. It is made of sandstone, nine and one-half inches in length, and pecked over the entire surface except the extreme ends which are smoothed, but not polished. This pestle, which is a good representative example of the common short specimens, was found on a well known Indian village site near Trenton, New Jersey, where hundreds have been collected.

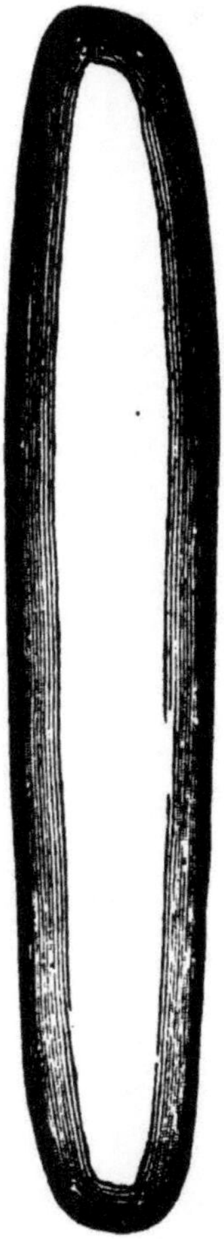

Fig. 141.—New Jersey. ¼.

Fig. 142.—New Jersey. ⅓.

The great abundance of pestles, in many localities where mortars are seldom found, has been frequently noted, and much surprise has been expressed that such should ever be the case. It is readily explained,

[41] Schoolcraft. History and Condition of Indian Tribes, pt. 1, p. 86, pl. 21, fig. 1.

I think, by supposing that there was a stationary mortar in some large rock near by, to which the women of the tribe resorted, and also, by the fact, that wooden mortars were more largely in use than has been

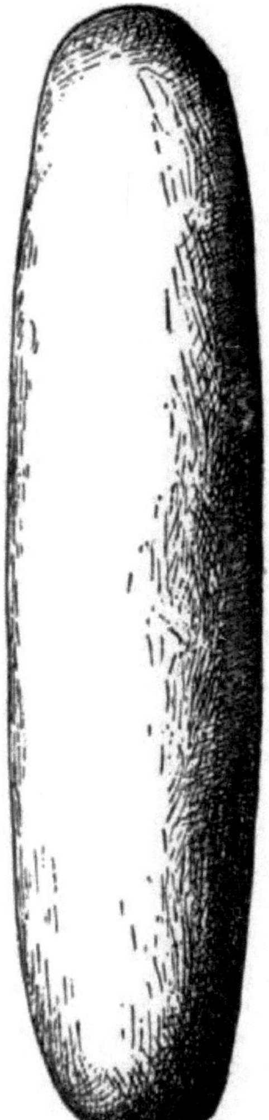

FIG. 143.—New Jersey. ½.

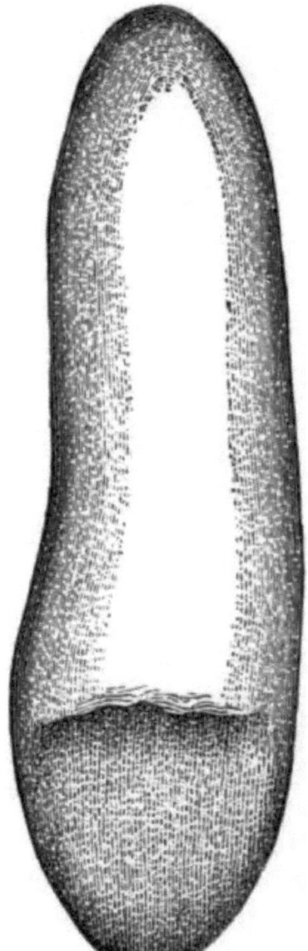

FIG. 144.—New Jersey.

supposed. These, of course, have generally long since decayed, but some two or three are known to have been found and are still preserved.

Fig. 144 represents a form of pestle, common in New Jersey. It is a cylindrical pebble that has apparently been split intentionally for three-fourths of its length. The unbroken end shows some traces of contact with stone, and is slightly battered. The split end is worn quite smooth. The evidence of intentional splitting of these pestle-pebbles is seen in the fact that so many have been found of this form. Either they are intentionally split, or naturally fractured pebbles were habitually chosen for pestles.

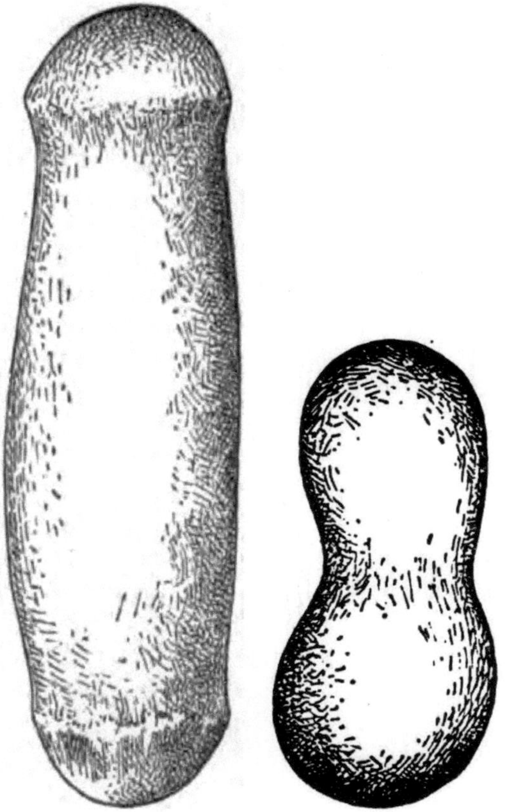

FIGS. 145 and 146—California. ½

Figs. 145 and 146 represent cylindrical pebbles of small size from California, which have been used as pestles. Fig. 145 has been slightly modified by pecking on both ends, and is somewhat smoother at the extremities than over the body of the implement. Fig. 146 shows but slight alteration of its surface by use, but the fact of its having been found in a grave, associated with a mortar of the smallest size, indicates that it has been used as a pestle. Small cylindrical pebbles like these are of common occurrence on the Atlantic seaboard.

Fig. 147 represents a magnificent example of a large pestle, remarkable, not only for its length, which is twenty-eight inches, but for the artistically carved serpent's (?) head at one end. This pestle is not cylindrical.

Its surface is formed by a series of flat planes of uniform width,

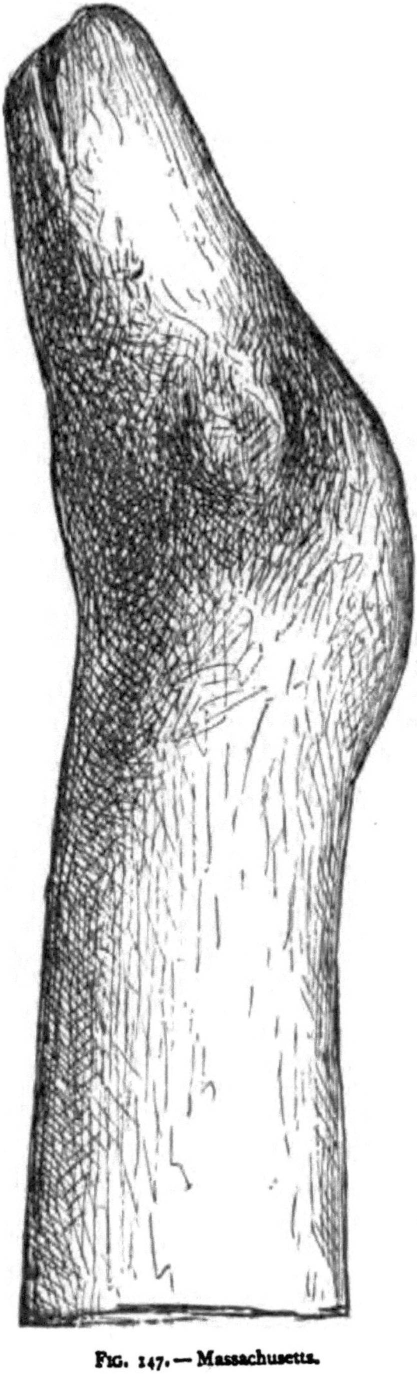

FIG. 147.—Massachusetts.

extending the entire length of the implement. Some of them are quite highly polished, while others are merely smooth. The circumference of the plain, or pounding end, is nearly five and a half inches, which is more than an inch larger than the circumference of the pestle immediately below the carved head. In the Archæological Museum, at Cambridge, Mass., are two others of these ornamented pestles, one of which is quite small, whilst the other is not so long, but of greater diameter than fig. 147. Both have more mammal-like heads carved at one end, and are noticeable from the fact that, in both, the ears of the head are prominently carved, while there is no trace of an eye.

Professor Perkins[42] describes an ornamented pestle over two feet in length, now in the museum of the University of Vermont, as "cylindrical, as usual, and is rounded at one end, while the other is carved to resemble the head of some animal—it may be a wolf."

[42] American Naturalist, vol. v, p. 12. 1871.

Mr. Putnam,[43] in a foot-note referring to the above, adds: "In the collection of the New York State museum, at Albany, there is a long 'pestle' of identical pattern and having the same rough carving as the one described from the Burlington museum. These are the only ones that have come under my observation having the handle carved to represent an animal, though most of the longer implements of this character have a knob at the handle, as if for the purpose of suspension."

A pestle with a carving of a serpent's head, found in Onondaga Co., New York, is of much interest, as the implement is not straight. The carved head is bent to a greater angle than in fig. 147, and the plain or pounding end is somewhat curved in the opposite direction.

This crooked specimen would not answer for the ordinary uses of a pestle, and suggests the idea that it may have been a club or batôn. That stone clubs were ever in use among the Iroquois or Algonkin tribes is, however, merely mentioned as possible.

The pestle, here described, is figured in a MS. on the Antiquities of Onondaga Co., New York, by Rev. W. M. Beauchamp of Baldwinsville, New York, in the library of the Archæological museum, at Cambridge, Mass.

A pestle, closely resembling that represented in figure 147, is described by the same writer, as "twenty-seven inches long, quite slender and uniformly cylindrical, its average diameter being about two inches. The diameter is nearly equal throughout; one end is somewhat contracted, and the opposite more so, to form a neck for the carved head which terminates it. This carving, though not elaborate, yet distinctly and strongly indicates a head, somewhat dog-like and somewhat fish-like, in some features resembling one animal, in others the other." Judging from the illustration that accompanies this description, it would seem more probable that a serpent's head was intended to be represented, as the head upon that here figured is certainly intended for that, and the two bear a marked resemblance. Professor Perkins also

---

[43] *L. c.*, vol. xiii, p. 739, fig. 1, of pl. 2. 1879.

refers to two stone pestles from Vermont, which have rude carvings on one end. Of one of these he remarks, "the carved end represents, rather rudely, but yet plainly, the head of a squirrel or some similar animal." Of the common forms of this implement, as found in Vermont (and the remarks apply equally well to all New England), Professor Perkins says they are of "all varieties. Some are fusiform, used evidently as rollers, since both ends are in some cases polished by the friction with the hands, others as obviously used for pounding, as the more or less rectangular form precludes the possibility of rolling them; one of these is three inches square and nearly a foot and a half long, flat on all sides; other specimens are flat on two sides and rounded on the other two; others are club-shaped, and so on."

Long pestles, with knobs on one end, judging from the series of these implements preserved in the several large museums, are of even rarer occurrence than those with animal heads; but very few are found even with a narrow groove near one end. This form, which is so characteristic of the long pestles found in California, is not abundant anywhere save on the Pacific coast. Of the hundreds of pestles from the Atlantic shores, none are so distinctly grooved and collared, as those from California. Figs. 148 and 149 represent specimens of these collared pestles, taken from graves at Dos Pueblos, southern California. It is somewhat strange when we consider the labor of carving an animal's head on one of these pestles, that so very few of them have even a groove about one end, that they might the more easily be suspended; though they appear very generally to have been so suspended. "The Pennacooks," according to Schoolcraft, "were accustomed to suspend a stone pestle from the limb of a tree, which acted as a spring and saved the squaw labour in lifting the pestle. These pestles were usually ornamented with a human head, or the head of a deer, or some other animal, sculptured at one end." While the conclusion, that they were *usually* sculptured, was an unwarranted inference on the part of Mr. Schoolcraft, the few carved pestles that are found show that such was sometimes the case. The absence of these carvings or of a groove or collar on certain specimens cannot, however, be considered as evidence that they were not also suspended. The distinction drawn between true pestles, so called, and the highly polished, finely finished but plain specimens, which have been called war-clubs, as has been already stated, is, in all probability, not warranted.

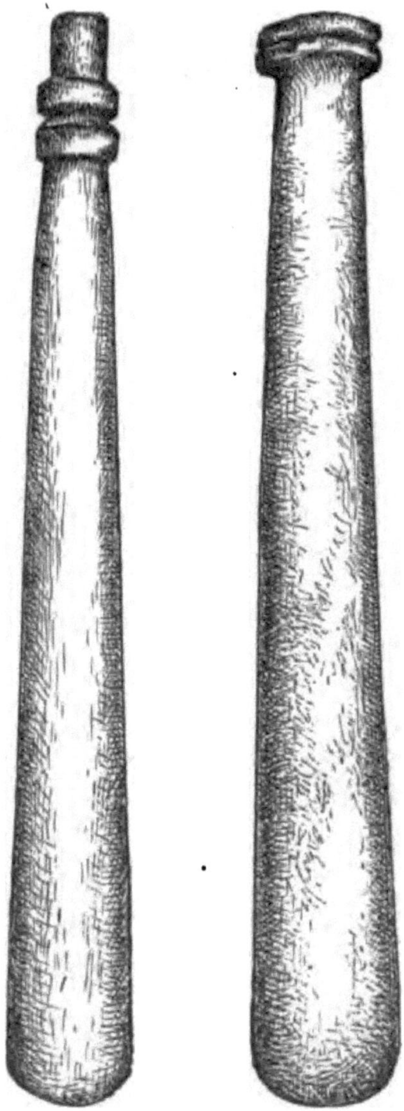

FIGS. 148 and 149. — California. ½.

Fig. 150 represents a small pestle, pounder or muller, such as is common in western and southern localities, but of rare occurrence in New Jersey or New England. This specimen has been carefully shaped by pecking, and on the ends, one of which is perfectly flat, are shallow cup-like depressions. These depressions are not uncommon in the short mullers found in Ohio, which are implements of known use, and therefore their presence, in this instance, is not an indication that fig. 150 is not a pestle, or muller. The purpose of the groove about the middle of this implement is difficult to determine. When such narrow grooves are near one end, as in the California specimens, it is evident that the pestles so marked, when in use, were suspended from the limb of a tree; but fig. 150 is too small to be used in this manner, and the groove being in the middle of the implement would prevent it from taking, if suspended, other than a horizontal position.

Fig. 151 represents an average specimen of the short flaring mullers, or crushers common in Ohio, but comparatively rare in New England and the middle states. In a series of about ninety pestles, there are but three of these mullers. While not pestles strictly speaking, yet their use is so similar as to warrant their being classed together.

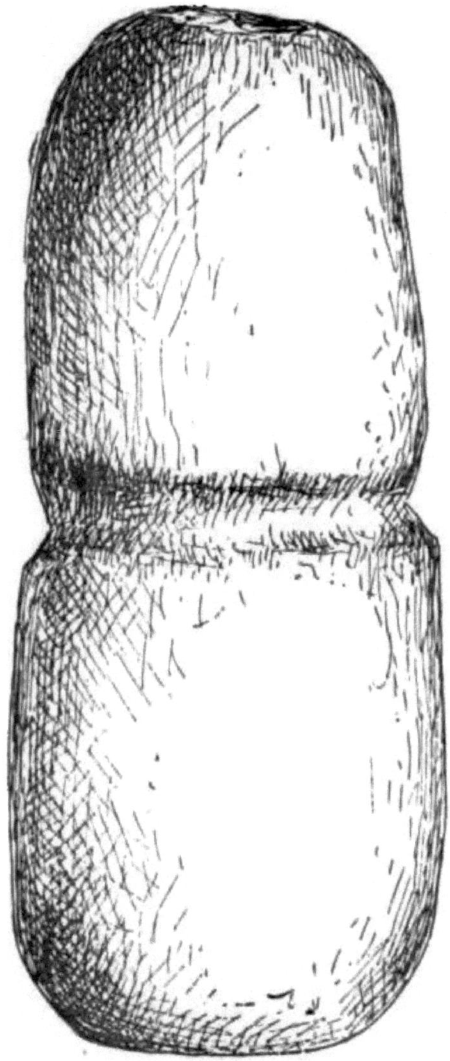

FIG. 150.—New Jersey. ½.

Fig. 151 is made of a very compact granitic rock, and is carefully worked over its entire surface. The crushing end, or base, is nearly level and more worn than battered, as though the implement had been used with a rubbing, rather than a pounding motion. The two other examples of typical mullers, to which reference has been made, are of much ruder finish, and the flaring base is less pronounced than in this instance.

Among the many shallow mortars that have been examined, none possessed any peculiarity which would make an upper stone, like fig. 151, more desirable than the flat oval pebbles that have been described, fig. 139.

Besides the large stone mortars and pestles, to which attention has been called, there are occasionally found small stone cups or miniature mortars, invariably, I believe, made by utilizing pebbles having natural hollows on one side, which have been deepened by more

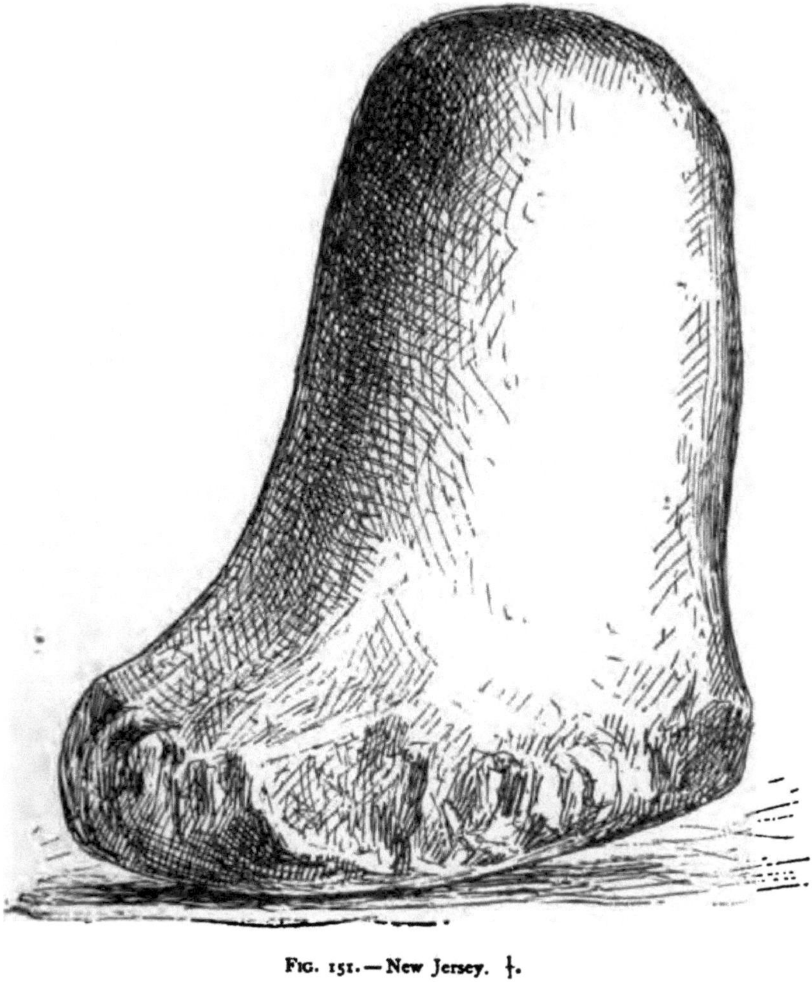

FIG. 151.—New Jersey. ½.

or less pecking, until a serviceable, though small stone vessel was produced. These were used for grinding the red paint, which, in the shape of coarsely powdered oxide of iron, is found in great abundance in the geodes that lie scattered over much of the territory of New Jersey. In these little mortars, this paint is readily reducible to an impalpable powder.

Loskiel[44] remarks of the Indians, "They bestow much time and labor in decorating their faces; laying on fresh paint every day, especially if they go out to dance. They suppose that it is very proper for brave men to paint, and always study a change of fashion. Vermilion is their favorite color, with which they frequently paint their whole head. Here and there black streaks are introduced, or they paint one-half of their face and head black and the other red."

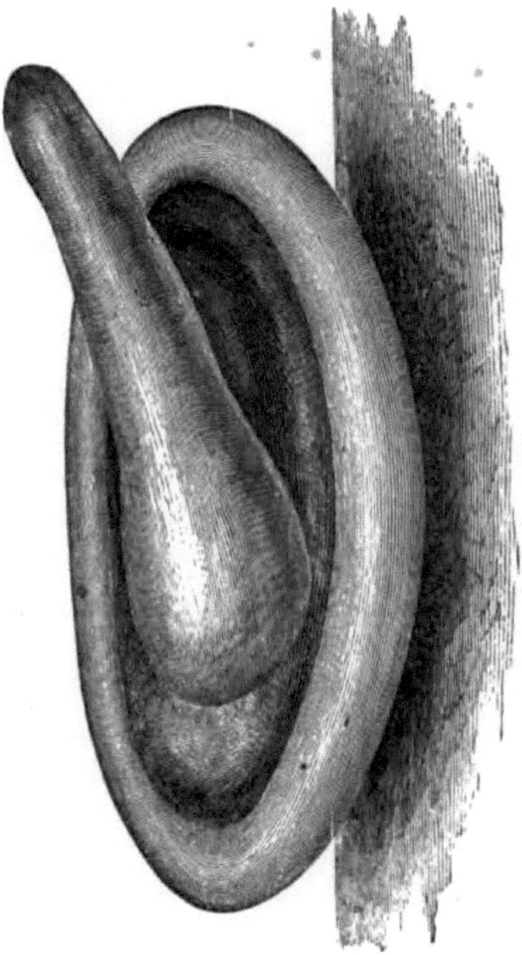

FIG. 152.—New Jersey. ½.

Fig. 152 represents a medium-sized paint-cup, made of a water-worn pebble. Three and three-fourths inches long by two and one-quarter inches wide, this specimen has sides and ends of a uniform width of half an inch, giving thereby a *large* cup-shaped depression for the total dimensions of the specimen. It may be objected that

44 Mission to North American Indians, p. 49. London, 1794.

the size of fig. 152 is too small for the supposed use. Undoubtedly many paint-mortars were larger, but this specimen is not as small as some we have found; and, in favor of the theory that it was so used, is the fact that it was found in a grave, with a series of arrowpoints, a celt, a knife, and some fragments of pottery. The locality and the evidently artificial character of the cup-shaped depression prove beyond a doubt that it is an Indian relic; and that its use was for paint-mixing seems more probable than any other that can be suggested. Associated with this delicate paint-cup was the little club-shaped pestle, drawn as resting in the cup. It is a pretty pebble, three inches in

FIG. 153.—California. ½.

length, slender and oval at one end, and flat, oval, and double the width at the opposite end. This pestle has probably been worn away considerably in its use with the accompanying cup. The width of the club-like end, and that of the slender stem, seem to agree perfectly with the width and depth of the cup's hollow, and the slender portion with the points of contact of the rim of the cup and the handle of the paint-crusher.

These small mortars, or paint-cups, fig. 152, are not confined to the Atlantic coast, but are of even more frequent occurrence on the Pacific coast. Fig. 153 represents one of these small stone cups made of serpentine, and differs from the eastern specimens only in being made from a mass of this stone, and is therefore wholly artificial; while very seldom, if ever, are they found of this character in New England or the middle states.

Fig. 154 represents a very small paint-cup made from a water-worn pebble, which has received its present hollow, or cup-shaped depression, wholly by pecking, after the manner of working the deep grooves on the common stone axe. This paint-cup is almost circular in shape, being a little flattened on one side. It is one and five-eighths inches in diameter, and has but a depth of three-eighths of an inch at the centre, or deepest part of the depression. There are still marks of the stone hammer in this hollow, which, however, feels perfectly smooth to the touch, and has the same color and amount of polish as the exterior surfaces of the specimen. This specimen, like the preceding, was found in the grave of a child, with a number of greatly decayed bone beads, and a highly polished black stone, with a number of small holes of natural origin, through it.

FIG. 154.—New Jersey. $\frac{1}{1}$.

As in the preceding example of paint-cups, so in this instance, a long, slender pebble, of considerable polish, more worn at one end than at the other, was found with the cup. The two certainly seem fitted for use with each other.

CHAPTER XI.

POTTERY.

Of all the traces of man's handiwork none are so unmistakable, and so imperishable, as fragments of pottery. Nature produces nothing with which it can be confounded. However minute the scattered sherds, they cannot escape recognition, and wherever found we can confidently point to them as evidences of man's former presence.

While primitive pottery, in a fragmentary condition, can be obtained by careful search in almost every locality, the frequency of its occurrence necessarily varies. On every former village site, it naturally occurs in greatest abundance; but it is only when we chance upon a burial place, that has escaped disturbance, that a perfect vessel can be obtained.

In the upper valley of the Delaware, in the vicinity of the beautiful Water Gap, "fragments of earthen-made articles are found in almost every field near the river."[45] Here "the material was prepared by pounding certain kinds of shells and mixing with suitable moistened clay; having dried this compound in the shade, it was then burned in the oven or kiln, made for the purpose, and became hard, and would stand exposure to the fire.

"The earthen pots are made of various sizes, holding from a pint to several gallons. The larger ones were used, among other purposes, for boiling the sap for maple sugar. Of the same material were made pitchers, vases, bowls, plates, etc.

"Unbroken articles of earthenware are now rarely met with, but

---

[45] Brodhead. Delaware Water Gap: Its Scenery and History, p. 107, 2d ed. Philadelphia, 1870.

fragments, sometimes in large pieces, are found in quantities, some of these showing a degree of taste and skill in ornamentation."

The above is true of all parts of the state, and a comparison of the pottery found about the Water Gap, which was a place of great importance to the Indians, with that of other localities, shows only such differences as arise from the quality of the clay used, or the proportions of clay and shell adopted by the different potters. About Trenton, New Jersey, which was also the site of an extensive and important Indian town, the pottery is of different colors, and of both mixed and unmixed clay. Here the raw materials are found in great abundance, and the character of the ware seems largely to have been determined by the character of the beds, from which the potter took his clay. A large portion of the pottery made by the Indians, however, was not made from pure clay, just as it came from the bed, but the clay-earths that overlie the others were utilized and made available by mixing with them quartz granules and pounded shell. Much of the pure clay, which in many places was accessible, would need far more manipulation than the Indian potters would care to give it, and as the mixture of clay and shell was simpler and would meet all their requirements, it was, very naturally, most frequently used. They nevertheless possessed the knowledge of successfully working in pure clay, as sherds are found so made, and their well-formed clay smoking pipes are a further proof of the fact.

In the immediate vicinity of the seacoast, pottery is found as frequently as along the river valleys.

In his report on the Artificial Shell-Deposits in New Jersey, Dr. Chas. Rau[46] mentions as evidence of the occupancy of such of these shellheaps, as were examined by him near Keyport, New Jersey, the occurrence of "numerous fragments of pottery," which he subsequently describes as consisting "of a dark clay, either mixed with coarse sand, or pure, for the most part rather slightly burned; some of the sherds still bear the ornamental lines and notches cut in the

---

[46] Rau. Smithson. Annual Report, 1864, p. 371.

surface of the vessels. The mixing of the clay with pounded shells does not seem to have been practised by the Indians of this region."

Throughout the New England states, the pottery presents no essential difference from that found in New Jersey. Like that found in the middle states, the material of which it is made, and the arrangement of the lines, dots and cord-marks vary indefinitely, but, as yet, no vessel has been obtained that is in any way peculiar either in form or ornamentation. Professor Perkins [47] records of Vermont, that "specimens of earthenware occur all over the state, chiefly in the form of fragments. * * * All of the earthenware was ornamented somewhat, some but little, some more, the decoration consisting of impressed figures of a great variety of form. * * * No decoration by the application of paint or any coloring material occurs. * * * The material of which the jars was made is essentially like that found elsewhere." Of pottery, as found in the New England shell-heaps, Professor Wyman has remarked, that "it is poorly represented, only small fragments having been found. Like those from other parts of the United States, the pots were made of clay, with or without the admixture of pounded shells, and were imperfectly burned so that the walls are both friable and porous. The ornamentation, when it exists, is of the rudest kind, consisting of indentations or tracings with a single point, or, as in some cases, with a series of points on one and the same instrument;" and in some cases, "the lines in the surface had been formed by impressing an evenly twisted cord into the soft clay, the cord being laid on in various positions."[48]

Two examples of medium sized vessels, each showing traces of handle-like projections, which, though broken, appear to have been the animal-like figures that characterize so great an amount of the well-known Missouri pottery, were found near Trenton, New Jersey, under such circumstances as indicated that they had been in use, and were buried there by the Indians of that region. The clay is appar-

---

47 Perkins. *L. c.*, vol. xii, p. 737.

48 Wyman. American Naturalist, vol. i, p. 581.

ently the same, and the general style of workmanship is so identical with the pottery found in the mounds of southeastern Missouri, and in the stone graves of the Cumberland valley, that there can be but little doubt that these vessels were brought from some far western locality. Suggestive as is this fact, it becomes more so, when it is remembered that this same river valley was at one time occupied by not only the resident Lenapès, but also by bands of the restless Shawnees. The latter dwelt in large numbers near where Philadelphia now stands; they had also a considerable village at the Delaware Water Gap, and are known to have occupied at one time the valley of the Cumberland. In fact, that stream was originally known as the river of the Chaouanons. Inasmuch as, mingled with the ordinary objects of stone, made by the resident Delawares, are found many that appear to be out of place, and characteristic of the southern and western Indians, it does not seem strange to find specimens of earthenware also, and yet the cases are very different. The spearpoints and ornaments are referable to southern Indians, and such as are of southern patterns or types may have been brought by the southern Shawnees; but the two black pots, in question, are not like the surface-found vessels of the later Indians, but are identical with those from the mounds. Can it be that the Shawnees were the ancient potters that fashioned that wonderful ware? But, if so, why did they not continue its manufacture in New Jersey? Here are found no animal-shaped vessels; and the clay images of the Delawares and the Iroquois are of a very different character of workmanship, and, besides, they were not designed as ornaments for vessels.

Fig. 155 represents the largest example of a clay pot, in the combined collections of the Museums of Cambridge and Salem, Mass. One, larger, from Vermont, will be subsequently referred to. This symmetrical and well-made vessel is of the shape and size that appears to have been most generally in use, judging from the numberless fragments that are found.

In fig. 155 the ornamentation is of modest character, and, as is almost always the case, it is almost entirely confined to the neck and rim. In all vessels of this pattern, the lower portion is perfectly plain.

This specimen, which is probably of Iroquois or Huron manufacture, was found near Wiarton, County of Bruce, Ontario, "under a cliff of limestone formation, upwards of one hundred feet high, and within a few feet of Colpay's bay."

Fig. 156 represents a beautiful, but fragmentary little vessel found by the late Prof. Jeffries Wyman, at Hingham, Mass. It is very thin,

FIG. 155.—Ontario. $\frac{1}{3}$.

and the clay of which it is made is nearly pure. The ornamentation is much more elaborate than that upon the preceding specimen. A noticeable feature of this pot is the almost pointed bottom. Larger vessels of the same pattern, used for cooking and holding food, were suspended by a cord passed around the narrowed portion or neck of the vessel. Fig. 156 is thin and small, and required very careful handling when in use. There is a marked difference between the ornamentation of this tastefully designed specimen and that of the following small vessel found in New Jersey, though many fragments of ornamented pottery are found.

Fig. 157 represents a nearly perfect specimen of a clay vessel of

FIG. 156. — Massachusetts. ¼.

small size, found near Trenton, New Jersey. Vessels of this kind appear to have been habitually placed in the graves of the Delaware Indians, but so imperfectly were most of them made, that, unless handled with the utmost care, they crumble on being removed from the earth.

This little vessel measures three and three-fourths inches in height, and is of the same width at the mouth, including the flaring of the rim. The clay used has a very slight admixture of shell, and is identical with much of the pottery found in fragments upon the surface of the ground. The ornamentation is of the most crude and meaningless character.

When found, this pot or vase was filled with a black powder, and covered with a square plate of mica, nearly half an inch in thickness. Mica is of common occurrence in Indian graves in New Jersey, and has occasionally been found in Indian graves in New England. Mr. F. W. Putnam[49] has described, as among the contents of Indian graves discovered in Beverly, Mass., several large plates of mica, which he states are not usually met with in this connection.

FIG. 157. — New Jersey. ½.

Fig. 158 represents a very handsome vessel of moderate size, "found in the town of Colchester, Vermont, in 1825, and is now preserved in the museum of the University of Vermont. It was found some distance below the surface and covered by a stone. * * * * The jar is made of a kind of clay made very coarse by small bits of mica, quartz and felspar, and obtained, it may be, by pulverizing granite."[50] The amount of ornamentation, which while simply com-

---

[49] Putnam. Bulletin of the Essex Institute, vol. iii, p. 123. Salem, Mass., 1871.

[50] Perkins. *L. c.*, vol. v, p. 15, figs. 1 and 2.

binations of lines and dots is really of tasteful design, is quite unusual; but the square top occurs quite frequently in the pottery of the entire Atlantic coast.

Fig. 159 represents a larger vessel found at Bolton, Vermont, many years ago. This specimen closely resembles the specimen figured on

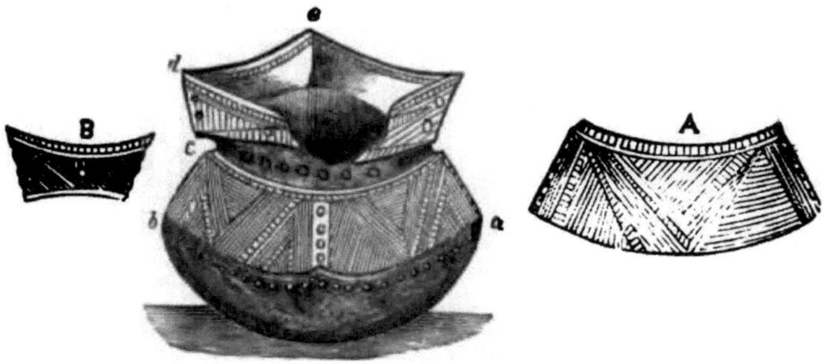

FIG. 158. — Vermont.

page 173, and is of about the same size. This specimen "is not ornamented except by a ring about the neck," about which are various oblique and notch-like incised lines. The depth and greatest diameter of this specimen are alike, nine and a half inches, and at the mouth, is seven and one-half inches in width. The capacity of this vessel is three gallons.

FIG. 159. — Vermont.

It is probable to vessels of this size, more particularly, that Kalm refers (Travels in North America), when he says "the old boilers or kettles of the *Indians* were either made of clay, or of different kinds of potstone (*Lapis ollaris*). The former consisted of a dark clay, mixed with grains of white sand or quartz, and burnt in the fire. Many of these kettles have two holes in the upper margin, on each side one, through which the *Indians* put a stick, and held the kettle over the fire, as long as it was to boil. Most of the kettles have no feet. It is remarkable that no pots of this kind have been found glazed, either on the outside or the inside. A few of the oldest *Swedes* could yet remember seeing the *Indians* boil their meat in these pots. They are very thin, and of different sizes."

While many fragments of pottery have been found, which had one or more perforations, they were not generally of a character to suggest that the object of the perforation was for suspending the vessel; but that fragments of pots had been utilized as trinkets, and drilled for suspension as rude ornaments, associated with other objects, or to be attached simply to the dress. This is further indicated by the fact, that many of these perforated pottery fragments have had the edges smoothed, and the more prominent angles worn or ground off.

Kalm further suggests indirectly, that some of the clay vessels, used by the New Jersey Indians, had feet. No specimens, I believe, have ever been found along the northern Atlantic coast, which had any foot-like projections, or were indeed sufficiently flat upon the bottom, to retain, unaided, an upright position. May not this careful observer have seen examples of the western pottery—possibly of Shawnee origin—such as have been found in New Jersey, among the then but recently discarded objects of Indian origin? Much of the so-called Missouri or black pottery, as we know, is provided with "feet," or foot-like knobs.

As further showing the character of the ornamentation common to the pottery of our Atlantic coast Indians, a series of fragments exhibiting various designs and combinations are also given. These specimens, however, must not be looked upon as in any way typical, for the combinations, possible, of the lines, dots and corn-cob marks are nearly countless, and it is probable that no two vessels were made exactly alike. Of the thousands of sherds often found in a few acres of newly ploughed ground no two can, by their ornamentation, be shown to have belonged to the same vessel.

While this rude pottery has been, by some, carefully classified as cord-marked, incised, stamped, thumb-nailed, and otherwise, it is here considered collectively, as these various methods of ornamentations are so frequently combined, that it is difficult to determine which method has the best claim to the specimen. Even if these processes were singly used, there does not seem to be sufficient difference in their results, to render careful classification necessary. Surely the difference resulting from the fact that the squaw, when moulding a pot, scratched it with her thumb-nail, rather than with a stick. is of little

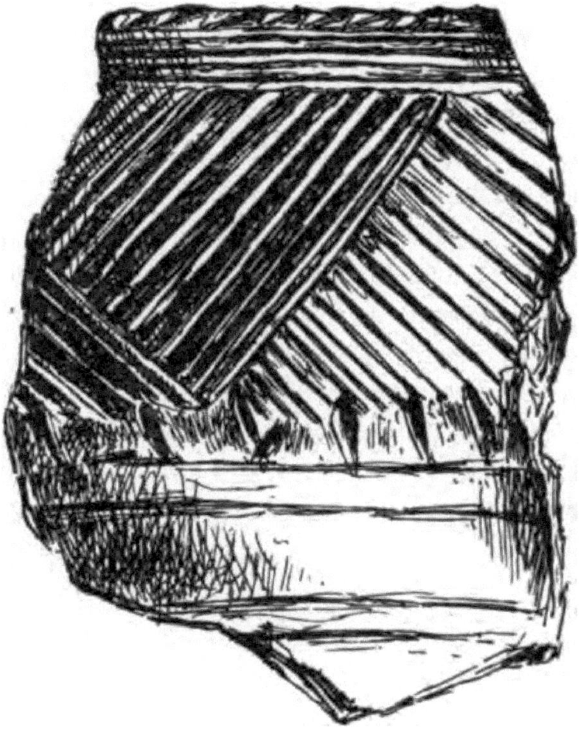

FIG. 160.—Pennsylvania. ½.

scientific interest, especially when the two methods were equally used by the same potter.

Fig. 160 represents a large fragment of a vessel, found on Shawnee island above the Delaware Water Gap, Penn. The lines are of very regular width, equidistant, and give a very pleasing effect, as they are here combined. The common practice of terminating the ornamentation by a few short perpendicular lines is not without merit, as it is desirable to prevent the abrupt junction of the plain and ornamented portions; but whether this was the intention of the ancient potter is questionable.

Figs. 161 to 164, inclusive, represent combinations of straight lines that are of common occurrence. These lines are in many cases carefully incised, and are of uniform width and depth, but the accuracy of all such incised lines depends largely upon the composition used in lieu of pure clay. When very much mixed with sand and shell, the lines, dots and other devices, are not clearly defined, neither are they of regular or uniform measurement.

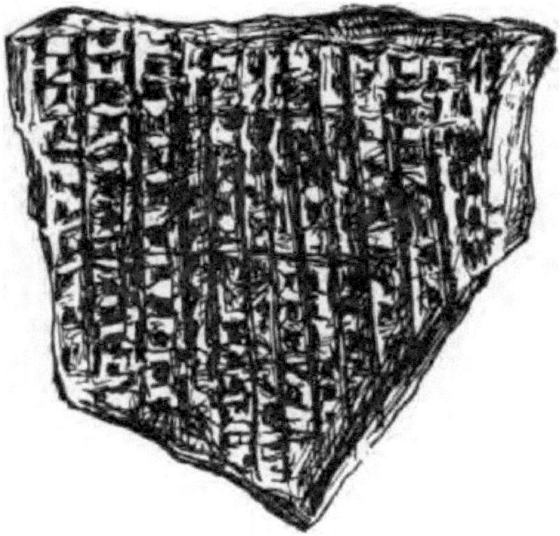

FIG. 161.—New Jersey. ½.

Fig. 164 represents an angular fragment of a thick and very small

FIG. 162.—New Jersey. ½.

FIG. 163.—New Jersey. ½.

vessel. The oblique lines in this example are pointed, as though made by the point of a sharp bone needle. The parallel lines suggest a highly conventionalized human face. This is the more probable, as the fragment may be from one of their large clay smoking pipes, which were often ornamented with various devices.

Figs. 165 and 166 represent the outer and inner sides of a fragment

Fig. 164.—New Jersey. ½.

Fig. 165.—New Jersey. ½.

of clay pot. The lines here are made of a series of square impressions which, being separated by a narrow septum, give a pleasing impression to the eye. This series of lines may have been made by a bone implement such as described by Professor Wyman, in his notice of New England Shellheaps, in Vol. I, of the American Naturalist, as having "a series of points on one and the same instrument." There does not seem to be any greater abundance of this pattern of linear decoration, than of the preceding.

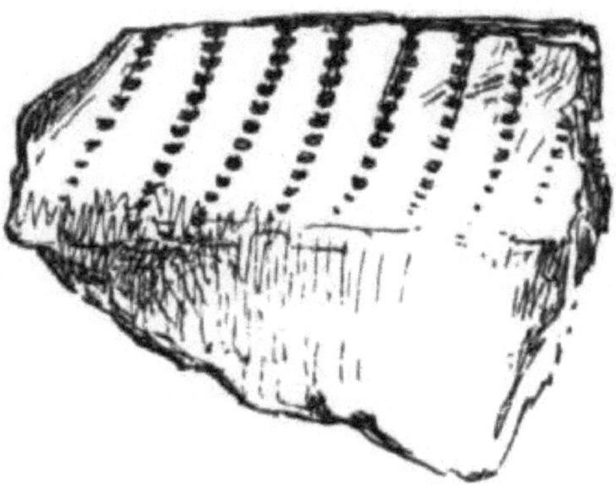

Fig. 166.—New Jersey. ½.

Fig. 167 represents a fragment with a still simpler combination of lines. In this instance it is made up of very irregular and somewhat indistinct scratches. This is the rudest attempt at ornamentation that we find, and is in broad contrast to the fragment, fig. 168, where the lines consist of delicate impressions made by the hollow stem of the bulrush. The bottom of each impress has a nipple-like protuberance, which is very distinctly seen. Fragments thus marked are not of common occurrence, either in combination with other markings or alone.

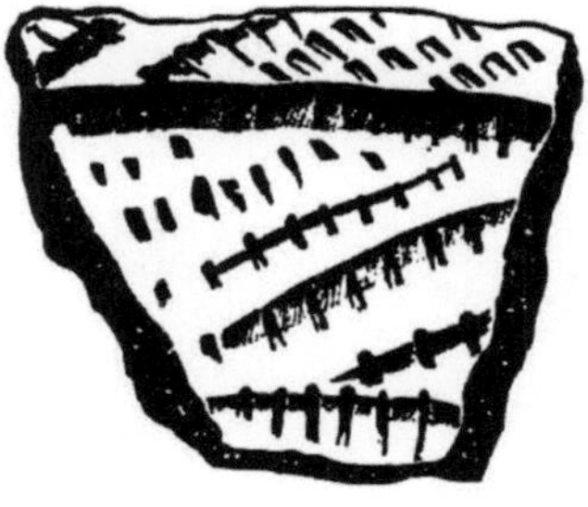

FIG. 167.—New Jersey. ½.

Figs. 169 and 170 represent two examples of linear decoration, where the lines consist of squares impressed in the soft clay by a wooden or bone implement. The clay has not been actually removed from these little pits, but pushed aside and subsequently smoothed off when the clay was soft. Such at least is the appearance of the fragments I have examined. The little fragment, fig. 171,

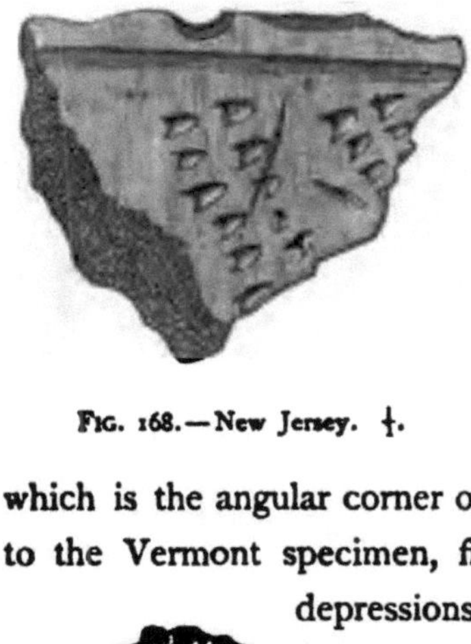

FIG. 168.—New Jersey. ½.

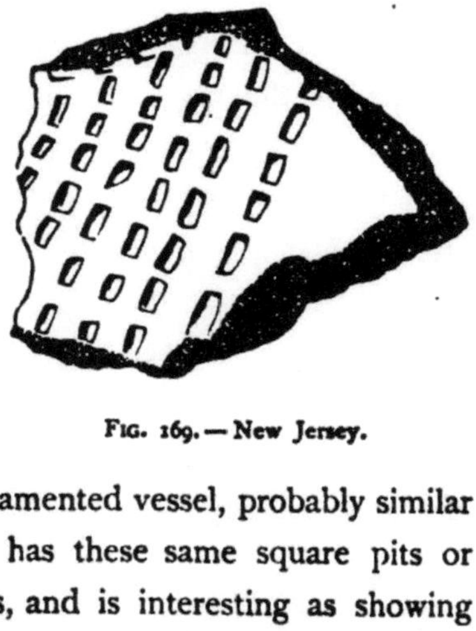

FIG. 169.—New Jersey.

which is the angular corner of an ornamented vessel, probably similar to the Vermont specimen, fig. 158, has these same square pits or depressions in rows, and is interesting as showing in how delicate a manner this style was occasionally executed.

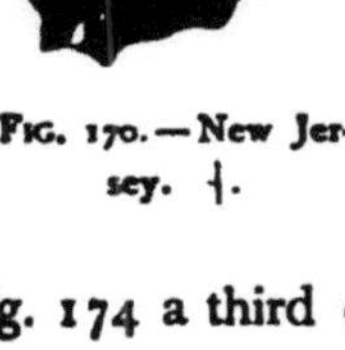

FIG. 170.—New Jersey. ½.

The character of ornamentation, so well shown in figs. 169 and 170, is by no means confined to the Atlantic coast of America. Figs. 172 and 173 represent sherds similarly marked from Wisconsin, and fig. 174 a third example from a shellheap in Massachusetts; which, however, is more nearly allied to fig. 167 than to the regularly marked example, fig. 169.

Fig. 175 represents a fragment of pottery, which is of interest as showing a style of decoration that is rarely seen in New Jersey, and is never found on any of the New

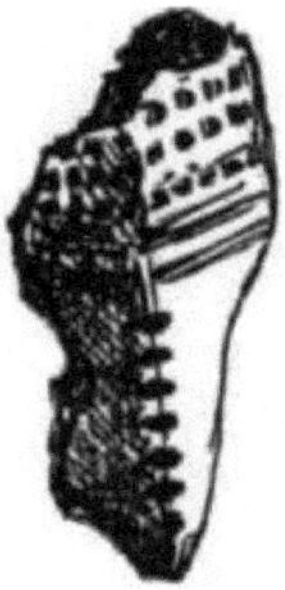

FIG. 171.—New Jersey. ½.

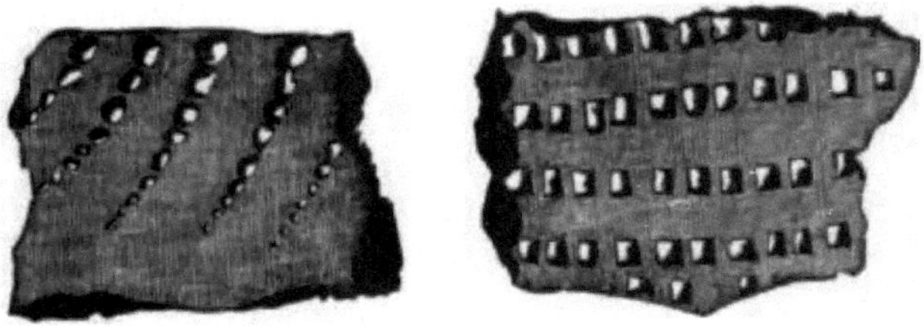

FIGS. 172 and 173. — Wisconsin.

England pottery. This is the row of large conical prominences near the rim of the vessel, three of which are seen on fig. 175. These prominences are made by punching from the inside, while the clay is yet soft; and the depressions there made correspond in depth to the knob-like elevations on the exterior surface. Were these prominences regularly placed and of more uniform size, they

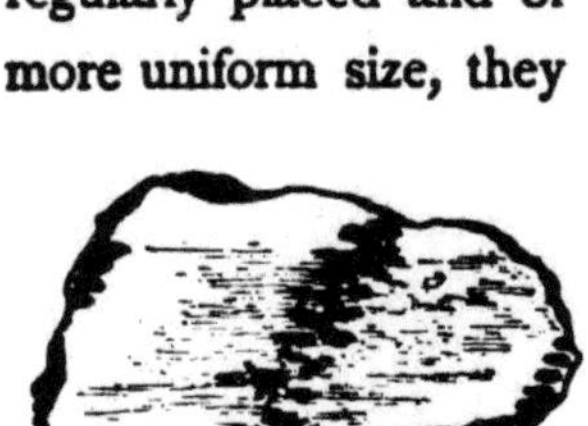

FIG. 174. — Massachusetts.

FIG. 175. — New Jersey.

would be really ornamental. Figs. 176 and 177 represent fragments from Illinois with the same character of marking. One of the prominences has been worn off, thus perforating the side of the vessel.

Fig. 178 represents a style of ornamentation of rare occurrence.

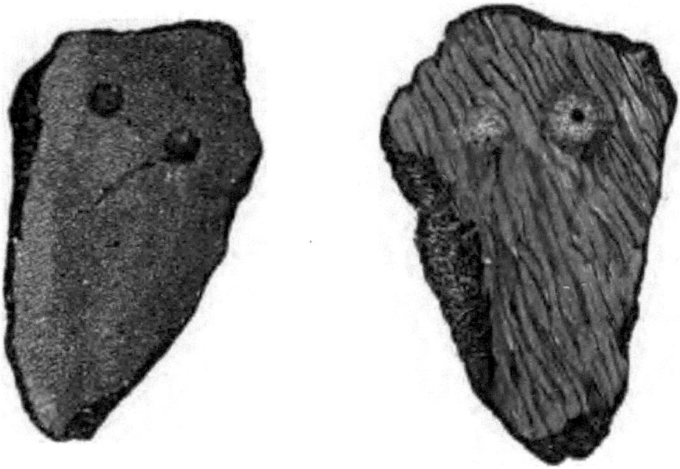

FIGS. 176 and 177.—Illinois.

In this case a hollow tube has been pushed deeply into the soft clay, and a mass removed equal to the calibre of the tube, and the depth to which it penetrated. This is nearly to one-half the thickness of

FIG. 178.—New Jersey.

FIG. 179.—New Jersey. ½.

the vessel, so that the sides were necessarily much weakened. The inside of the vessel has not been bulged or at all affected by this method of marking the exterior surface; but the spaces are too thin where the depressions are opposite to be durable, and possibly for this reason, vessels were so rarely ornamented in this manner.

Fig. 179 represents a very plain and prosaic manner of decorating pottery. The semilunar lines are merely the impress of the thumb-nail upon the clay while it was yet soft. It would, indeed, be possible to make such lines by other methods, but the general impression this style of ornamentation gives us, is doubtless the true one; that in lieu

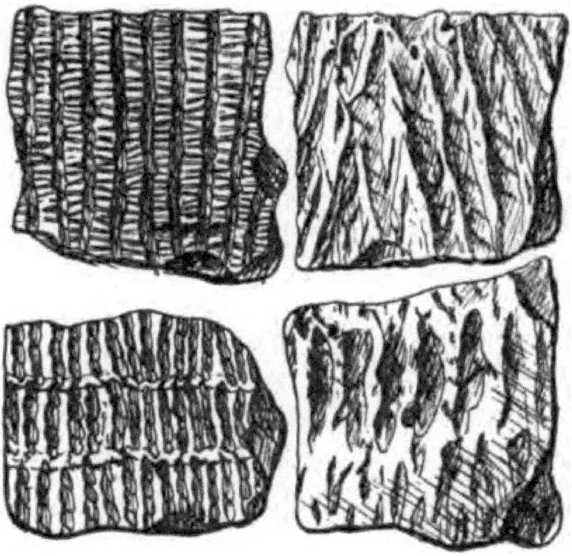

FIG. 180.—New Jersey. ¼.

of a cord or pointed stick, the potter utilized her thumb-nail, and gave a multiplicity of "new moons" to the dish or pot that she had fashioned.

Another and very common method of relieving the plain surfaces of brown earthenware, was by wrapping a cord about the vessel before it was burnt, which gave linear impressions of a rude pattern, as is shown in the two right hand specimens of the group forming fig. 180; or by covering the vessel with coarse cloth, which produced the more elaborate marking, seen in the other specimens of the group. These styles of ornamentation, known as "cord-marked" and "cloth-marked," are found wherever pottery occurs on the Atlantic seaboard.

CHAPTER XII.

STEATITE FOOD-VESSELS.

THROUGHOUT the eastern and middle states, wherever other traces of the former occupants of this country are found, there will usually be among them fragments of steatite or soapstone that have been shaped and smoothed artificially. These fragments, which are as readily recognized as the sherds that often cover the ground on the former sites of Indian villages, are portions of steatite vessels that were in common use among the Indians, from Maine to Maryland. "They are, generally, more or less oblong in shape, rather shallow, and provided with two knobs, or handles. In fact, the term *dish* would probably convey a better idea of their shape than the term *pot*, though the latter is applicable, as they often bear evidence of having been in contact with fire and were undoubtedly used for the preparation and cooking of food. The accompanying illustration, fig. 181, representing a portion of one of these vessels from Massachusetts, illustrates the common form of these pots, though there are numerous variations in size and shape." (F. W. Putnam in the Eleventh Annual Report of the Museum of Archæology, at Cambridge, Mass.) Peter Kalm, the Swede, who visited New Jersey in 1748, refers to the dishes in use by the Indians as "made sometimes of a greenish, and sometimes of a grey pot-stone, and some are made of another species of apyrous stone; the bottom and the margin are frequently above an inch thick. The *Indians*, notwithstanding their being unacquainted with iron, steel, and other metals, have learnt to hollow out very ingeniously these pots or kettles of pot-stone." The number of perfect vessels that have been found, and the immense quantities of fragments, bear evidence that almost as many vessels of steatite, as of clay, were in use.

The method of working steatite, both in procuring suitable masses of the mineral from the quarry, and in subsequently shaping it into vessels of the general character found along the Atlantic coast, has been so fully described by Messrs. Schumacher and Putnam, that the sum of their observations will be here given, as the best explanation of this early industry of our native races.

Mr. Schumacher[51] discovered on Santa Catalina island, off the coast of California, "pits and quarries" where soapstone vessels had been made, and also the tools used in the several processes. The vessels (*ollas*), which in California are globular or somewhat pear-shaped,

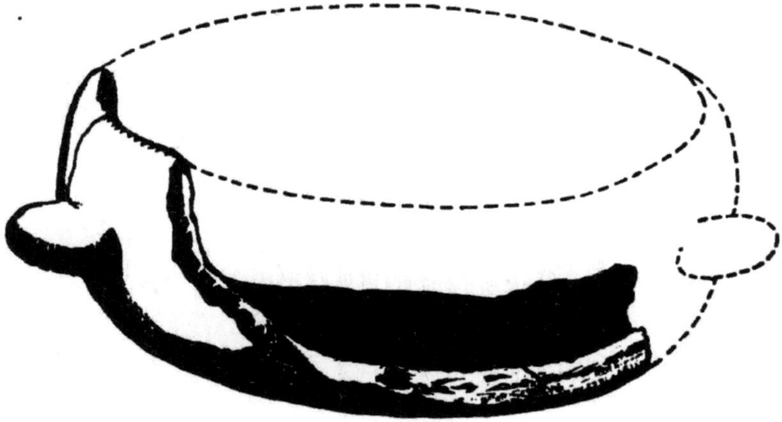

FIG. 181.—Massachusetts. ⅓.

are roughly blocked out in the living rock by the use of rude slate chisels, some of which bear a marked resemblance to the European palæolithic implements. "After the pot-form had been worked out, it was broken from the living rock by working under it and by gradual pressure of the chisel around the base. The detached pot-bowlder was next rounded into proper form; it was then hollowed out until a certain thickness of the pot was reached; and finally, carefully finished with the scraper. As the thickness of the *olla* increases towards the bottom—it usually thickens from about half an inch at the rim to one

[51] Schumacher. Eleventh Annual Report of the Peabody Museum of Archæology, p. 258, Cambridge, Mass.

and a half at the bottom—it requires skill to attain this evenly. No mechanical apparatus was used for this purpose (as shown by certain irregularities in the form of the pot) but simply the touch of both hands in anteposition, one gliding outside the already finished surface while the other worked inside towards the guiding hand. In this wise, with some practical experience, a greater accuracy is attainable than at first might be supposed, especially if the work proceeds from a known thickness to which reference can be taken, which is here the case as it progressed from the rim.

"A new pot is without polish, and has only the smooth surface imparted by the scraper; while those which had been in use attained frequently a polished surface by wear, which the soft and greasy nature of the potstone is inclined to adopt."

As the forms of pots or food-vessels found on the Atlantic coast are flatter and more dish-like, the same skill was not required in their production; but the finish and occasionally the ornamentation make the ware, that was produced by the Indians of the Atlantic coast, no less artistic and desirable for all culinary purposes. The same method of working the rock, while in place, was practised by the Atlantic coast tribes, as was adopted by the former occupants of Santa Catalina island. A ledge of steatite has been examined and is thus described by Mr. Putnam;[52] "At the time of my visit, many * * * * rude chisels (roughly pointed stones that were lying about and bearing evidence of having been used) had been carted a few hundred yards distant to fill up a low piece of ground and others had been thrown in a pile on the ledge. A careful estimate of the number convinced me that at least two thousand of these rude stone chisels had been found on the ledge, or in the immediate vicinity. They were all nearly the same size, rudely chipped to a blunt point at one end, and roughly rounded to fit the hand at the other. Those brought to the museum vary in length from five to eight and one-half inches, and in weight from one to four pounds; the majority being about seven inches in length and from

---

[52] Putnam, *l. c.*, p. 273.

two to three pounds in weight. These chisels were made from the hard stone of adjoining ledges, and their manufacture must have required considerable labor. A short trial of the chisels upon the soapstone showed the facility with which the steatite could be pecked by these rough implements, and what patience combined with muscle would accomplish.

"Associated with the stone picks, or chisels, were between seventy-five and a hundred large rounded stones, weighing from twenty-five to a hundred or more pounds each, which might have been used as hammers for the purpose of breaking off large masses of the soapstone.

"The bed of steatite had been excavated its full width, and nearly all its length and depth as far as at present exposed. The remains ot the circular and oval masses, that had been broken off from the sides of the ledge, showed that the seam of steatite was formerly from six to twelve feet deep; the whole of this mass of rock having been worked out and probably made into utensils.

"Several fragments of pots were found in the débris of the ledge, evidently broken during manufacture, and also several unfinished pot-forms just as detached from the matrix; while on the ledge itself the pot-forms could be followed out through their various stages of development.

"The fact that soapstone vessels, of the peculiar shape and character of those made at this ancient New England manufactory, are widely distributed east of the Mississippi river, though more common in the New England states than elsewhere, may be one of the many indications of aboriginal trade."

Fig. 182 represents a second and more nearly perfect example of the average soapstone vessels that are common to the Atlantic seaboard. Judging from the character of the innumerable fragments that have been gathered, the great majority of these vessels were less than one foot in length and comparatively few were circular or globular in shape. Of a large series in the Gilbert museum at Amherst College, the majority have a capacity of several quarts, and one, of unusual size, a capacity of about sixteen quarts; but these are exceptional.

While the New England examples of steatite vessels as a rule, have perfectly plain exterior surfaces, those found in New Jersey were frequently quite elaborately ornamented with deeply incised lines, similar to much of the aboriginal pottery found in that state, and this resemblance was increased by exposing them to the fire.

Unfinished steatite vessels have frequently been found in New Jersey, at considerable distances from any known ledge of soapstone, worked by the Indians, and it is supposed that they were carried to tribes not having access to the mineral, and bartered in this unfinished condition, in order that the purchaser might finish them to suit himself, or rather herself, as it is probable that this kind of work fell to the lot of the women. However this may be, it is difficult otherwise to

FIG. 182.—New Jersey. ⅓.

explain the occurrence of these scarcely more than blocked-out specimens of steatite pots and dishes.

It has been frequently noticed by collectors and others that, in limited localities, many fragments of steatite vessels are found, and while ordinary forms of stone implements may be abundant, there is no trace of pottery. This absence of pottery, where steatite was abundant, has been commented upon, and some very questionable assertions made concerning the use of the two materials. The absence of fragments of pottery is, of itself, but negative evidence that the people who used stone vessels did not use pottery also. When we consider that pottery occurs in the middle shellheap period, and no steatite is found referable to so early a period, it is more than probable that the use of soapstone is of a comparatively recent date, and in no locality east of the Mississippi river had it wholly replaced the more convenient and more fragile vessels of clay.

It can scarcely be held, on the other hand, that where there occurs an abundance of pottery, no steatite vessels were ever used, because no traces of them are found. Stone vessels of the character of the average soapstone pots of the Atlantic coast Indians were not readily broken, and scores of them might have been in use for weeks or months, and yet all escaped destruction, and be carried away by the people occupying the site whereon the pottery fragments were found. In New Jersey, no village site that has been carefully examined has failed to produce a few fragments of soapstone, and many of pottery; but often, so worn, discolored and pebble-like were the fragments of steatite, that it is very possible that pieces of vessels made of this material have been overlooked, and there is abundant evidence to show that fragments of these soapstone vessels were frequently used by the Indians as available crude material for small pendants and other objects. In some cases, the fragments have been perforated in one corner, and the broken edges simply worn smooth.

Early unpublished records of the customs of the Indians that lingered in New Jersey, after the English settlers had firmly established the city of Philadelphia, refer to the cooking vessels of stone made by the savages, and mention the great abundance of them found by the farmers when ploughing, for the first few times, their newly occupied fields. These steatite pots, that the Indians had discarded, were generally preserved and were found "exceedingly useful in the kitchen."

CHAPTER XIII.

PITTED STONES.

Among the many stone objects that are to be classed strictly as domestic utensils, left by the former occupants of the Atlantic seaboard of North America, are certain slabs of hard stone, that have been deeply pitted in one or more places. While no one shape or size can be considered as typical, the majority are slabs measuring about one foot square, with from three to ten pits, usually upon one side only, though occasionally they are found upon both. Like the slabs themselves, these hollows or pits vary considerably in size, the largest measuring nearly two inches in diameter, the smallest about half an inch.

As is usually the case with domestic utensils, these implements are of common occurrence on village sites, but are rarely found singly in out-of-the-way places. So far as the series gathered in New Jersey bears upon this matter, it may be stated that nearly one hundred were found where the ground was literally covered with fragments of pottery and steatite pots, mixed with charcoal and other evidences of fires. Whatever may have been the purpose of these pitted stones, it is evident that they were closely connected with household, and probably, culinary occupations.

Fig. 183 represents an excellent example of these stones, although it exhibits more traces of artificial shaping in the slab itself, than is usual. This specimen is nearly square in section. On one side is a series of the pits which characterize these implements. They have been carefully pecked out, and are quite rough. In no specimen found in New Jersey is there any trace of rubbing, or of polish. This characteristic roughness in the pits seems to show that they were not used in any way connected with a revolving object, as has been suggested with reference to similar stones found in Ohio, and which have been considered as *spindle-socket stones*.[53] These pitted slabs are equally abundant in the southern states, and Col. C. C. Jones, jr.,[54] has given many reasons for believing that they were used as nut-cracking stones. He remarks, "their cavities are so located that one, two, three, four, five, and sometimes more nuts could be cracked at a single blow;" and as it is well known that our walnuts, chestnuts, shell-barks and even acorns were largely used as food, it is more than probable that this suggestion correctly explains the use of these implements.

FIG. 183.—New Jersey. ½.

Unless they have been overlooked by collectors generally, throughout New England, these implements are not common, or it may be that they are entirely wanting. None appear to have been received at the museum at Cambridge from any New England locality.

If used as slabs upon which to crack nuts, may it not be that in some localities, level surfaces of rocks in place were used, instead of portable slabs, and thus the absence of the latter may be accounted for? The depressions in these slabs of stone vary greatly in outline, some being oval; and also in the degree of smoothness of the surface both of

[53] Whittlesey, Ancient Earth Forts of the Cuyahoga Valley, Ohio, pp. 33-35, pl. viii. Cleveland, 1871.

[54] Jones, *l. c.*, p. 319.

the depressions and the surrounding level portion of the slab. Dr. Chas. Rau (Smithson. Contributions to Knowledge, No. 287) considers it improbable that those of fine finish, with almost polished cup-like depressions should have been used merely for nut-cracking, and suggests that they were used as receptacles for paint. A fact that has some bearing upon this point is that of the occurrence occasionally of small specimens of these pitted stones in graves. It is quite in accordance with what we know of the burial customs of the Indians, to find among the various articles inhumed with the dead, such objects as related to their toilet, which during life, were almost as valuable as their weapons. On the other hand, it is extremely improbable that a mere household article, used, it may be, only by the women, should be placed in a grave. The division, therefore, of these pitted slabs into nut-stones and paint cups is warranted; but it must be borne in mind, that unquestionable stone cups were in common use, and the use of these for toilet purposes was the rule, that of the small pitted slabs, the exception.

CHAPTER XIV.

CHIPPED FLINT IMPLEMENTS.

ALTHOUGH the term "flint implement" has acquired a meaning too comprehensive to justify its employment as a descriptive term, yet it still seems necessary, with our present limited knowledge of the various uses of certain objects of Indian make, to apply it to some forms of stone objects. We can feel well assured, on examining the specimens referred to, that they are finished implements, and that they have been made by chipping. Beyond this, all is conjecture.

Guided by the study of other forms, there is but little doubt that these roughly chipped, and indefinitely shaped flint implements, of the several different patterns and sizes, were intended for widely different purposes. As the use or uses of any one of them is not known, and as they have many points in common, it is better to treat them collectively, rather than to consider them solely with reference to such differences as are suggestive of varied, but undetermined uses.

Of the larger objects treated in the present chapter, there has been much written; and some evidence has been adduced to show that they were agricultural implements, or "spades;" but of those found in New Jersey, but very few bear trace of use, such as digging in sandy soil would inevitably produce, and spades of a different pattern, that do bear the marks of such use, are quite common. Of the smaller sizes, little can be said, as their form is essentially indefinite. There are possibly a score of uses to which they are adapted, but there is no evidence that any one of them was really that of the implement.

Fig. 184 represents a good typical specimen of one of the indefinite implements of chipped flint, to which reference has been made. The character of the chipping shows that the speci-

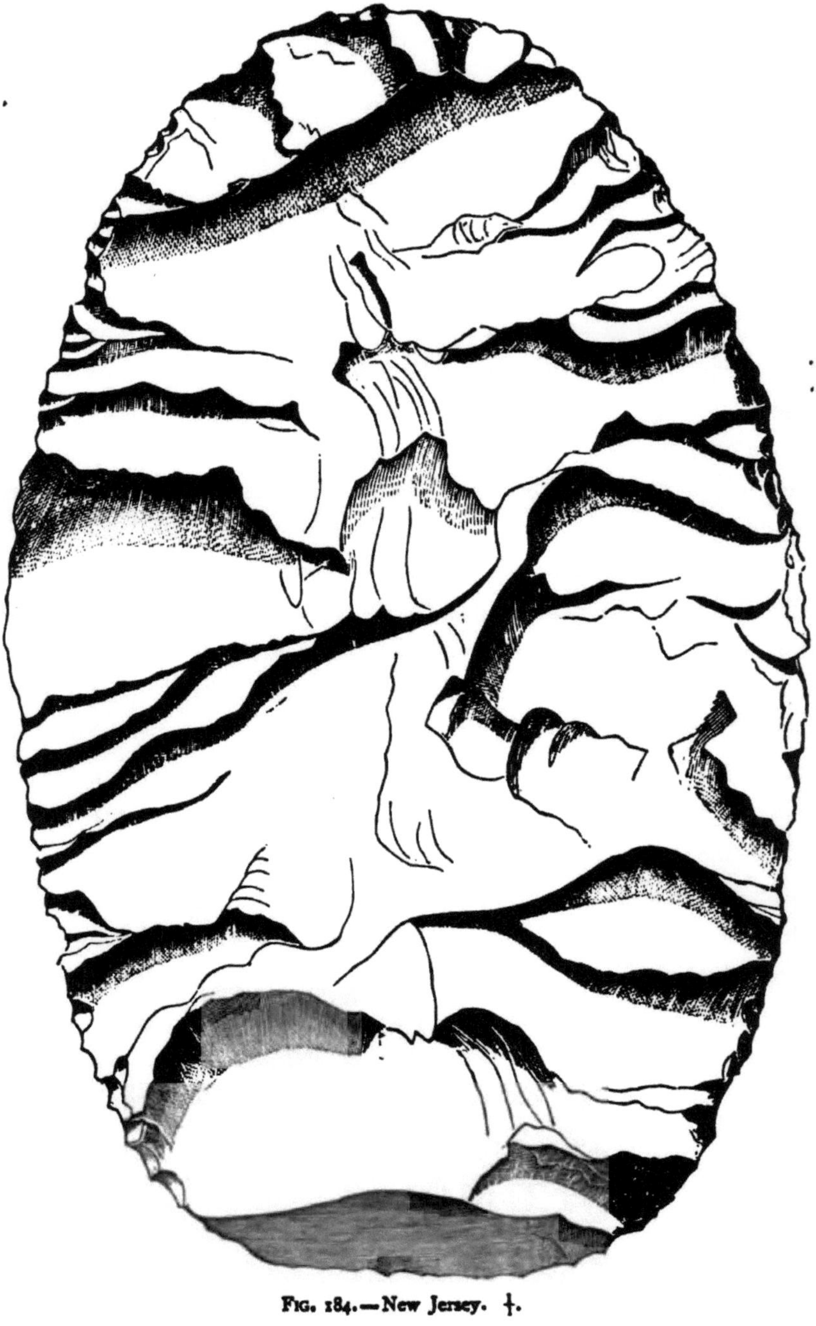

FIG. 184.—New Jersey. ½.

men was unquestionably finished, and not a "blocked-out" example of some other form.

The implement most nearly allied to this specimen is the "shovel" described by Dr. Chas. Rau,[55] as an "oval plate of flint, flat on one side and slightly convex on the other, the outline being chipped to a sharp edge." Fig. 184, on the contrary, is identically chipped upon both sides, and this is true of nearly two hundred that have been found. As the evidence that the Illinois specimens were spades or shovels, found in "the peculiar traces of wear which they exhibit," or the "glaze" and striæ extending "in the direction in which the implement penetrated the ground," is almost wholly absent in all cases that have come under my observation, there is of course no reason for pronouncing these New Jersey specimens to be spades.

Fig. 185 represents a second example of these large, broad implements of undetermined use. This specimen varies but little from the preceding. Both belong to a series of one hundred and fifty which were discovered in ploughing. They were carefully packed together in the smallest possible compass, about two-thirds of the number being placed on one end, and walled about with the remaining third, lying on their sides, and overlapping each other. Had these specimens been alike, or had they even resembled each other as much as do figs. 184 and 185, it might have been maintained that they were a deposit of unused shovels; but many of the series were considerably shorter, some being scarcely more than half the size, and quite acutely pointed. As there was a gradation in size, there was evidently a mixture of different implements, and how are we to determine the limits of shovel measurements and the maximum size of the nearest allied implement? Setting aside the fact, that unquestionable shovels and hoes of very different patterns are of common occurrence, there is nothing to warrant the conclusion that these implements, so essentially unlike the undoubted western shovels described by Dr. Rau, belong to that class of agricultural implements.

---

[55] Rau. Smithsonian Annual Report for 1868, p. 401, fig. 1, Washington, D. C.

While it is undoubtedly true that the majority of "deposits" of flint implements have consisted of these large forms of uncertain uses, there

FIG. 185.—New Jersey. $\frac{1}{1}$.

seems to be nothing in this fact, that throws any light upon the nature of the implements themselves. Dr. J. F. Snyder, of Virginia, Illinois, has given us a very interesting account of a number of these deposits, discovered in the west, but as he refers to the deposits of other forms also, it is evident that the fact of their concealment in the ground has nothing to do with their character. Already reference has been made to extensive deposits of spearpoints and arrowheads in New Jersey. Two other instances, one of grooved axes, and the other of polished celts, were mentioned when these implements were described. Dr. Snyder,[56] in his article on "Deposits of Flint Implements," already mentioned, quotes from Strachey, as follows, and it appears to explain the whole subject of buried implements—"Their corne, and (indeed) their copper, hatchetts, howses, beades, perle, and most things with them of value, according to their estymacion, they hide, one from the knowledge of another, in the grownd within the woods, and so keepe them all the yeare, or untill they have fitt use for them."

Fig. 186 represents a chipped implement similar to the preceding, but much smaller. While the larger examples of this pattern are but very rarely found singly upon the surface, those of this reduced size are by no means uncommon. This would seem to afford some reason for supposing that they had been used in an essentially different manner, or why should we not find the larger size also? Fig. 186 bears a very marked resemblance to one of a large series figured by Dr. Chas. Rau, and described as "roughly edged fragments, which were destined to be made into arrow and spearheads at some future time." As such objects have not been found in the chips on the sites of arrowmakers' workshops, but are scattered singly about, as is the case with other forms, and as some of this size were associated with the larger specimens here figured, the assumption that they are raw material, from which other forms of implements were chipped, is hardly warranted by the facts.

Assuming that they are finished implements, it has been suggested

---

[56] Snyder. Annual Report, Smithsonian Inst. for 1876, p. 433. Washington, D. C. The History of Travaile in Virginia; by William Strachey, printed for the Hakluyt Society.

that they were used to some extent as blades or points to war-clubs. In describing the several forms of war-clubs formerly used by the Iroquois, mention is made of one armed with deer's horn."[57] It was made of hard wood, elaborately carved, painted and ornamented with feathers at the ends. In the lower edge, a sharp pointed deer's horn, about four inches in length, was inserted. It was thus rendered a dangerous weapon in close combat, and would inflict a deeper wound than the former (one with a ball). They wore it in the girdle. At a later period, they used the same species of club, substituting a steel or iron blade resembling a spearhead." May it not be, that many of these short, thick, pointed jasper objects, such as fig. 186, were used, in the manner described by Mr. Morgan? Stone, as a ball, was used to weight a club, and Schoolcraft[58] has figured a series of war-clubs of various finish, some of which are mounted with distinctly pointed hatchets of stone. Strongly made as all such specimens as fig. 186 are, they are now frequently found with their points broken off; showing they had been exposed to violent

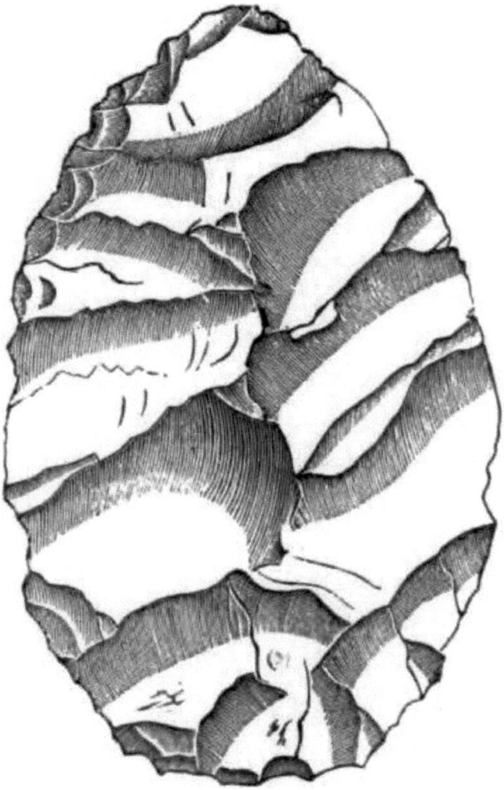

FIG. 186.—New Jersey. ½.

[57] Morgan. League of the Iroquois, p. 363. New York, 1850.

[58] Schoolcraft. Hist. and Condit. Ind. Tribes, pt. II, plates 73–74.

usage. While the purpose of these "flints" may be very different from that suggested, there is evidence that, occasionally, such use was made of them. In the autumn of 1875,[59] three chipped jasper specimens of short flints were found in three graves, each of which had been inserted in one end of the femur of an elk (?). The bones had so far decayed, that they were destroyed utterly, in attempting to remove them. Two of these flints were similar to fig. 186, but the other was apparently the pointed half of such an implement as fig. 185.

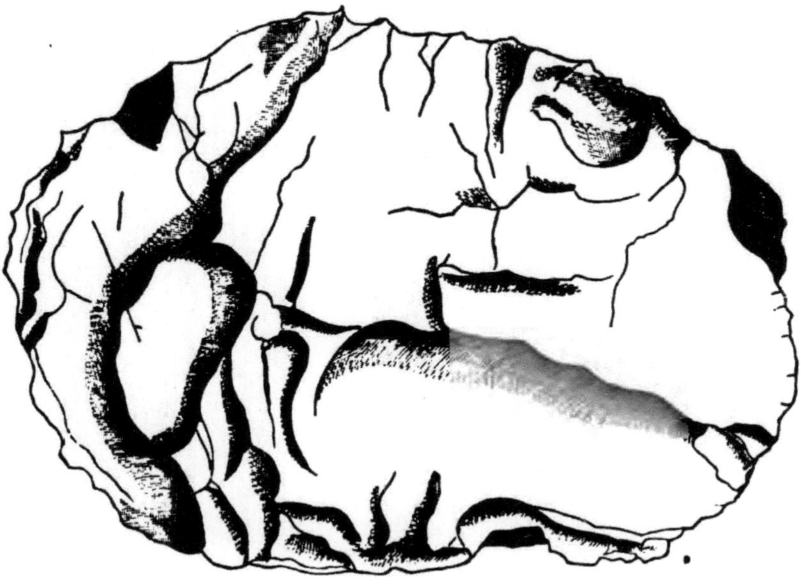

FIG. 187.—New Jersey. ½.

Figs. 187 and 188 represent specimens of very indefinite patterns, having no feature that is suggestive of use. They are almost too blunt to be considered as possible points to a war-club. Certainly this is the case with the former. The edges appear to be in their original chipped condition, and if so, they are not sufficiently sharp to warrant us in classing the implement as a cutting tool. Neither do they bear any resemblance to the larger scrapers.

59 Abbott. American Naturalist, vol. x, p. 116. Boston, Mass., 1876.

The association of these forms, with all the others, especially on village sites, gives us evidence that they are finished implements, and beyond this, it seems difficult to proceed. They bear considerable resemblance to many of the palæolithic implements found in Europe, but as the purpose of these latter is unknown, the similarity throws no light upon the use of those found in New Jersey.

Fig. 189 represents a beautiful specimen of a chipped jasper implement, from Massachusetts, so marked in every feature, that its association with the preceding forms is apparently an error. It is, however,

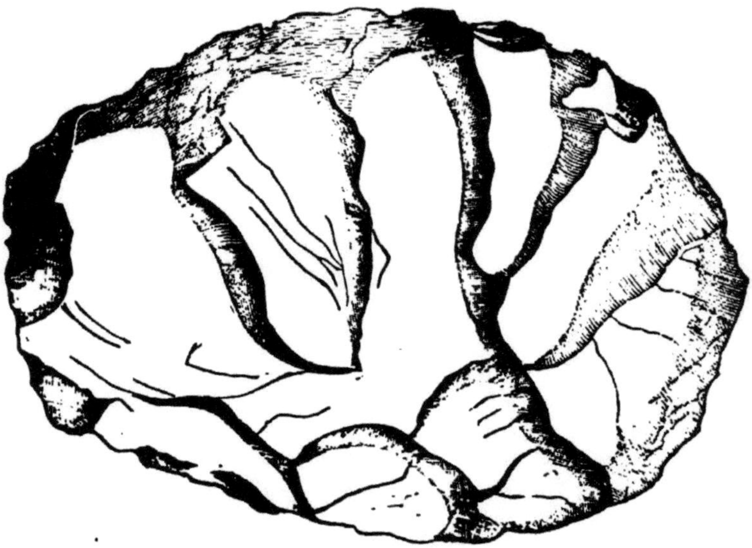

FIG. 188.—New Jersey. ½.

simply the maximum size of the narrow and pointed kind of implements, and from it a series without a break can be made down to those with such an indefinite outline, that no use can be imagined for them. Fig. 189 may or may not be a spearpoint, lancehead, or dagger. Undoubted objects of this character, of a very different pattern, are of frequent occurrence, but it is very seldom that a specimen of chipped jasper implement of this form is met with.

This specimen measures eight imches in length, by two and five-eighths inches in width; is very thin and acutely pointed. That it was not an implement of every-day use is shown by their great scarcity, and also by the fact that it is so delicate that the slightest rough usage would break it. In this, it closely resembles the long, thin, dagger-like implements found in southern California and Tennessee.

Fig. 190 is an example of flint implement which, in size

FIG. 189.—Massachusetts. ½.

FIG. 190.—New Jersey. 1/1.

and character of chipping, might readily be used as a knife, spear or dagger; although, as it has neither a notched nor a stemmed base, the difficulty of attaching it to a handle, renders it somewhat doubtful, whether it has been used as such an implement or as a weapon.

Other forms of these chipped jasper implements are found occasionally, but none that differ in any important manner from those that have been here figured; unless we except such as are distinctly pointed at both ends. These are not so common as those that have a blunt base.

Mr. S. L. Frey[60] has figured a specimen of this kind, found in New York; one of three taken from a grave. When of sufficient length to be used as the "head" of a war-club, might not these doubly pointed implements have been passed through the club handle, and so make a doubly armed weapon? The edges of all these implements are too sharp or jagged to have been held in the unprotected hand, even if only used as knives. This fact suggests the probability of a handle of some kind.

In a MS. notice of the antiquities of Onondaga County, New York, to which reference has already been made, there are drawings of several flint "lance-heads" of large sizes. Fig. 8 of the MS. measures nine and one-half inches in length, by two and one-half inches in width for nearly two-thirds of the length, when the blade rapidly narrows to an acute point. This specimen, while of unusual length, is not finely worked nor symmetrical in its outline.

---

[60] Frey. American Naturalist, vol. xiii, p. 641.

CHAPTER XV.

BONE IMPLEMENTS.

To a certain extent, bone, instead of stone was habitually used for marking many forms of domestic utensils and weapons; but exactly how far it replaced stone is questionable. Still, it must be remembered that bone, being far more perishable than stone, may have been used in the manufacture of many kinds of implements that have long since crumbled into dust. Hence, the absence of implements of this material cannot always be taken as evidence that they were not used. After making due allowance for all possible conditions, it is probable that implements of bone are exceptional. In no known inland locality do they outnumber those of stone, and only in the New England shellheaps can they be considered as more common.

In New Jersey the occurrence of single objects made of bone, with here and there an occasional bead, is quite unusual. Less than half a dozen specimens of handiwork in this material are among a series of nearly twenty thousand stone implements from this state.

In the Mohawk valley, New York, Mr. Frey[61] found bone implements to be "much more rare than those of stone," and also remarks that "the sites of villages that are uncleared and uncultivated, and where these bone tools alone are found, are very few."

Prof. Wyman,[62] on the other hand, speaks of bone implements in the Massachusetts shellheaps as "quite abundant."

Fig. 191 represents a bone spoon, of a peculiar pattern. It is care-

---

[61] Frey. Amer. Nat., vol. xii, p. 752, figs. 15 and 16. Philad., 1878.

[62] Wyman. Amer. Nat., vol. i, p. 581, plates 14 and 15. Salem, Mass., 1868.

fully shaped from a portion of the bone of the jaw of a porpoise. The handle has been quite elaborately ornamented with incised lines, placed at regular distances, which produce a pleasing effect.

This specimen is one of four, now preserved in the museum of the Academy of Science, at Salem, Mass. The other specimens vary somewhat from this in shape, being all of them shorter, and in two instances much broader. The four specimens were found in an Indian grave in Lagrange street, Salem, Mass.

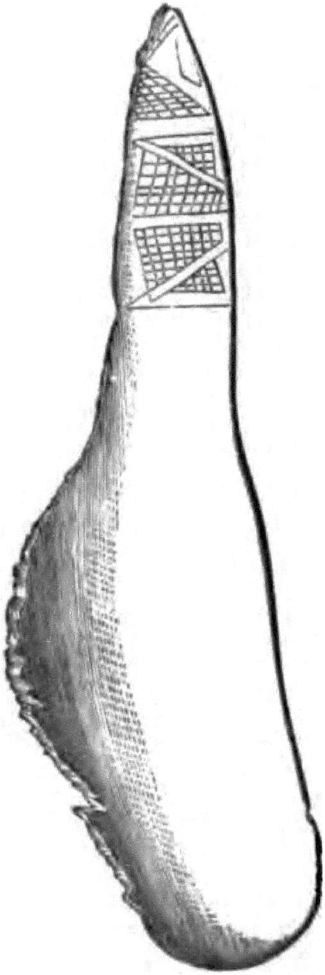

FIG. 191. — Massachusetts. ½.

In the same museum is a fifth example, varying in no important particular, which was found also in an Indian grave at Eagle Hill, Ipswich, Essex Co., Mass.

It is not probable that many bone spoons of this pattern were ever in use, as they certainly were in nowise as serviceable as many natural productions; especially portions of many of the larger marine shells, and of some of the fresh water bivalves, which, without any alteration, could readily be used for the same purpose. Indeed, Holm[63] speaks of such shells as being in use among the Delaware Indians; remarking that "their spoons were muscle shells," which shells they also used in boat building. (See Chapter XVIII.)

Fig. 192 represents a spoon or paddle-shaped bone implement, made from a portion of a rib of some large mammal, probably an elk. To what extent the specimen is fragmentary cannot be determined, but while

[63] Holm, *l. c.* p. 124.

apparently a portion of some implement, it is not evident what was its character. It certainly bears no resemblance to the bone spoons made of the jaw of the porpoise, as represented in fig. 191. The narrower portion has been cut or ground away to some extent, and the edges are quite smoothly polished. Near the end of this handle-like portion, there is a countersunk perforation, and upon the concave side of the wider part there are rudely outlined the heads of two birds.

FIG. 192.—New Jersey. ½.

In the collections of domestic implements and weapons of the Alaskan and northwest coast Indians, in the museum at Cambridge, are large series of wooden implements, many of which are known to be wooden spoons, while others are considered as "models of paddles." Many of the former have no greater resemblance to a spoon than the example of supposed bone spoon, here figured, but all are highly decorated, either by painting or elaborate carving. A series of seventeen, in a cylindrical basket, of Haidah Indian manufacture (P. M. No. 17,021), are all ornamented in either one or both ways, as mentioned. If used as domestic implements, as is probable, they are no doubt represented, in the implements of the Atlantic coast tribes, by objects of the general character of fig. 192.

It is of interest to note the resemblance between these birds' heads, and those that occur on the semilunar slate knife (fig. 43, chap. v) and upon the shell disks found in the stone graves of Tennessee.

Fig. 193 represents an interesting specimen of a bone fish-hook, from Long Island. The illustration so clearly shows the character of the implement, that a detailed description of it is unnecessary. Objects of this character are exceedingly rare, either as found on the surface, or in shellheaps. While of so simple a form, bone fish-hooks of this pattern do not appear to be common in any locality in eastern North America, although Peter Kalm, in his "Travels in North America," writes of the Delaware Indians, that they "employ hooks made of bone, or birds' claws, instead of *fishing hooks*. Some of the oldest *Swedes* here told me, that when they were young, a great number of *Indians* had been in this part of the country, which was then called *New Sweden*, and had caught fishes in the river *Delaware*, with these hooks."

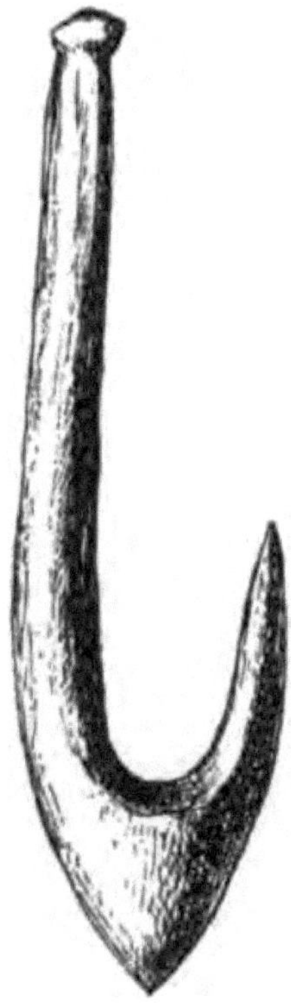

FIG. 193.—New York. ½.

Col. C. C. Jones, jr., remarks of the southern Indians, that with them "fishing with hook and line seems to have obtained to a very limited extent, if we may judge from the remarkable absence of anything like bone, flint and shell hooks in the mounds and refuse piles. Very few hooks have been found, so far as our information extends, and they were of bone."

On the Pacific coast, bone hooks of a more complicated pattern are not uncommon. They are made of bone and shell, and differ from the plain hooks, like fig. 193, in having the stem short and curved towards the point of the hook, and also, in having a well defined barb, on the outer side of the hook, some distance from the point. Generally they are notched on the outer edge of the stem, for the more secure attachment of the line, which, after being wrapped about the hook, was covered with asphalt. (See pl. xi, of vol. vii, Archæology, U. S. Geog. Survey West of 100th Meridian.)

In northern Europe, this pattern of bone fish-hook is more common. Nilsson, in his admirable Stone Age in Scandinavia, figures a specimen differing only in having a very slight barb-like projection, very near the point.

Fig. 193 was found in a shellheap, near Sag Harbor, Long Island, New York, by Mr. W. W. Tooker of that place, to whom I am indebted for an opportunity of describing it. The shellheap from which this specimen was taken has yielded "many objects of Indian workmanship," but no other specimen, I believe, of similar fish-hooks.

Fig. 194 represents an interesting bone implement, which, it has been suggested by the late Professor Haldeman, was used for ornamenting pottery. This implement is made from one of the long bones of a deer, and its entire surface has been worked over.

The broad sides of this specimen are not uniform, level surfaces, but are cut or ground off, so that they slope irregularly towards the edges. As seen in the illustration, fig. 194*b*, the implement is slightly curved, and on the convex face of the lower and narrower end, fig. 194*c*, are the two grooves, which have been used, it is believed, to make the parallel lines so commonly found on the pottery of the Atlantic seaboard.

The upper, broader end, fig. 194, it will be noticed, has five similar grooves; these being on the opposite side of the implement, and thus on the convex face of this end.

This decorating stick or "ébauchoir" measures seven and three-fourths inches in length, three-fourths of an inch in width and one-fourth in thickness.

An examination of thousands of fragments of pottery, on which were series of lines, such as this implement would produce, clearly showed that some such object as this was generally used to make these linear impressions, as there was such a uniformity in the width of these lines, and in their distances apart, that it can scarcely be conceived that each line was separately produced. Experiment with soft or raw clay showed further that an implement of this kind produced the identical lines in earthen vessels, that are found to be characteristic of the majority of the fragments that bestrew the ground in so many localities, along the northern Atlantic seaboard.

This interesting bone instrument, which is supposed to be a unique specimen, was discovered by Mr. F. G. Galbraith, in the valley of the Susquehanna, in Lancaster Co., Penn., and by him presented to the

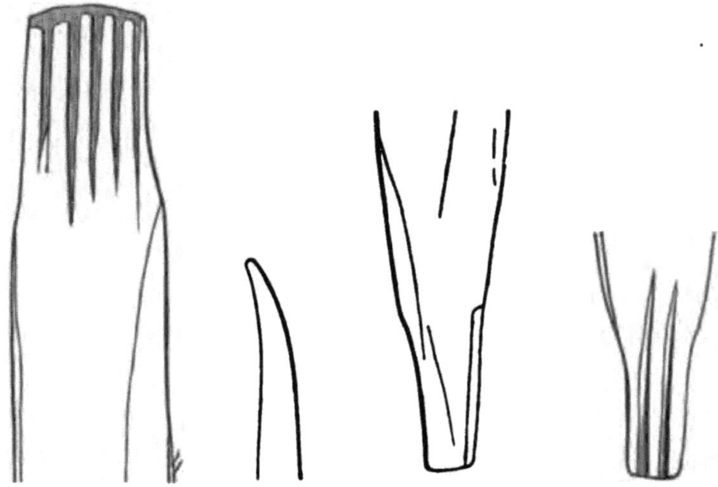

FIG. 194.—Penna. ½. *b.* *a.* *c.*

late Prof. S. S. Haldeman, to whom the author is indebted for the excellent wood-cuts representing it.

Professor Haldeman, believing this specimen to have been used in decorating pottery, has suggested the name of *ébauchoir* for it, which is a proper one to adopt, if it is necessary to go beyond our own language for terms that shall correctly convey a lucid idea of the purposes of such objects as were in ordinary use among the Indians.

Fig. 195 represents an implement which "is ten inches long, two and a half broad at the top, and one at the point. It is made of a branch of the antler of a moose or elk. The breadth of the upper portion is not seen in the figure, as the piece is represented edgewise.

It is obliquely truncated at the lower end, so as to give it a chisel-shaped edge, and shows the effect of having been hacked by some dull tool. Attached to a handle it might be used to dig with, or might serve as a head-breaker, or 'casse-tête,' as described by Father Rasles.[64] From Frenchman's Bay."

Fig. 196 represents "a flat pointed instrument, three and three-quarters inches long, and one and one-quarter wide. It is made of the dense exterior portion of an antler and the lower end has been ground down to a thin sharp edge as in fig. 196*a*."[65]

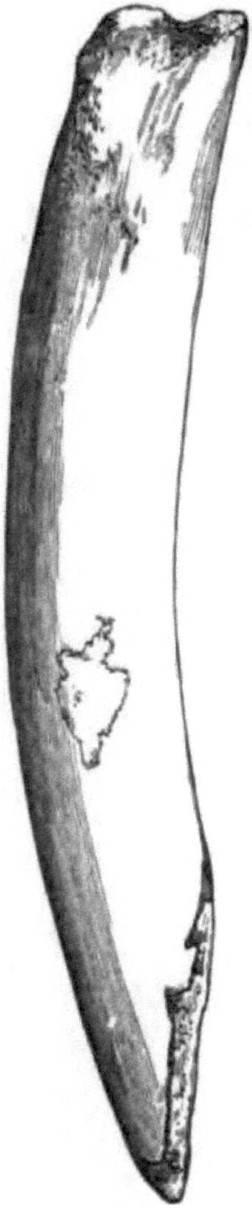

FIG. 195.—Mass. 1/2.

FIG. 196.—Massachusetts. 1/1.

196*a*.

Fig. 197 represents "a piece of one of the branches of the antler

[64] Lettres Edifiantes et Curieuse. Paris, 1838. Vol. i, p. 670.

[65] Wyman. Amer. Nat., vol. i, p. 580. The cuts here given, figs. 195 to 203 inclusive, with the descriptions, are from this article.

of a deer, from which the tip has been cut off. The sides near the pointed end have been worked down so as to present four faces, two of the angles uniting them being quite acute. The detached piece having a deep notch would be provided with two points or barbs and would be adapted to serve as the point of an arrow. Such points were used by the aborigines, and we are informed by Winslow,[66] that when the Pilgrims were making their first explorations on the shore at Cape Cod, previously to landing at Plymouth, some of the arrows shot at them had the kind of point just described. From Cotuit Port."

FIG. 197.—Mass. ½.

Fig. 198 represents "an artificially pointed fragment. From Crouch's Cove."

Fig. 199 represents "an artificially pointed fragment of bone, suitable for use as an awl. From Crouch's Cove."

While bone awls, of the same general pattern as fig. 199, are only of occasional occurrence in some inland localities, of New Jersey and New York, they are comparatively common in Ohio and other western and southern states. Inasmuch as prehistoric human bones are frequently found in a good state of preservation, it is not improbable, to say the least, that bone awls also might have escaped destruction in many instances, and hence that they would be far more common than they now are, had they ever been in general use among the Indians of the Delaware valley.

FIG. 198.—Mass. ½.

Fig. 200 represents "one of the lower incisors of a beaver, ground

[66] Young's Chronicles of the Pilgrims, p. 158. Boston, 1841.

to a thin, sharp edge. From Crouch's Cove." Mr. Frey[67] has

Fig. 199.—Mass. $\frac{1}{1}$.

Fig. 200.—Mass. $\frac{1}{1}$.

Fig. 201.—Mass. $\frac{1}{1}$.

figured a somewhat similar implement made from a beaver's tooth, found in a grave in the Mohawk valley, New York.

Figs. 202 and 203.—Mass. $\frac{1}{1}$.

Fig. 201 represents "a fragment of a bone of a bird, obliquely truncated and artificially sharpened. From Crouch's Cove."

Fig. 202 represents "a well wrought and polished spindle-shaped instrument, the lower end of which is flattened, and has a sharp edge: the upper portion is rounded with the end broken off, but appears to have been worked to a sharp point. From Frenchman's Bay."

Fig. 203 represents "a slender piece of bone, smoothly wrought and pointed. From Frenchman's Bay."

Figs. 204 and 205 represent other forms of bone implements, both of which are "made of flattened pieces, each being cut from the walls of one of the long

[67] Frey, *l. c.*, p. 782, fig. 13.

bones, and showing the cancellated structure on one of the sides. From Frenchman's Bay."

Figs. 206 and 207 represent two other examples of bone implements made in the same manner as the preceding. From Crouch's Cove.

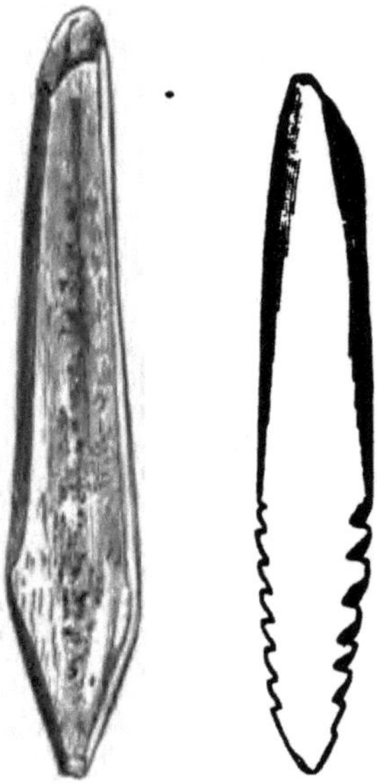

Figs. 204 and 205.—Mass. 1/1.

Those bone implements here described, with lateral notches, and barb-like projections, bear a marked resemblance to the bone "dartheads" figured and described by Major W. H. Dall,[68] in his admirable work on the shellheaps of the Aleutian Islands; and also to harpoons figured by Nilsson,[69] in his volume on Scandinavia. It is evident that these bone implements, from the Massachusetts shellheaps, were put to the same use as those from the Pacific coast of this continent and from other parts of the globe.

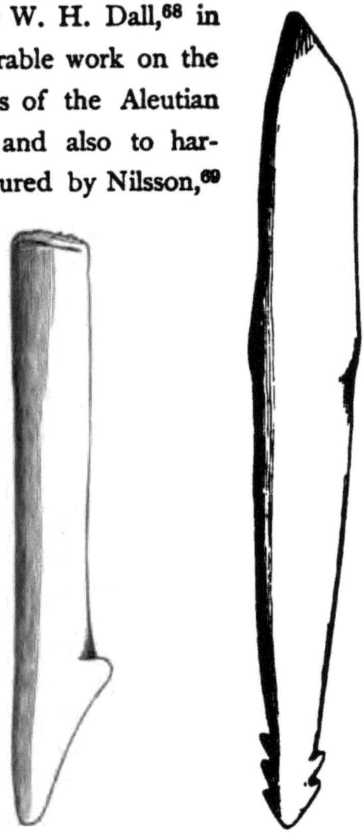

Figs. 206 and 207.—Mass. 1/1.

Fig. 208 represents a split fragment of a long bone, pointed at one end. There is nothing in the specimen in its present condition, to give any intimation of its use. Exact locality not recorded.

---

[68] Dall. Tribes of the Extreme Northwest, p. 76 and plate opposite (no number). Washington, D. C., 1877.

[69] Nilsson. Stone Age in Scandinavia, pl. iv, London, 1868.

Fig. 209 represents a somewhat similarly shaped object, but it has no point, and there is no indication that it ever had one. One edge is serrated, and "is quite sharp, but from this [serration] the bone rapidly increases to one-third of an inch in thickness, so as to render it wholly unsuitable to be used as a saw. From Eagle Hill."

Figs. 210 and 211 represent implements that are quite like the bone "sewing-awls" figured by Mr. Dall,[70] in the volume already quoted.

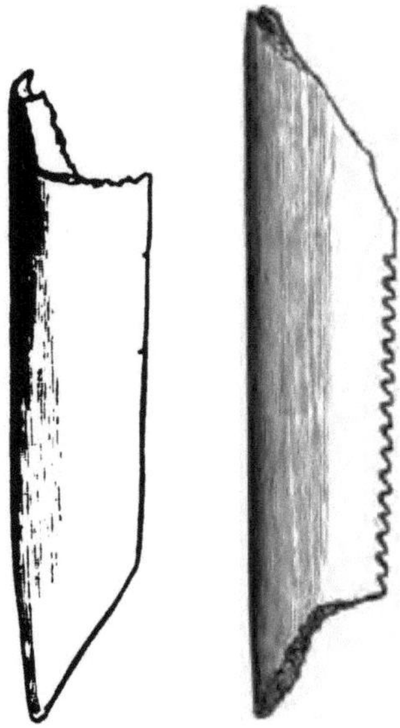

FIGS. 208 and 209.—Mass. $\frac{1}{1}$.

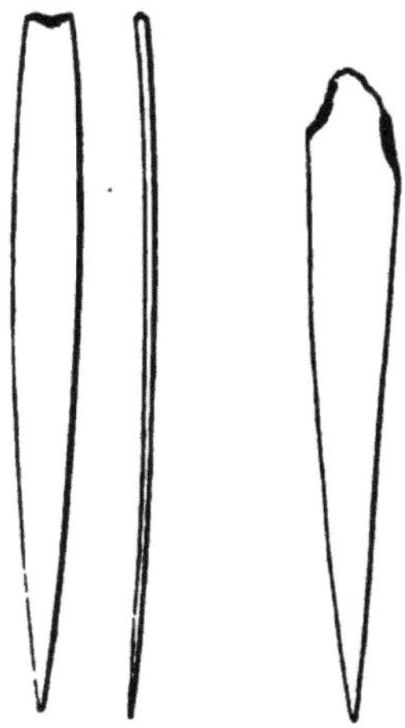

FIGS. 210, 210*a*. and 211.—Mass. $\frac{1}{1}$.

These specimens "are flat, scraped very thin, as seen in fig. 210*a*. One of these is made from the bone of a bird. From Eagle Hill."

The implements here described will probably cover the range of patterns in common use among the Indians, who formed the shell-heaps along the entire Atlantic coast, as even the forms of bone implements from the fresh water shellheaps of Florida[71] do not vary in any important particular.

---

[70] Dall, *l. c.*, plate opposite p. 79.

[71] Wyman. Fresh-Water Shell Mounds of Florida, p. 51; plates 3 and 4. Salem, Mass., 1875.

CHAPTER XVI.

AGRICULTURAL IMPLEMENTS.

ALTHOUGH the method of cultivating maize, the plant most intimately associated with the native races of America, was practically the same, there yet appears to be a considerable difference in the patterns of the implements, used in preparing the soil for the seed, and in its subsequent care and culture.

Over a limited extent of western territory, chipped flint hoes of a peculiar pattern are frequently met with. They are broad blades, with deep; lateral notches, near one end, by which they can be readily fastened to a wooden handle. Of these, it has been remarked by Dr. Rau,[72] who first described them, that "if the shape of the described implements (shovels and hoes) did not indicate their original use, the peculiar traces of wear which they exhibit would furnish almost conclusive evidence of the manner in which they have been employed; for that part with which the digging was done appears, notwithstanding the hardness of the material, perfectly smooth, as if glazed, and slightly striated in the direction in which the implement penetrated the ground."

This form is not found in New Jersey, but its place is taken by a "pecked" and grooved implement, in many respects similar to the gouges, already described. That they were ever very numerous is doubtful, as comparatively few have been found; but their widespread use is indicated by their occurrence throughout the middle and southern states.[73] They have been found in Connecticut, and are described among the implements met with in Georgia.

[72] Rau. Smithson. Annual Report for 1868, p. 379, fig. 1. Washington, D. C.

[73] Jones. Antiquities of Southern Indians, p. 301, pl. xvii, fig. 1. New York, 1873.

Implements known as "hoes," both of stone and elk-horn, not altogether unlike these in the more important features, have been figured and described by Professor Nilsson.[74] They differ principally from the New Jersey specimens in being perforated for the insertion of a handle. Of the use of such "hoes," Professor Nilsson says, "It must be acknowledged that if agriculture, as seems most probable, consisted originally in burning tracts of forest, and then sowing among the ashes, these rude hoes must have been very suitable for such operations."

Hoe-like implements, however, even those used by the Atlantic coast Indians, were not always of this pattern. In numerous references to the cultivation of maize, in the early histories, mention is made of bone implements, and of certain bones, which were thus used without alteration.

Of the Delaware and Iroquois Indians Loskiel[75] mentions that "they used formerly the shoulderblade of a deer, or a tortoise-shell sharpened upon a stone, and fastened to a thick stick, instead of a hoe;" and we readily see, on comparing such "hoes" with those of stone, here figured, that the latter are fully capable of doing the same work, and of doing it well.

A second class of agricultural implements, common in the west and south, but of rare occurrence in New Jersey and almost unknown in New England, are known as "spades." They are "oval plates of flint, flat on one side and slightly convex on the other, the outline being chipped into a sharp edge." These implements are of such a marked character that they never can be mistaken for any other form. The convex "backs" of these flint spades, and their large size and thickness, make them a very distinct and well-designed implement. As they are generally marked by fine striæ, and often polished as if by digging in loose soil, their use as spades seems to be beyond doubt. Very often

---

74 Nilsson. Stone Age in Scandinavia, p. 74, pl. viii, figs. 180 and 181. London, 1868.

75 Loskiel. Mission to N. A. Indians, p. 66. London, 1794.

objects are found associated with these true spades, which bear a general resemblance to them, and hence have been improperly classed with them. That all the large, disk-shaped and oval chipped jasper implements were spades or even hoe-blades, is not probable. What they really were will, in all probability, never be known.

The agricultural implements found on the Atlantic seaboard, about which there can be no reasonable doubt, are hoes and spades, of certain well-defined shapes. Of greater abundance than these, is a series of slate implements, that may or may not be hoe-blades. There are many reasons for considering these latter as agricultural implements of the character named, and yet if they are so, there is no reason why the jasper specimens of the same size and shape should not be included in the same class. To call one an agricultural implement and the other something else is certainly unwarranted.

FIG. 212. — New Jersey. ½.

Fig. 212 represents an average specimen of these long, slender, fine-edged slates, which we have designated lance-heads. They vary little from five to seven and one-half inches in length by from two to three and one-half inches in width; the longer specimens are usually the more slender ones, suggesting the possibility of the broader and shorter specimens having had a different use from the others. Of course, the same degree of probability obtains in this case, as in all others, that a difference in size, and slight variation in outline, may be indicative of different purposes. From the very nature of the case, it is obviously impossible to do more than conjecture, and wild guesses are neither scientific nor commendable.

Of a series of fifty-eight of these possible hoe-blades,[76] about twenty-five are made of argillite, and in their weathered surfaces, have all the appearance of age that characterizes the true palæolithic implements. The others are of slate and slate-like material, and the surface appears as if comparatively freshly chipped. In size, they vary from four and one-half to seven inches in length, by two and one-half to three inches in greatest width. One half of them have distinctly straight bases, and are obtusely pointed at the opposite end. In but one instance is there a really sharp point. While in some specimens there has been a wearing away of the pointed end, the general appearance of a large series gives the impression that the points were never sufficiently sharp, to have been used as a lance or spearpoint. A few specimens are the same at each end, and look as if they had been originally chipped in this shape. An interesting feature is seen in the well-defined notch that is found only on one side. This occurs in fully twenty per cent. of all found. In some specimens, this single notch on one side is very broad and deep, measuring in one example half an inch in depth, and an inch in width. This is, of course, exceptional, but from one of this size there is a gradation down to the slight, but readily discernible notch, an eighth of an inch deep. That these were in some way utilized in fastening the implement to a handle is probable.

Chipped slate implements of this pattern are found scattered along the whole Atlantic seaboard, but in varying numbers, in the several states. As they are found singly more frequently than the jasper implements of the same general pattern, it would appear as if they outnumbered the latter; but if we take into consideration the great

---

[76] In the museum of the Academy of Science, at Salem, Mass., is a large series of these objects, all gathered from a few adjoining fields. So numerous were they, and being found associated with an even greater number of fragments, it was supposed that they were weapons, possibly used and broken, here, in some great battle. The supposition that they are lance-heads, and indicative, through their numbers, of a battle field, is now believed to be an error. (See Smithsonian Annual Report, 1875, p. 269).

numbers of the jasper specimens found in "deposits," they really are less abundant.

A few facts that seem to have some bearing upon the question of the use of these slate implements will now be briefly mentioned. There are still to be seen, in various parts of New Jersey, certain barren, weed-grown tracts, or "clearings," as they are called, if still surrounded by a forest growth, known as old Indian fields. At the time of the settlement of the country by the Europeans, these tracts were under native cultivation. In many, the cropping seems to have been so persistent, year after year, that the fertility of the soil was finally exhausted; and to this day, it shows the ill effects of improper treatment. On such spots, there are found a larger number of these rude "hoe-blades" than elsewhere; and associated with them, are the true hoes, which will be subsequently described. It must not, of course, be understood that these implements are really rare in localities where maize could not have been cultivated. Such is not the case. All agricultural tools of native make are found in essentially *un*-agricultural localities. Their abundance on these old Indian fields is certainly suggestive, although why they should have been mostly left in the corn-fields does not appear.

If it were true that these slate implements are "hoe-blades," ought we not to find upon them the peculiar scratches that are supposed to mark all agricultural implements that have been used? The unquestioned hoes, and the "spades" that are found in New Jersey do present these traces of use; though the *jasper* implements, such as figs. 184 and 185, never do. The few flint spades we possess of the western form show them, and so, too, does a small proportion of the slate hoes. On the others it is believed that they have been obliterated by the weathering of two centuries. Occasionally, also, there occurs a specimen which is chipped flat upon the under side, convex upon the upper, and with a slight curve extending the whole length of the implement. These invariably have a single notch upon one side. In such specimens, we have a "link" between the western flint spades as described by Dr. Rau,[77] and the "hoe-blades" found in such great abundance in New Jersey. No similar implement of jasper has as yet been found.

Fig. 213 represents[78] an excellent average specimen of a hoe, such as is found in New Jersey. It has been "pecked" into shape, and subsequently ground until nearly all trace of the original pecking has been obliterated. The outer or upper surface (that shown in the illustration) is ridged, the height decreasing as it approaches the edge of the implement, thus making the greatest thickness at the head, where it measures an inch and a half. This head, oval in shape, is separated from the blade of the implement by two deep grooves or notches, which do not extend over any portion of the under surface, and scarcely meet upon the upper. This specimen now measures five and one-fourth inches in length, by a little over two and one-half inches in width. It evidently was originally much longer, and has been worn away by long continued use.

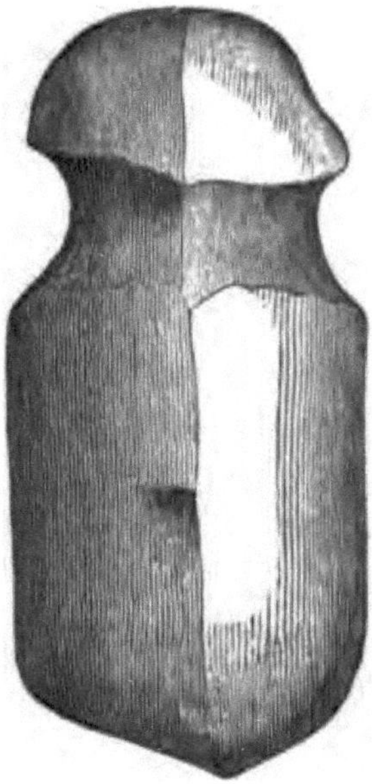

FIG. 213.—New Jersey. ½.

The fact that these hoes generally show such evident marks of use, and that the supposed "hoe-blades" do not, may be explained on the supposition that the latter do not possess the strength and durability of the former, and readily break when smartly struck upon a stone. They would probably be broken long before they would become striated, and worn by use. Probably the enormous number of halves of these blades, found scattered over small areas, is due to this cause.

---

[77] Rau. Smithsonian Annual Report for 1868, p. 379.

[78] The illustration is defective in the representation of these short lateral, and partially dorsal grooves.

From experiments made with these blades as they *now* are, it is probable that a new lot would have to be made at every seedtime.

Fig. 214 represents a second example of these heavy stone hoes, This specimen, like the preceding one, was originally much longer. but seems to have been worn down so much, that it was probably discarded. In one particular, it varies from the preceding: the under side is concave, while in fig. 213 it is flat. This curvature of the blade might be supposed to indicate that implements of this character were really gouges, and as such referable to the class of objects described in Chapter III; but the presence of the striæ and the general appearance of the surface, show clearly that, like the preceding, this specimen also is a hoe.

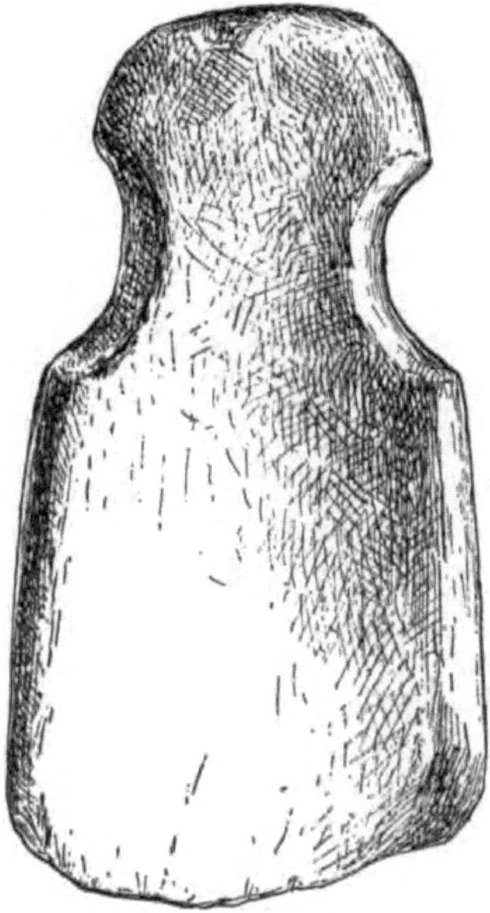

FIG. 214.—New Jersey. ½.

Fig. 215 represents a stone spade, such as is found in considerable numbers in Mercer Co., New Jersey, but which is not known in other localities. These spades are all of the general shape and character of chipping shown in fig. 215, which sufficiently explains itself. All of those collected show distinctly the striæ, and polish amounting, in some, to a glaze, which characterize the western spades, of an elongated oval, or quadrangular outline. The material of which these New

FIG. 215.—New Jersey. ½.

Jersey spades are made varies, but none are of jasper. Slate, argillite, and other material of this character, have been used. While the narrowed "handle," and broad blade are well-marked features in all these specimens, they merge into each other, and are not distinctly separate, as in the spade-like implement figured by Col. C. C. Jones.[79] In some, however, the junction of the two portions is more abrupt than in the specimen figured.

All of these chipped spades were found on a limited area of some three or four hundred acres; and were associated with the supposed "hoe-blades" and the grooved hoes, previously described. It is possible that they may be a local form, but the pattern is so simple in design, and so well adapted for digging in loose, sandy soils, that similar objects may be confidently looked for in other localities.

When the extent to which maize was raised is considered, it is evident, that even with the supposed "hoe-blades," the number of stone agricultural implements is too small to have met the requirements of that cultivation, and hence it is more than probable, that the hoes made of the shoulderblades of deer, and of tortoise shell, mentioned by several early writers, greatly outnumbered those of stone.

---

[79] Jones. Antiquities of the Southern Indians, p. 302, plate xvii, fig. 2. New York, 1873.

CHAPTER XVII.

PLUMMETS.

As has been done in previous chapters, a series of objects, uniform in the more characteristic features, but infinitely varied in minor details, have been grouped together under one title. While the general name given to the group is convenient, and in a measure descriptive of their appearance, it has but little reference to their uses. It has been remarked,[80] that "their principal use * * * * as 'plummets' may be * * * * questioned, as there are far too many of them found, and of too great a variation in size, to lead us to infer that they were used mainly for that purpose. Though if it were necessary, in ancient architecture, to establish a perpendicular line, the implements were at hand, as 'weights' with lines attached." The same author has also remarked of these plummets, that they "are of quite common occurrence in the vicinity of Salem, Mass.," and there are "in the collection of the Peabody Academy of Science, a large number of specimens varying in size from an ounce or two to several pounds in weight, but all made on the general pear-shaped pattern, though they exhibit about as many modifications within that shape as are shown by the hundreds of varieties of the pear itself. Local archæologists in New England generally consider them as 'sinkers,' from their shape and from the fact that they are more often found along the seashore than in the interior."

"The very large size of some specimens would perhaps indicate some different use from any proposed" for the western specimens, which, as a whole, are smaller. "In fact, some of them run so decid-

[80] Putnam. American Naturalist, vol. vi, p. 649, foot-notes. Salem, Mass., 1872.

edly into the class of 'pestles,' that it is almost impossible to draw the line between the two, though the extremes are well marked. The peculiar shape of these implements has also caused them to be regarded as weights, used to stretch the thread in spinning. This supposition is rendered very probable by the fact that stone weights have been used in spinning, and from the statement (made to me in conversation by Dr. Edward Palmer), that stones are still in use among the Indians of the northwest for a similar purpose. As it is generally admitted that the moundbuilders understood spinning different kinds of fibre, and *twisting* certain materials into threads which they, by some process akin to weaving, subsequently manufactured into a kind of cloth, the use of these implements as weights seems very probable, and as household implements they would often be more or less elaborately finished or carved. For my own part, I have for some time considered them as representing, according to size, material, shape and finish, either 1st, pestles, 2nd, sinkers, 3rd, spinning weights, or 4th, ornaments."

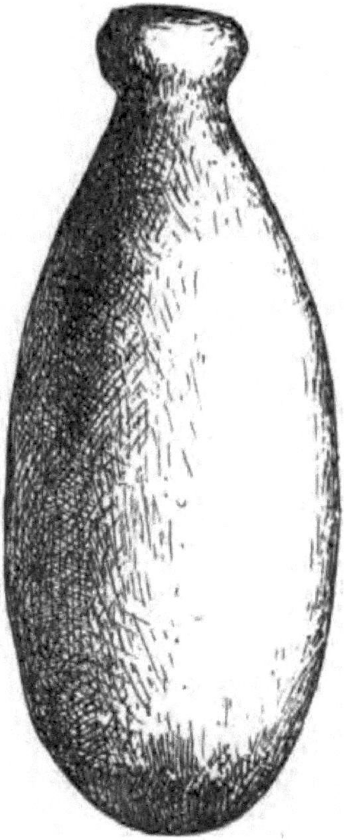

FIG. 216. — Massachusetts. ½.

Fig. 216 represents a characteristic specimen of the New England plummets of about the maximum size. Whether it should be classed as a "pestle" or not, cannot be readily determined; but the fact that unquestionable pestles are by no means infrequent in the same localities, would seem to throw doubt upon their use as pestles, although "many of them * * * * would serve well for use as such, provided grit was no objection as a component of 'Indian cake;' though the grit would be avoided if such pestles were used in wooden mortars similar to those in use by the early white settlers in this country. The extreme softness of the stone of which these large pear-shaped implements were made, combined with the fact that they seldom exhibit signs of use at their rounded end, was * * * * (one) argument against their use as pestles," but not the *only* one, as that author considered.[81] There does not seem to be any objection to the use of these larger specimens as sinkers, from the mere fact of their size and weight. In comparatively deep waters, or in strong currents, specimens as heavy as this would often be required. Bearing on the subject of their use as pestles, however, it should be stated that there is a specimen in the Salem collection that is said to have been found in a stone mortar, and another in the cabinet of the Amesbury Nat. History Society. Fig. 216 was found near Amesbury, Mass.

FIG. 217.—Maine. 1/1.

Fig. 217 represents a much smaller specimen of a plummet, which was probably used as a sinker for a fishing line; at least, it is well adapted for this purpose. The original was long ago described and figured by Schoolcraft, as "a fisherman's sinker, of the Penacook tribe, accurately wrought in stone." Objects of this size are not confined to fishing stations, however, and it is quite as possible, that similar specimens had different uses in different localities. Of a very large series of these plummets in the collections of the Museum at Cambridge, Mass., but few are highly polished and possessed of that

[81] Putnam. Bulletin of Essex Institute, vol. v, June, 1873. Salem, Mass.

accuracy of outline so characteristic of many found in Ohio and other western states. Of the larger examples found so frequently in the neighborhood of Salem, Mass., a large majority are merely pecked into shape and have no other finish. These have an added interest, from the fact that in the same neighborhood, are found in like profusion, the stone gouges that are supposed to have been principally used in making the "dug-out" canoes. If such was the ordinary use of these implements, their occurrence in unusual numbers, associated with a like abundance of plummets, would certainly seem to indicate that the latter were used as "sinkers," as we know the canoes were made more for fishing purposes, than for the mere convenience of travelling by water.

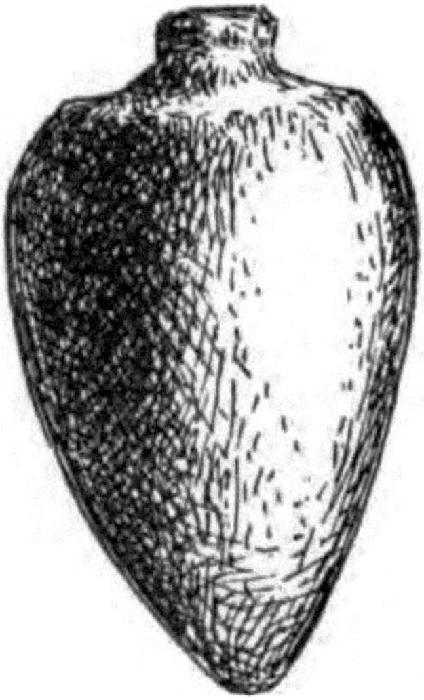

FIG. 218. — Massachusetts. 1/1.

Fig. 218 represents a still smaller specimen of these plummets, which certainly is most admirably designed as a sinker, for fishing in comparatively quiet waters. It does not weigh so much, and is but little longer than lead sinkers that are now used in trolling for rock-fish; and it is a matter of astonishment that such an admirable implement should have been found so very seldom in New Jersey. As in the southern states, an occasional specimen, like fig. 218, may be found; but all such as have been examined, purporting to be from the valley of the Delaware river, are ingenious frauds, copies in steatite of New England specimens, manufactured for "the trade" by unscrupulous dealers. The simpler form of the grooved globular pebble may, in New Jersey, replace the plummet, as it is believed by Col. C. C. Jones,[82] to do, in Georgia. That author groups notched and perforated pebbles, and grooved

[82] Jones. Antiquities of Southern Indians, p. 339, pl. xix, fig. 12. New York, 1873.

globular pebbles as alike, sinkers, and figures a rudely fashioned plummet, which is of rare occurrence, and suggests that it was "employed to weight the hand-line in fishing with a hook."

Fig. 219 represents a plummet-shaped implement, four and one-half inches in length, and one inch and three-quarters in its greatest breadth. The material is a very fine-grained sandstone, and the whole surface has been worked down, until its smoothness almost amounts to a polish.

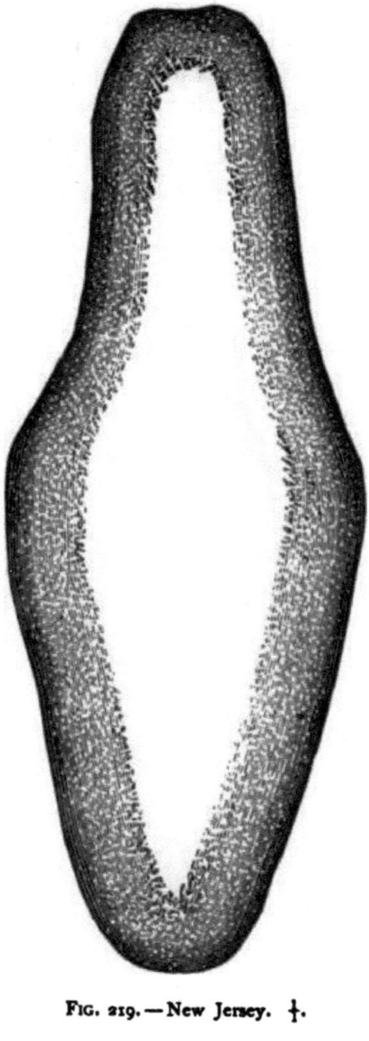

FIG. 219.—New Jersey. 1/1.

While the resemblance of this specimen to the common forms of plummets is marked, it varies from all the New England specimens that have been examined, in not having a groove near the upper or smaller end, for its suspension. This, however, does not indicate that the purpose of the implement may not have been one of the several suggested by Professor Putnam, on page 228. Plummets of both stone and hematite are frequently found in Ohio. Whether used as plummets, sinkers, or ornaments, this want of a groove or other means of fastening it to a cord, renders the object of the Ohio specimens, as well as of fig. 219, a difficult matter to determine. Mr. Henderson,[83] from whose interesting account of plummets, much

[83] Henderson. American Naturalist, vol. vi, p. 644, figs. 132–138. Salem, Mass., 1872.

information has been derived, considers the grooves upon such specimens even as figs. 220 and 221, too small to have insured a secure attachment to a fishing line.

We are therefore quite in the dark as to the precise belongings of this New Jersey specimen, as it, like so many objects that are found of Indian make, is so unlike anything in use at the present time, that it is quite impossible to identify it.

As an interesting instance of the identity of form of certain patterns

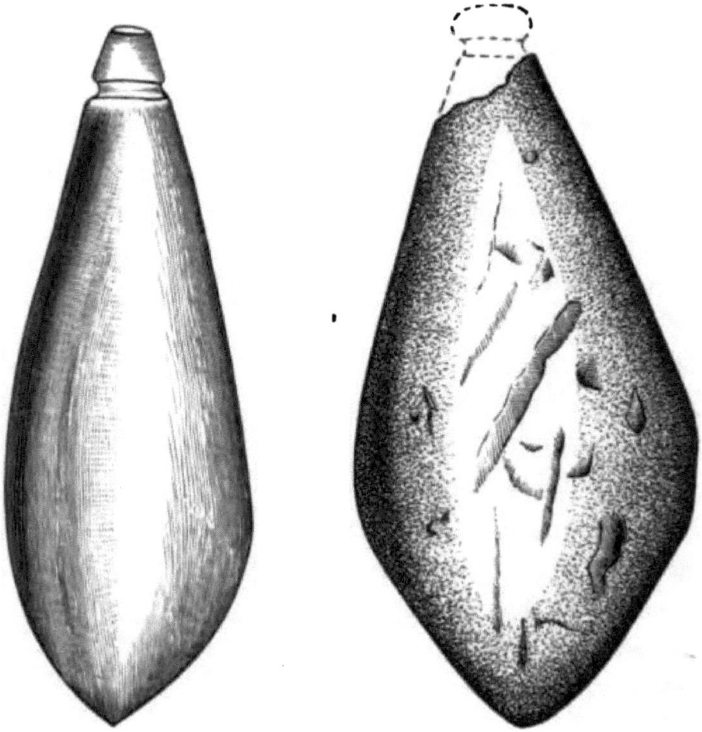

FIGS. 220 and 221. — Illinois. ½.

of stone implements that are found in all parts of the country, and used by people as dissimilar in their modes of life as the native races of the Pacific and of the Atlantic coasts, and also the moundbuilders, of whom it may be said that their separate origin and distinct racial characteristics have not yet been proven, notwithstanding the laborious investigations of many laborers in American archæology, and the ingenious theorizing that swells the pages of so many volumes; illustrations of a series of the western forms are here introduced which, while made of many different materials, are practically the same as the specimens from Maine and Massachusetts, figs. 217 and 218.[84]

Fig. 220 "represents what may be styled the typical form of these implements. It is made of iron ore, ground down and polished, until it is almost as smooth as glass." It is one of eight found near Quincy, Illinois, embedded in solid clay.

Fig. 221 represents a second example, and varies from the preceding, in being much broader at its widest part. This specimen is also from Illinois; and is "made of a whitish limestone containing numerous small joints of crinoids."

FIG. 222.—Ohio. 1/1.

FIG. 223.—Illinois. 1/1.

Fig. 222 represents a much smaller specimen, of the same class of implements, and is of unusual interest from the fact that it was found at a depth of "sixteen feet below the surface of the earth." "It is made of green stone. With it were found a small stone scraper and a stone disk."

Fig. 223 represents still another, made of copper. "It is composed of small pieces of native copper pounded together; and in the cracks between the pieces, are stuck several pieces of silver, one nearly the size of a half dime."[85] This specimen is from a mound near Marietta, Ohio.

---

[84] Henderson, *l. c.*, pp. 642–649.

[85] Squier. Aboriginal Monuments of New York, p. 187. Washington, D. C. 1856 (Smithsonian Contributions to Knowledge, vol. ii).

Considering the small size of some of these specimens, particularly fig. 222, it may be thought that the small trinkets, or cylindrical pebbles with an encircling groove, described in Chapter XXVII, belong to the same class. They certainly approach very near to them, though they are so small that they would be nearly worthless as sinkers. This, however, is not true of fig. 222, which is probably a sinker of the minimum size.

While plummets, as a class, seem to have been polished and sym-

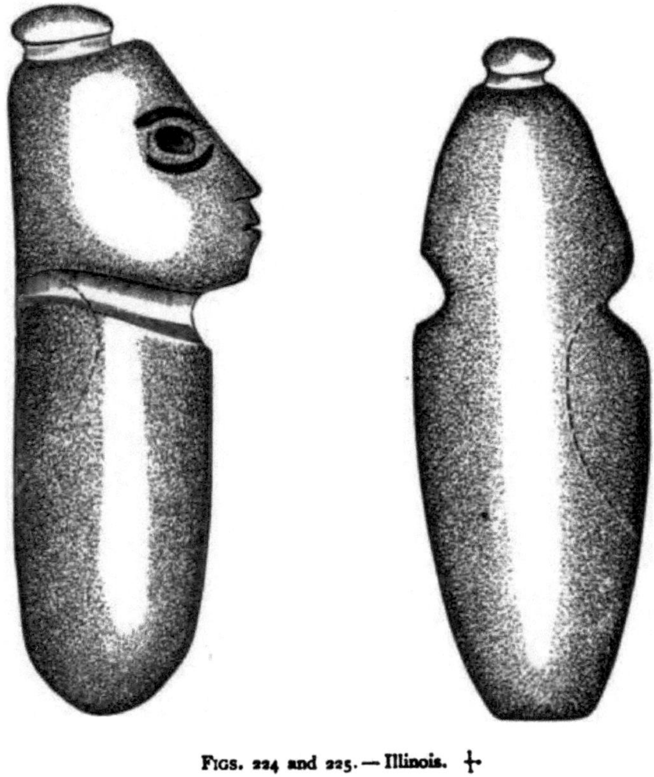

FIGS. 224 and 225.—Illinois. ½

metrically shaped, they were not generally ornamented in any other manner. Figs. 224 and 225, however, represent one of a small series, that is a marked exception in this respect. This specimen has already been described and figured by Mr. Henderson, and the illustrations, as well as several of the preceding, are those given in his valuable paper, already referred to. Fig. 224 is made of dark limestone, and the top of the implement has been carved to represent an Indian's head. The carving is successful certainly, so far as "presenting the characteristic features of the Red Indian." "The streaks of black paint above and below the eyes, the black eye-balls and scalp-lock, give it a hideous appearance which, perhaps, caused it to be looked upon with reverential awe by its superstitious aboriginal owner." "A glance at fig. 225, which is a back view of fig. 224, shows that the primitive artist preserved, as far as his design would admit, the general plummet form shown in the other implements figured, while the slight groove around the small end shows plainly that it was to be suspended by a string, and I think fully warrants the conclusion that this implement is but a modification of the plummet."

There is in this carved specimen an instance of that taste for realistic carvings, which is more artistically exhibited on some of the New England pestles, than on this plummet; and considering also that "the characteristic features of the Red Indian" are shown on this plummet, it is evident that the Indians used and made a large number of these implements in the west, as well as along the Atlantic seaboard. If it could be shown that this carving on fig. 224 was the handiwork of the moundbuilders, then why have we not an indication that they and the Indians were closely related? Certainly there is little that is peculiar in the representations of the human face, as exhibited on stone carvings taken from mounds; and in that little, there is a resemblance to the later (?) Red Indian. Until the origin of the known races of North America is unquestionably determined, it will clearly be unsafe to designate this or that implement as the production of another people, of a supposed different origin.

CHAPTER XVIII.

NET-SINKERS.

AMONG the many familiar forms of primitive stone implements common to the valleys of the Delaware and Susquehanna rivers, are those small, flat, notched pebbles, usually nearly square, but not unfrequently of the most irregular shapes, which are universally known as net-sinkers. The fact that the use implied by this name is suggested at once by the very appearance of this implement is not perhaps a safe guide in determining such questions, although, in this case, it is almost certainly true. If used as net-sinkers, then, it is evident, that occasionally a series of these objects should be found together, as numbers of them, placed at short distances along the lower margin of the net would be required to keep it in place when in the water. If they could be found so situated, that their position was evidence that they had been attached to a single net, all doubt as to this form of fishing tackle having been used by the Indians would be dispelled, even though they alone remained to testify to the fact. While no direct reference to nets is made by Holm,[86] in his history of the Delaware Indians, he does allude to other productions of a character that, at least, renders the weaving of a net a probability. In detailing the several duties of the women, he remarks: "they make much use of painted feathers, with which they adorn skins and bed-covers, binding them with *a kind of net work* which is very handsome, and fastens the feathers very well." Attention has also been called to the fact that they made "thread and yarn, from nettles and hemp." Peter Kalm also refers to fishing tackle made of hempen cord. (See

[86] Holm, *l. c.* p. 129.

p. 147.) These facts are sufficient to identify the notched pebbles as net-weights or sinkers, but other evidence is not wanting.

In the summer of 1878, a series of these notched pebbles was found in the wasting northern shore of Crosswick's creek, about two miles from its mouth, at Bordentown, New Jersey. They were in an irregular heap, in some instances one just above the other, but in contact. They were twenty-two inches below the surface of the meadow, which is composed of a fine sandy mud, that has been slowly accumulating, at this point, for centuries. There were seventy-three in the series, and supposing them to have been placed at a distance of a foot apart, they would have supplied a net just long enough to stretch across the creek at this point. It is not improbable, however, that these notched pebbles were left long anterior to the formation of the present channel of the creek, and hence are of a remote antiquity; for recent as are the alluvial deposits in our river valleys, some of them are not to be counted by scores of years more or less.[87]

"Fishing-nets may be counted among the utensils invented at very early periods, on the spur of necessity, by men in various parts of the world. That they were already in use in Europe at a remote antiquity is proved by their remnants preserved in an almost marvellous manner in the Swiss pile-constructions of the stone age, as, for instance, those of Robenhausen and Wangen. In the earliest works on North America the fishing-nets of the Indians are mentioned but not described. Cabeça de Vaca,[88] the first European who gave an account of the interior of North America, refers in various places, though in a transient manner, to the nets of the natives whom he met during his long wanderings. Garcilasso de la Vega[89] and the anonymous Portuguese gentleman, called the Knight of Elvas, the two principal authors who have left accounts of De Soto's expedition (1539–43) are likewise

[87] Abbott. American Naturalist, vol. x, p. 71.

[88] Relation et Naufrages d'Alvar Nuñez Cabeça de Vaca (Ternaux Compans), Paris, 1837, pp. 24, 142, 177, 179. Original printed at Valladolid in 1555.

[89] Narratives of the Career of Hernando de Soto, etc., translated by Buckingham Smith, New York, 1866, p. 112.

deficient in all such details as might serve to illustrate the original character of Indian nets. The latter relates, however, that the Spaniards, while at a place near the Mississippi, called Pacaha (Garcilasso has it 'Capaha'), caught fish in a lake with nets furnished by the Indians. This establishes at least the fact that the tribes of the Mississippi valley employed fishing-nets, when first seen by Europeans. The Indians of the present New England states made strong nets of hemp. For this we have the authority of Roger Williams,[90] who gives also the word ashòp, which signifies a net in the language of the Narragansetts. Yet it appears that the Indians of the Atlantic coast (and others) were rather in the habit of 'spearing' fish than taking them in nets. Some were also killed by arrow-shots. According to Van der Donck, the Indians in the neighborhood of New Amsterdam (now New York) employed, during the middle of the seventeenth century, various kinds of nets; but this author does not state whether these nets were original Indian inventions, or adopted from the Dutch colonists. The Natchez, on the lower Mississippi, made their nets from the bark of the linden tree, and knitted them quite in the European fashion." (Chas. Rau, in American Naturalist, vol. vii, pp. 145 and 146.)

Fig. 226 represents an ordinary specimen, such as occurs by the hundreds in the valley of every creek, and along the river shores, in New Jersey, and other eastern states.

Beyond determining their use, there is nothing in their variety, shape, or method of manufacture, but is so simple and evident, that there is little interest attached to them; unless it be to wonder why it was, if these primitive sinkers were used in the manner described, as their numbers and circumstances under which many are found cer-

---

[90] Roger Williams. A Key into the Language of America, London, 1643: Providence, R. I., 1827, p. 102. The practice likewise prevailed of erecting in the water large labyrinth-like enclosures of lattice-work, flanked by long weirs, the whole forming a sort of gigantic trap, into which the fish were driven. Such a contrivance of the Virginia Indians is figured and described in the first volume of De Bry's "Peregrinationes" (Frankfort on the Main, 1590).

Beschryvinge Van Nieuw-Nederlandt. Amsterdam, 1656, p. 70.

Du Pratz. Histoire de la Louisiane. Paris, 1758, vol. ii, p. 179.

tainly indicate was the case, that so carefully wrought an implement as the "plummet" of New England, should have been used for the same purpose. If they were used only as sinkers for fishing lines, it is not a matter of surprise; but certainly if these small flat pebbles were all that were required to weight the nets used in the rivers, then larger pebbles of the same shape would have answered the needs of the fishermen of the New England coast just as well, and it is strange that so simple a form of weight should not have been used, if indeed, it was not.

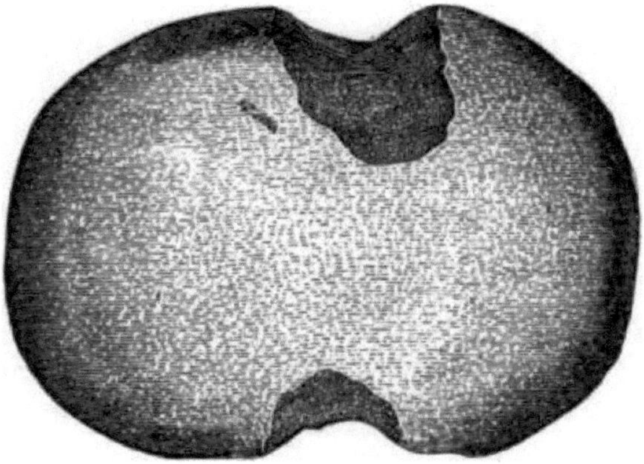

FIG. 226. — New Jersey. ½.

Figs. 227 and 228 represent examples of these sinkers from the shores of the Susquehanna river. They are of the same pattern as those found in all other river valleys of that region; as a class, however, those found in the Susquehanna valley are somewhat larger than those of the Delaware. Dr. Rau[91] writes of the specimens, above figured, "the material

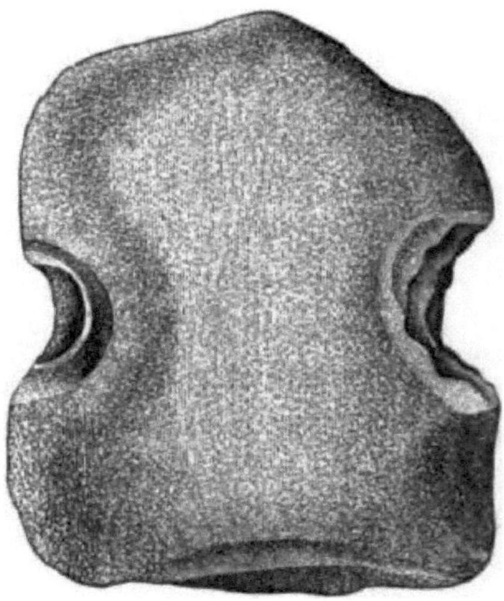

FIG. 227. — Pennsylvania. ½.

[91] Rau. American Naturalist, vol. vii, p. 140, figs. 30 and 31.

is almost exclusively a flat-breaking, silico-argillaceous stone of gray or brownish color, sometimes containing diminutive particles of mica, and consequently bearing the distinct character of graywacke."

The net-sinkers found in the Delaware valley are made of every sort of stone, and even thin pieces of coral rock have been utilized. Some are even halves of "hoe-blades." No one form of implement presents a much greater range of minerals, than these sinkers, although the majority are made of flat river pebbles of sandstone.

Fig. 229 represents an average specimen of net-sinker from the shores of the Delaware river. But few are of this pointed shape, however, and a great many do not have the notches so deeply cut.

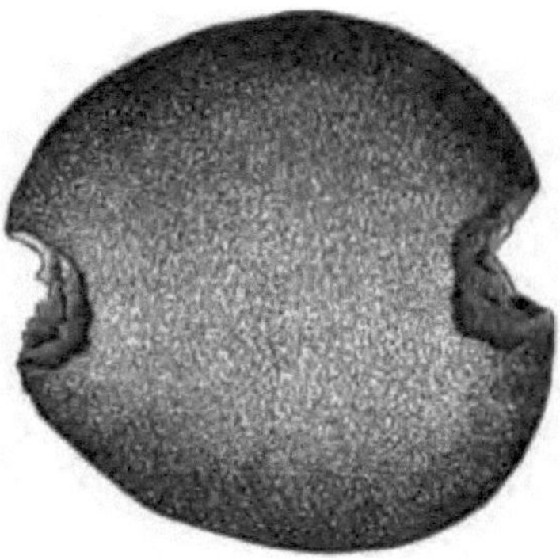

FIG. 228.—Pennsylvania. $\frac{1}{1}$.

Not all of these notched pebbles need necessarily be classed as net-sinkers, in the ordinary acceptation of that term. Some are found that are too cumbersome to have been used in that way, but were probably anchors or set-weights for fishing apparatus of a different kind. In June, 1879, while relic hunting in the Delaware valley, with Prof. F. W. Putnam, of the Museum at Cambridge, Mass., the author found a very large notched pebble on the shore of the river, a short distance above the Water Gap, in Monroe Co., Pennsylvania, which, judging from the size, and the fact of its having four notches, was used as an anchor or set-weight. This example measures eight inches square, and weighs nearly five pounds. To secure a net, which was placed in the stream, as gilling nets and fykes are now set, such a weight would have been frequently a necessity, especially where there was a swift current, as there is in the river, at the point where this specimen was found; but it is evidently impossible that such a stone could have been used, as one of a hundred or more, in dragging a sweep net through the water. Aside from their weight, stones of such size would constantly be caught by obstructions in the bed of the stream, and thus render the free movement of a net impracticable.

On the other hand, the fact of finding an implement of practically the same character, but of much greater size, in a fishing locality, associated with hundreds of smaller ones, suggests that it, too, was used by the Indians, in fishing, and is corroborative of the belief, that the smaller specimens were used in the same manner.

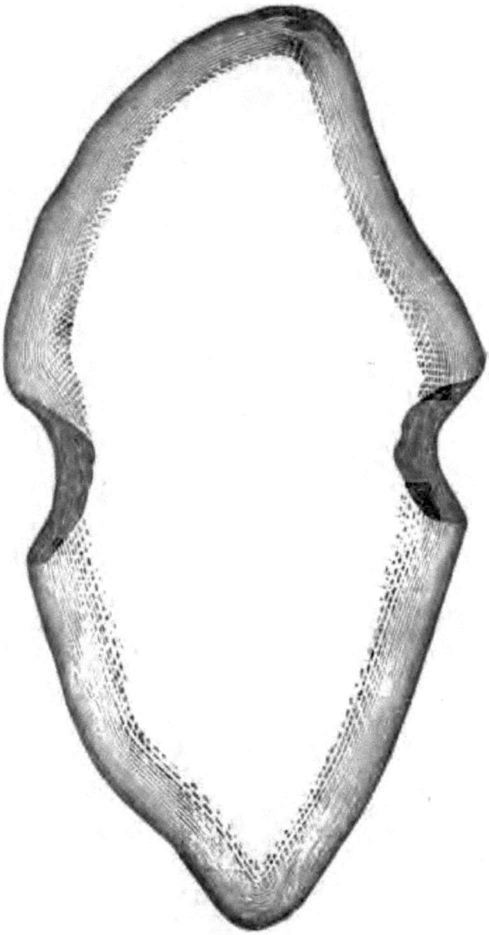

FIG. 229.—New Jersey. ½.

These large notched stones may have been used also as anchors.

Large angular pebbles or bowlders, with deep encircling grooves, have also been frequently found in the Delaware river as well as in many of the larger creeks flowing into it. These grooved bowlders, I believe, were used also as anchoring stones.

One of these so-called anchors, found in the bed of Crosswicks Creek, near Bordentown, New Jersey, is a compact sandstone bowlder, nearly a cube in shape, and weighs forty pounds. The groove divides the stone into equal parts, is evenly worked, and measures uniformly one inch in width and three-fourths of an inch in depth.

This specimen was found embedded in mud, at a depth of nearly three feet from the present surface. Near it were found a dozen notched pebbles, such as fig. 228, a grooved stone axe, and several fragments of pottery.

The circumstances under which this grooved bowlder was found clearly indicate that it was used as an anchor; and its being associated with a small series of the notched pebbles, described in the present chapter, is as interesting as it is suggestive. Unlike the large notched pebbles, referred to from the Water Gap, this specimen could not have been used as an attachment to a net; but at once suggests the use of a boat, and as we know that these boats were in almost daily use, it is not probable that they were always drawn from the water, when not in use. Holm[92] remarks of the Delaware Indians, that "their boats are made of the bark of cedar, and birch trees, bound together and lashed very strongly; they carry them along wherever they go; and when they come to some creek that they want to get over, they launch them and go whither they please. They also used to make boats out of cedar trees which they burnt inside, and scraped off the coals with sharp stones, bones, or muscle shells."

Such a stone "anchor," as the one mentioned, would readily hold, even in a swift current, the canoes and dug-outs, Holm describes.

Fig. 230 represents a large, oval, flat pebble, with a perforation, that has been drilled with great accuracy. Objects of this kind, so far as met with in New Jersey, are comparatively rare, and their purpose is not positively known. They are here classed as sinkers, because the localities where most of them are found suggest that they were used as net-weights or, in some manner, were connected with the occupation of fishing. Fig. 230, which is the largest of a series of fourteen specimens collected, is nearly one and one-half inches in thickness, and was found in a field, within a short distance of a navigable creek, and one much frequented by Indians, even so lately as historic times. The other thirteen were found in the immediate vicinity of Big

[92] Holm, *l. c.*, p. 130.

Timber Creek, Gloucester Co., New Jersey, and are but a small portion

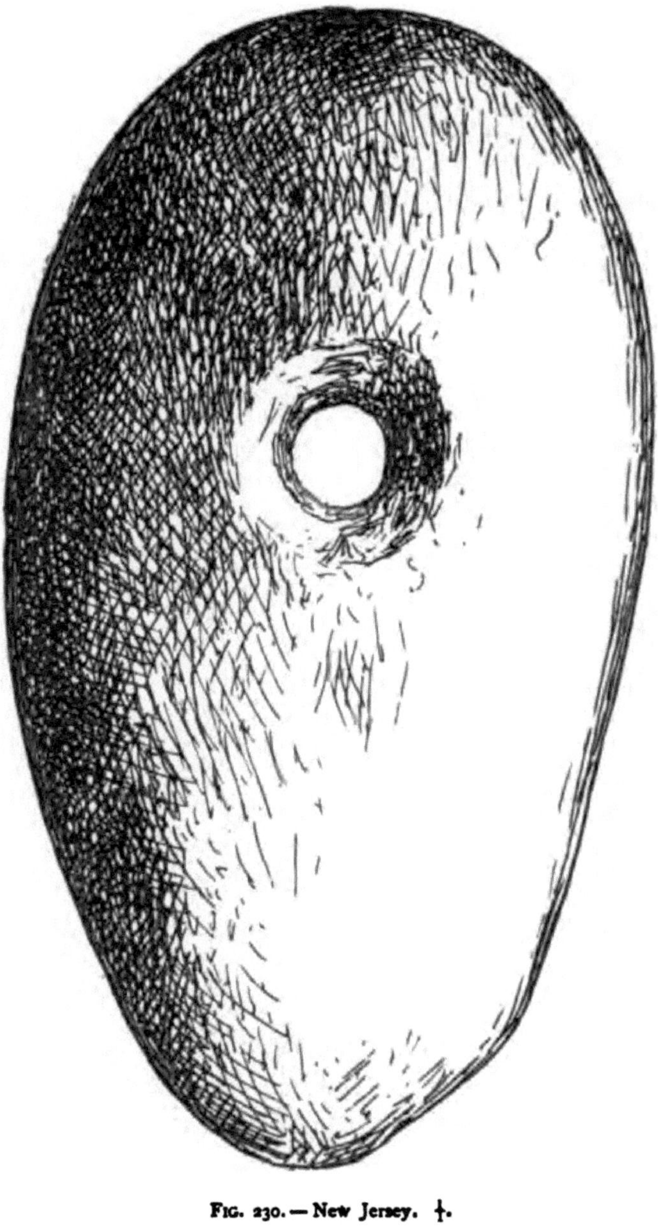

FIG. 230.—New Jersey. $\frac{1}{1}$.

of a large number that were gathered at one time. All that have been examined are made of compact sandstone, and are very heavy. Most of them have been drilled from each side, and the opening, near the middle of the stone, is much smaller than the orifice at the surface. Fig. 230 has probably been perforated with a hollow drill; the sides of the perforation being as even, smooth, and regularly striated, as the finest examples of drilled ceremonial objects.

A number of these perforated flat pebbles have been recently found on the shores and in the bed, of Lake Hopatcong, Morris Co., New Jersey. Those that I have seen varied in no essential features from fig. 230, except in the manner of the drilling. This was similar to that of the specimens from Gloucester Co., New Jersey.

Col. C. C. Jones[93] has figured and described a series of perforated net-sinkers, of which, he remarks, "all of the perforated sort that I have seen, with one exception, were formed either of soapstone or of clay. Consisting generally of flat or rounded pieces of soapstone, irregular in shape, they vary in weight from scarcely more than an ounce to a pound and upward. The perforations are from a quarter of an inch to an inch in diameter, and are indifferently located, either in the centre or near the edge of the stone."

None of soapstone have been noticed, among the New Jersey examples, nor any so small as those of but an ounce in weight. While in all probability used as net-sinkers, the New Jersey specimens were doubtlessly limited to particular kinds of nets or traps.

---

[93] Jones, *l. c.*, p. 337, pl. xix, figs. 1 to 6 inclusive.

CHAPTER XIX.

SPEARPOINTS AND ARROWHEADS.

When we come to examine any considerable series of chipped stone implements, and notice how varied are the patterns of what is practically the same object, we are forced to the conclusion that these several patterns were designed not for one, but for many purposes.

In the study of these varied forms, we cannot rest content with the knowledge that they are spearpoints. When, where, how were they used? Have we no clew to their meaning? The meagre records of those early voyagers, who first chanced upon our shores, tell us but little more than the discarded implements themselves.

Perhaps the efforts to determine the object of various stone implements by the character of the localities where they are usually found have not been altogether vain. With reference to some of the simpler forms, this is not an important matter, as their very simplicity and uniformity bespeak the use, as in the notched pebbles used as net-weights; and yet even here, the fact that they are found in abundance along our rivers and larger creeks, and that often scores are discovered associated together, in the very beds of the streams, is certainly an additional assurance that their purpose is known.

While the objects treated of in the present chapter are perhaps without warrant classified as spearpoints and arrowheads, it is not intended to convey the impression, that all the larger specimens belonged to the former class, and that the entire series of smaller flints were used to tip the shafts of arrows. This cannot be proven, though the shape, size and the relative abundance of the two series render it evident that such was usually the case.

Whether the one type or pattern that is here considered as having been used principally for capturing fish was generally so used, future investigation may possibly determine. At present, there is considerable evidence in favor of the view here expressed.

Schoolcraft[94] has referred to the larger of these spearpoints as "Antique javelins, or Indian Shemagon or spear." "This antique implement was one of the most efficacious in close encounters, before the introduction of iron weapons. A fine specimen is seven inches long and one and a half wide at the lower end, which is chipped thin to admit the splints by which it was fastened to the lower end. The length of the pole or staff could only be conjectured, and was probably five feet. The chief said, on presenting it, that it was one of the old implements of his ancestors."

When we consider how prominent and comparatively abundant are these large spearpoints among the relics of the Indians, it is not a little strange that the early writers, who refer to the Indians before they had wholly discarded stone implements, or very soon afterwards, should so generally have overlooked this form, while they frequently mention their axes and arrowheads. Neither Holm nor Kalm refer to the large spearpoints as a weapon of the Delaware Indians, or refer to the use of the spear or lance, in describing their methods of warfare; yet the number of these objects found is, of itself, sufficient to indicate that, at one time, they were in very common use. Is it probable that they had been discarded in great measure, at some remote period, and were veritable relics of a distant past, when the European settlers first reached our shores? The absence of direct reference to these characteristic implements seems indicative of this.

Fig. 231 represents a perfect specimen of what may be considered a typical spearpoint. The chipping is successful, so far as preserving a uniform thickness of the blade, and the edges are straight, and taper gradually to the moderately acute point. The stem is a perfectly straight projection from the base of the blade, of a little more than one-half its width. The material is a bluish-gray jasper, very com-

---

[94] Schoolcraft. Hist. and Cond. of Indian Tribes, pt. 1, p. 87, pl. 26.

monly used by the Delaware Indians for making implements of this character. While occasional specimens of this and allied patterns of spearpoints are found that are considerably longer, it may be said of them as a class, that they vary in size, from those that are as large as fig. 231 down to those that are on the boundary line between spearpoints and arrowheads. So far as spearpoints occur in New Jersey, less than one per cent. exceed six inches in length, and of these, but very few exceed that measurement by more than an inch.

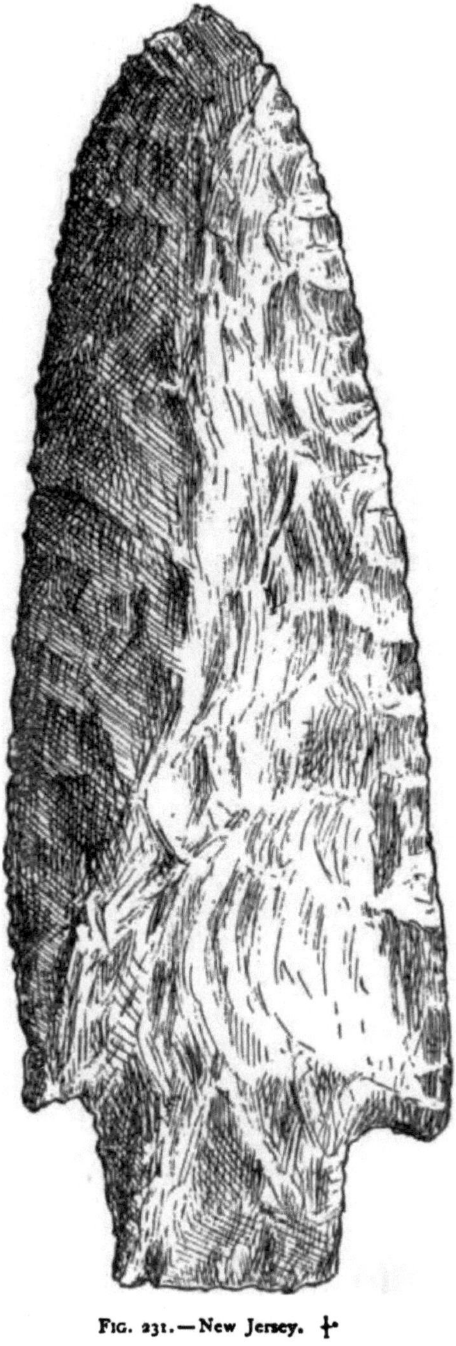

FIG. 231.—New Jersey. ½.

In the archæological collections of the American Museum of Natural History, at Central Park, New York, there is an example of large spearpoint, somewhat rudely chipped, which greatly exceeds in length any known specimens from New Jersey, or the New England states. This spearpoint measures eleven inches in length, and three inches in greatest width. It has a notched base and short stem, of about one-half the width of the blade. The material of which it is made is green jasper. It was found near lake Luzerne, in New York. A somewhat larger specimen, fourteen inches in length, is mentioned by Col. C. C. Jones, jr., as taken from a grave mound in Georgia. He remarks of this unusually large specimen, "no spearhead of such magnitude, so far as my knowledge extends, has been found within the limits of the southern states."

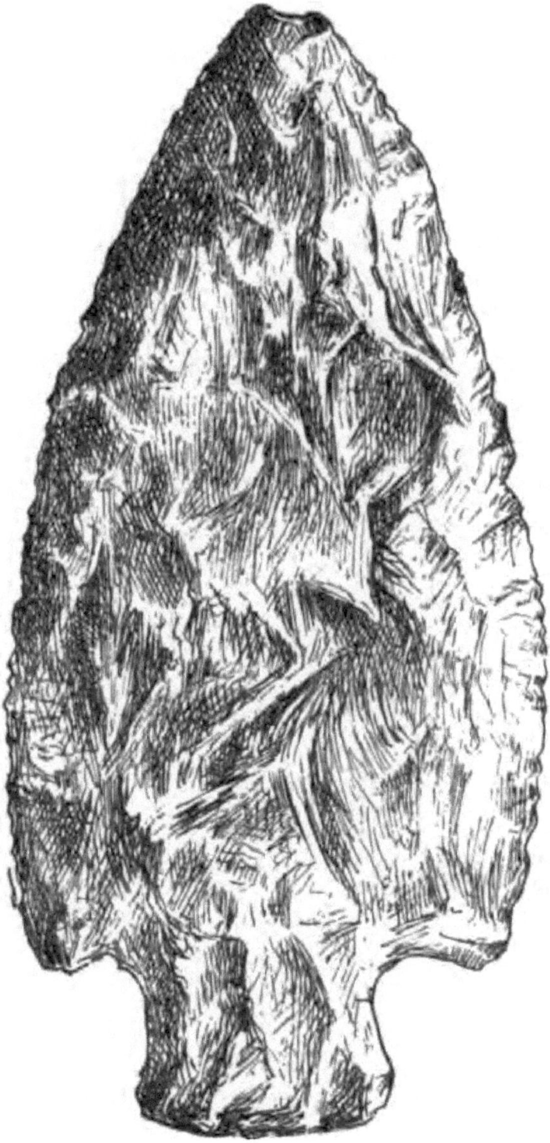

FIG. 232.—New Jersey. 1/1.

While it is thus shown that spear-points of a foot in length or longer do occasionally occur, they can scarcely be considered as examples of such spears as were in common use, and therefore the statement of the maximum size being about six inches is substantially correct.

This pattern of spearpoint is found in considerable numbers in the valley of the Susquehanna, Pa. Those in the cabinet of the late Professor Haldeman are made principally of limestone, and are of neat workmanship. In the western states, east of the Mississippi river, these implements are found frequently, and, as a rule, exhibit a higher degree of finish than similar objects found along the Atlantic seaboard.

In many localities in New Jersey, there are found great numbers of halves of these implements. Of these fragments, the great majority are bases. Why so many were overlooked, if the custom ever prevailed of gathering any of them for conversion into stemmed scrapers, does not appear. Indeed, it seems more probable that the points of these spears were gathered, and not the bases. In all cases within my own collecting experience, I have noticed a marked absence of points of spears and arrowheads.

Fig. 232 represents a second example of these large spearpoints, differing only in having a more convex outline, and in being a trifle shorter. This specimen is made of chert, and, considering the material, is handsomely worked. It has so far been the case in New Jersey, that these large spearpoints, usually broken, have occurred in numbers, in very limited areas, and no others have been found within several miles. This fact has been so frequently noticed as to give rise to the impression, that where these were found, a battle had been fought, and these broken weapons were lost and destroyed during the contest. How far this may be true is a matter of opinion. That it possesses an element of probability is undeniable.

A large number of these implements of this pattern and material have been found in the valley of the Delaware, from Easton, Pa., as far south as Salem Co., New Jersey. They are less common in the Susquehanna valley, I judge, than the narrower examples, like fig. 231. In the Connecticut valley many have been found, some of them even longer and more delicately chipped. In eastern Massachusetts there are but few found. The local collections made in New York, of which I have knowledge, contain comparatively few examples of flint spearpoints, as large as this. I am inclined to think, however, that the rarity of these implements is more apparent than real, and has arisen from the unfortunate habit of not preserving the halves, or even smaller fragments of them, when found. Spearpoints could nowhere be considered as common, if their numbers were judged only by the perfect specimens which have been preserved.

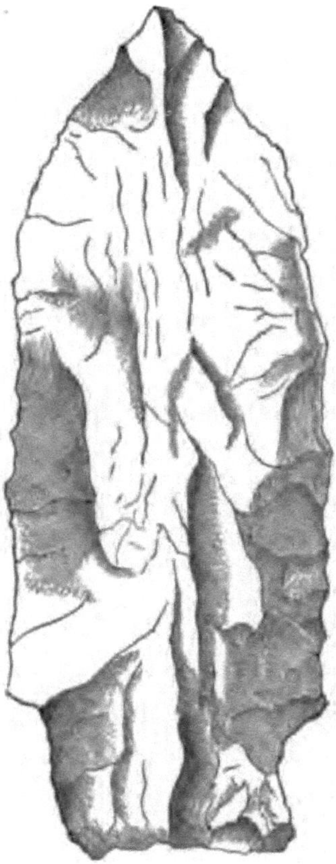

FIG. 233.—New Jersey. ½.

Fig. 233 represents a good example of a very common spearpoint, such as is found in every field, and wherever the ordinary arrowheads and other objects of Indian manufacture occur.

This specimen is chipped from yellow, quartz-veined jasper, a mineral that is shown to have been in great demand with the arrowmakers, by the large masses of it found on their workshop sites. Fig. 233 is a well-wrought implement, and has been finished more with reference to strength and durability, than appearance. The point and sides are still quite sharp, and the weapon is a good one whether used as a spearpoint or a knife.

Deposits of from twenty to one hundred spears identical with this have occasionally been found. Especially was this the case in southern New Jersey. What the object could have been of thus concealing implements supposed to be in constant use is, and probably must remain, a mystery.

Spearpoints of this size are comparatively common throughout the New England states. Professor Haldeman received many specimens, from several careful collectors in the Susquehanna valley, and a few were found in the rock-retreat discovered by him, in the Chickies Rock, near Columbia, Pa.

Fig. 234 represents what is probably the minimum size of spearpoints. Certainly objects so large as this could not have been arrowheads, and their use as knives is very problematical. The interesting feature of such specimens as fig. 234 is that they are made of argillite, and in the amount of weathering and rude workmanship they exhibit all the evidence of age that characterizes the palæolithic implements of the river-drift gravel. What relationship they may bear, if any, to those implements, has been discussed elsewhere. It is only necessary to remark, in this connection, that the evidence, which is very varied, of the general use of argillite prior to that of jasper and quartz, is almost unquestionable.

FIG. 234.—New Jersey. ½.

This spearpoint measures three inches in length, and one and a half inches in width. It may be taken as a fair representative of a class of objects that are found in extraordinary abundance in central and southern New Jersey. As many as one thousand have been found in an area of fifty acres. In the northern, hilly portion of the state, I have no knowledge of their abundance, and have seen but few specimens in local collections made in that section of the state. In Pennsylvania, they are not uncommon, and in the Chickies rock-retreat, discovered by Professor Haldeman, numerous specimens have been found. A peculiarity in their distribution is their frequent occurrence in the most unexpected localities, and often at a depth that suggests that they were lost when the face of the country was different from what it now is; and possibly that they were weapons used at the same time, and by the same people who fashioned the rude argillite fishspears that occur in the alluvial deposits of the river valley.

Argillite spearpoints, like fig. 234, must not be confounded with a class of similarly shaped implements made of slaty rock, which has undergone considerable weathering, and so has the same appearance as the argillite. These slate spearpoints and arrowheads are rudely made and usually of large size. In many localities they are quite abundant. Especially is this the case in the neighborhood of the Delaware Water Gap, where thousands of slate spearpoints and arrowheads have been found. At this locality, the slate of which they are made is the characteristic rock.

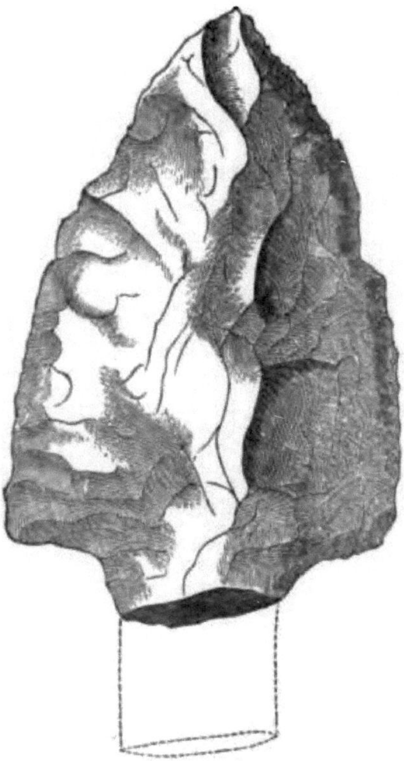

FIG. 235.—New Jersey. ½.

Fig. 235 represents a broad and rather short specimen of a spearpoint, differing but little from fig. 234. It is, however, made of jasper, is thicker and more carefully chipped, so that the edges are considerably straighter and sharper. This specimen is supposed to have had a straight stem, as other specimens identical in size and shape have been gathered, which were furnished with such a stem as indicated by the dotted lines, in the illustration.

Jasper and quartz spearpoints of this size and even larger, when found otherwise than singly, are associated with entire or fragmentary specimens of the various other patterns, thus showing that, however used, these several forms, and particularly those that vary so immaterially as do this and fig. 233, do not occur under circumstances that would suggest that they had been put to different uses.

In this connection it may be well to say that spearpoints, like the above, are not commonly found on village sites, or localities where the Indians were known to have congregated; but, except under the circumstances already mentioned, are found singly in what is still forest-grown land, or are ploughed up in fields which, when the Indians possessed the land, were covered with a dense forest growth. If we can judge from present appearances, it is probable that these implements were used principally in warfare, and, to a less extent, in hunting.

Fig. 236 represents a second example of the black, chert spearpoints, similar to fig. 232, but differing in the base, which is notched, instead of plain. The apparently trivial variation in the finish of the base very naturally suggests the possibility that these differences may indicate various methods of attaching handles or shafts, which, if long, would convert the weapon into a spear or lance; if short, into a dagger, as the case might be. There does not appear to be any evidence that the natives of the Atlantic seaboard used daggers of this character; yet it may be that, in the several patterns with varying bases, we have similar objects that were used for dissimilar purposes.

Fig. 236 was found near Salem, Salem Co., New Jersey, in a neighborhood remarkable for the number and beauty of the implements that have been left there by the ancient inhabitants.

Spearpoints of the above and allied patterns, with broad blades and short stems, do not appear to occur in Europe. No specimens of this character are given by Nilsson, as found in Scandinavia; and nothing similar to fig. 236 is described by Evans, as an English pattern of these implements. The javelin heads mentioned by him are usually smaller, and many are more nearly allied to the long oval and triangular flints, that have already been described as knives. The longest stemmed spear figured by Mr. Evans is one with long, curved barbs, and in this respect very different from the specimens found in America.

Fig. 237 represents a carefully chipped spearpoint of admirable proportions, with a notched base. Implements of this pattern, and size, are quite common, but they have been so generally broken either by use or subsequent exposure, that it is very seldom that a perfect specimen can be found.

Every variety of flint has been utilized in making these spear-

FIG. 236.—New Jersey. ½.

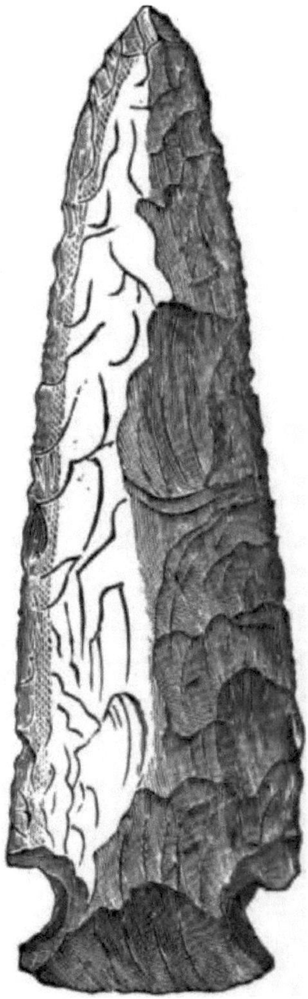

FIG. 237.—Indiana. ½.

points; quartz perhaps having been least frequently used. From Maine to Maryland, implements like fig. 237 have been gathered, and it is a form which is found abundantly in both the southern and western states.

In some localities, quite a number of argillite spearpoints of this pattern are found, which vary uniformly from fig. 237. They are thicker in the middle of the blade, and are more rudely outlined, in consequence of the detached flakes being of larger size. These peculiarities do not arise solely from the character of the material, for in the hands of a skilful workman argillite can be very evenly chipped: but they have every appearance of being the result of that want of skill which is now known to be one characteristic of the people who antedate the neolithic, or polished stone period.

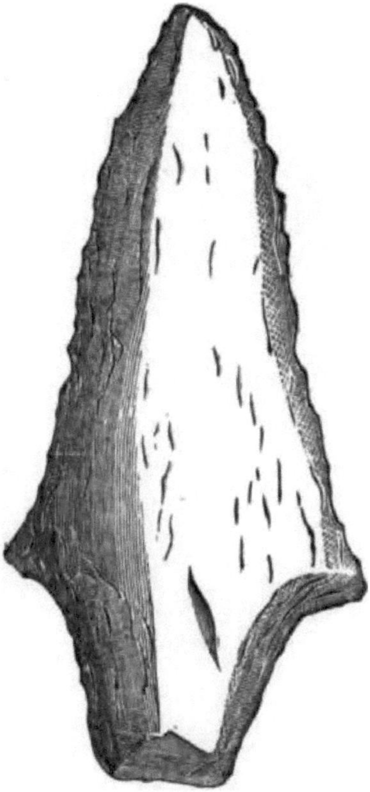

FIG. 238.—New Jersey. 1/1.

Fig. 238 represents a spearpoint which is well designed and carefully worked. The material is a tough, micaceous, quartzose rock, which, as bowlders, is frequent in the glacial drift of the central portion of the state. Whether from the peculiarity of the mineral, or design, is uncertain, but all similar weapons have the boundary lines of the flakes, detached in the making, nearly obliterated, and the specimen thus appears much like a *polished* spearpoint, an implement as yet scarcely known in the Atlantic states, for but few examples of slate spearpoints have been found which have been *ground* into shape, or smoothed subsequently to chipping. The slight "twist" in this specimen is due to the natural direction of the cleavage and not to design; and the same is in all probability true of the very distinctly twisted spearpoints and arrowheads that are made of flint and jasper.

Spearpoints of this pattern are not of frequent occurrence in any locality. I have seen none from the New England states, and but few from Pennsylvania. In Mercer Co., New Jersey, in which locality this specimen was found, a few spearpoints of this pattern, but made of a different mineral, were recently (1879) found lying together at the foot of a large elm tree, in a swamp, of many acres in extent. They were but a few inches below the surface.

FIG. 239.—New Jersey. ½.

It is not improbable that spearpoints of this pattern, and of the several forms described, were largely used for hunting deer and other large mammals. Josselyn (Account of two voyages to New England; London, 1674) has given a long description of the method of hunting the moose practised by the Massachusetts Indians, and says, that after a long pursuit over snow-covered ground, "at last they get up to him on each side and transpierce him with their Lances which formerly were no other but a staff of a yard and a half pointed with a fishe's bone made sharp at the end." Having authority for the statement, then, that lances were used for hunting by the New England tribes, it is warrantable to assume the Indians of the middle states, as having a like custom, varied only in that spearpoints of stone were principally used, instead of bone.

Fig. 239 represents a very beautiful pattern of supposed spearpoint which is comparatively frequent in the Ohio valley, but is quite rare in New Jersey or the New England states. A few specimens, however, are known from every state from Maine to Maryland. They appear to have been made in some one locality, and subsequently distributed by barter or otherwise over the seaboard states; as they are all made, so far as examined, of the same dull, bluish-gray jasper or hornstone. They are all alike, and cannot be distinguished from those found in Ohio and Indiana. The shape of the base and stem, of itself, gives no idea how these supposed spearpoints were hafted. It is not improbable, however, that a short handle and not a shaft was attached, and the specimen, therefore, is not a spearpoint but a dagger. This, however, is wholly conjectural, though its probability is increased by the fact that there are occasionally found flint implements of such design, as to prove that sometimes daggers, as we now understand that term, were made and used by the Indians of the Atlantic seaboard.

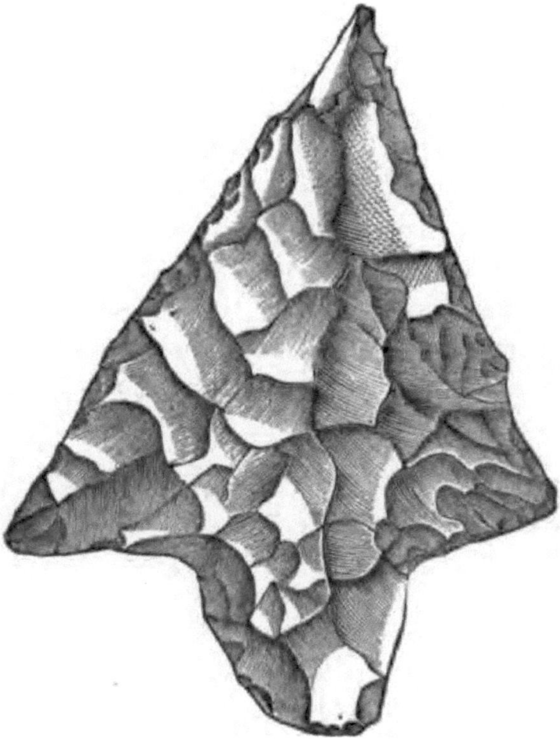

Fig. 240.—New Jersey. 1/1.

Fig. 240 represents an unusually large example of a form of spearpoint which is not very common, particularly of so large a size. These stemmed triangular spearpoints vary somewhat in shape. Sometimes the sides are slightly convex, and again, the short stems are notched, and not plain, as in fig. 240. They are always made of jasper and quartz; and the smaller sizes are, as a class, more delicately chipped than most of the other patterns of these implements.

This form of spearpoint is so different from any of the preceding, and is so far a widely distributed pattern, that it seems probable that it was used in some particular manner, whether for hunting or in warfare. The late Professor Haldeman found them quite common in the valley of the Susquehanna, and called them "fish-gigs" in the MS. catalogue of his collection from that locality. That they may have been used for spearing fish is not improbable; and, indeed, for capturing fish so large as the sturgeon, they are not poorly adapted. That they were also used in spearing turtles is also probable, from the fact that a series of eleven of these spearpoints were recently found in Gloucester Co., New Jersey, associated with an enormous quantity of the bones of the several species of water turtles, especially the snapper (*Chelydra serpentina*).

While spearpoints of this pattern are so well known to collectors in New Jersey and Pennsylvania, they do not appear to be common throughout New England.

Fig. 241 represents a rude argillite implement which is so similar to the spearpoints, that it is also classed as such, although it is not certain that it was so used. Of a very large series of this pattern, not one seems to have been acutely pointed, although it is possible that the point may have been worn away, or broken, and that the implement was subsequently used as a knife. The chipping is of the rudest character, even more carelessly done, than in many of the palæolithic implements of the river drift.

Rude as these specimens are, it must be remembered that the material of which they are made is very hard and susceptible of being brought to a very sharp edge, and therefore, although carelessly shaped, were not the less available either as spearpoints or knives. The entire series of these argillite implements are now much decomposed upon the surface, through weathering, but when made were quite sharp.

As in the case of the smaller spearpoint, fig. 234, it is uncertain as to the precise relationship these large implements bear to the palæolithic implements of the river drift, and the later fish-spears of the alluvial deposits. The degree of weathering of the surface of itself

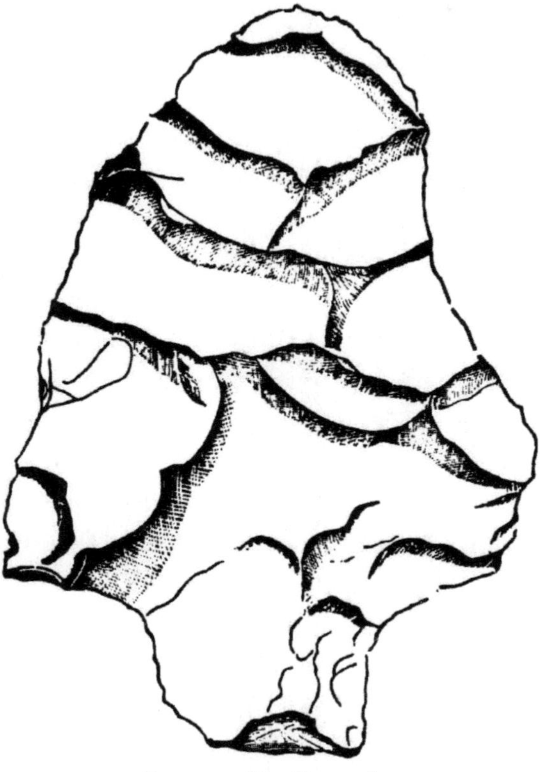

FIG. 241.—New Jersey. ½.

indicates considerable antiquity, but as this weathering may be comparatively rapid or very gradual, under certain circumstances, it does not afford, alone, sufficient basis for making an estimate of the age of these implements; sufficient at least, to carry them back to so remote a time as the date of the rude implements of the river gravels, or possibly even that of the fish-spears. These specimens have, thus far, been found usually in the uplands on and near the surface, mostly singly, and not always associated with jasper and quartz implements. While but little in the character of the localities where found tends to separate them, from the later handiwork of the Indians; it must be remembered that, like the smaller specimens, they occur frequently in localities where ordinary jasper implements are never found, and are brought to light, through landslides and the uprooting of trees, from depths greater than it is usual to find jasper implements,—circumstances which strongly suggest, if they do not prove, their greater antiquity.

While the implements of argillite and those of the same pattern of jasper and other silicious minerals are here considered collectively, because they are now so frequently found associated wherever relics of the early races occur, it is not intended to convey the impression that they are all necessarily of the same age, or origin even. As will be mentioned in subsequent pages, while the relative age or pre-Indian origin of any single specimen of argillite implement cannot be positively determined, except in the case of the palæolithic implements of the glacial drift, there is an amount of evidence in the circumstances under which many, if not most, of the argillite spearpoints and arrowheads occur to warrant us in referring them to an earlier people than the jasper-chipping Indians. When we come to consider the class of fish-spears, so called, made of argillite, the racial belongings of this supposed pre-Indian people will be considered.

Fig. 242 represents a well-known form of short, broad spearpoint, not abundant in any locality, but found occasionally over a great extent of territory. Investigations thus far show them to be more common in Kentucky than elsewhere. Like fig. 239, they seem to be all made of the peculiar blue-gray hornstone, so much used by the Indians of that locality. In some instances, the peculiar, deep, narrow notches are twice the length of those of fig. 242, although the specimen is no larger. The chipping of these spearpoints is always of the most finished character, and the effect is as artistic, as it is possible to produce with this material. As the edges are worked with all the skill and precision that mark the best examples of typical knives, it is not improbable that these objects were cutting rather than piercing implements; but, as was clearly shown, when the former implements were treated of, there was no possibility of discriminating, in many cases, between knives and spears.

No specimens of this pattern have been found in New England, of which I am aware, except a few examples from the Connecticut valley. In New York, they appear to be of rare occurrence. In Pennsylvania, they are more common, although none were found by the late Professor

FIG. 242. — New Jersey. ½.

Haldeman, during the several years he collected, with such success, in the valley of the Susquehanna.

In fig. 243 we have a second example of a carefully chipped implement, which varies but little from the preceding. The notches at the sides are not so deep nor are the barbs so well defined, but it can scarcely be doubted, that the uses of the two were the same. Fig. 243 is chipped from black hornstone, which is common in the shape of bowlders and pebbles in the river gravels; but it is not as delicately worked as the former. It is very probable that this specimen was copied from the more elaborately finished specimen, fig. 242, and that it was brought from some western or southern locality. This would seem to be true not only of a great many spearpoints, but also of other objects, which were more artistically designed and finished in the southwestern than in the seaboard states.

Fig. 244 represents a plainer example of these broad triangular points, and one that in size comes very near the largest size of arrowheads. These spearpoints are not very abundant, but are usually represented by one or more specimens, in every local collection. In Massachu-

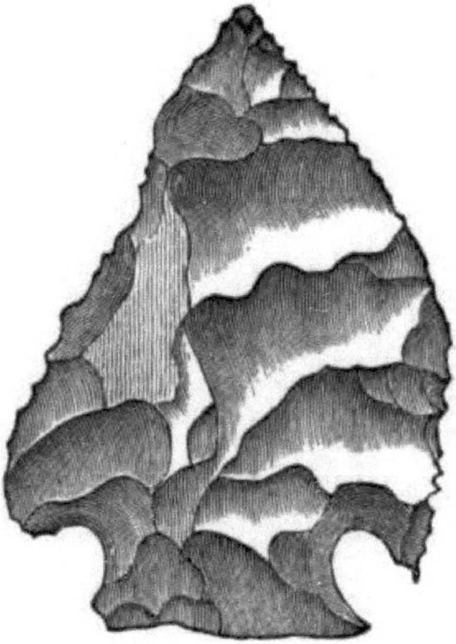

FIG. 243.—New Jersey. 1/1.

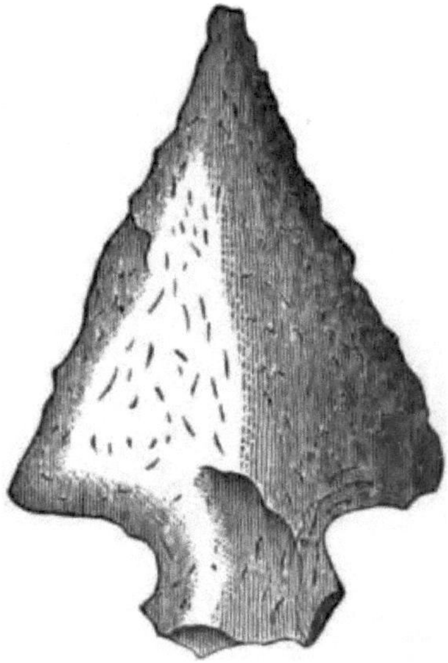

FIG. 244.—New Jersey. 1/1.

setts, they are occasionally found. In Connecticut, they appear to be more abundant. In New Jersey, they are found in some limited localities in considerable numbers, but never, apparently, in other districts of wide area. The late Professor Haldeman found but few specimens in the Susquehanna valley, and considered them as the same implement as the large stemmed triangular specimens, which he called "fish-gigs."

In concluding the subject of spearpoints, it is well to call attention to a class of specimens, which, though considered separately in consequence of a peculiarity in their finish, should not in reality be so treated, as there is no reason to believe that the peculiarity has anything whatever to do with the purpose or object of the implement. This feature is the twist or bend in the blade, which has been supposed to have been intentionally produced for the purpose of giving to the arrow or spear a rotary motion, during its flight through the air; this motion being increased by the manner in which the shaft of the weapon was feathered. In cases like fig. 245, where the entire blade is affected, this twist is wholly due to the nature of the mineral, and is a decided objection to the implement if intended as an arrowhead. If, however, it was designed for use as a dagger, it is rather an advantage, as the wound made by such an implement would be more jagged and severe, than that caused by a similar implement with smooth, straight edges.

FIG. 245.—New Jersey. ½.

Fig. 246 represents a smaller and much more rudely finished example of these "twisted" spearpoints from Indiana. The serrated edges, in this instance, add to the efficacy of this implement, as a deadly thrusting weapon, and its size, although somewhat smaller than the preceding, suggests its use as a spear or dagger rather than an arrowhead.

That the peculiar "twist" of the smaller examples of these implements can have no bearing upon the supposed rotary motion of the arrows armed with them, is shown in the fact that chipped knives are frequently found, that have the same peculiarity; and certainly these cutting tools had no need of a twist to aid them in a movement not required of them. In the few New Jersey specimens that I have collected, it is evident in many cases, that the original flake had assumed this bent shape when detached, and that it had been subsequently made into a spear or knife, notwithstanding the disadvantage of being crooked. In other cases, the chipping along the edge had been intentional, and the result was an arrowhead or spearpoint with bevelled edges ; the slope of the two sides being in opposite directions. The object of this is, of course, unknown.

The proportion of bent to straight spearpoints and arrowheads, as found in New Jersey, and throughout New England is very small, and far less than obtains in Ohio and the southwest. This is probably largely due to the absence of jasper which flakes in the peculiar manner that has been described. The supply of jasper and quartz used by the Indians was derived from the river-drift, to a great extent, and probably most of the arrowheads were made from pebbles which only yielded sufficient material to make two or three implements. Large bowlders were also brought from the river and used, as will be mentioned in a subsequent chapter. These bowlders are generally of a comparatively straight fracture, and few flakes are found which, to the maker of flint implements, would suggest the twisted spearpoint or arrowhead.

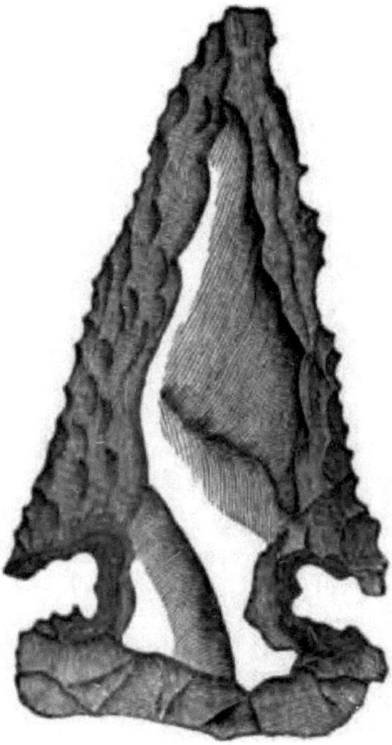

FIG. 246.—Indiana. $\frac{1}{1}$.

We have now to consider a series of implements of uniform pattern, which, from their similarity and apparent inadequacy to meet other purposes, may be supposed to be typical spears. Fig. 247 represents one, and is an excellent average example of the class. These slender spearpoints, very long in proportion to their width, are found in great abundance in many localities, while in others they are wholly wanting.

Although not unknown throughout the New England states, they are, except in the Connecticut valley, relatively scarce as compared with the numbers found in the valley of the Delaware river, south of Trenton. In that portion of this valley, they are exceedingly abundant. Often a dozen or more have been found in a very limited space along the shores of that stream. From the fact that they are very numerous along the banks of the river and the larger creeks, and are found but seldom at a distance from such streams, it has been supposed, not without reason, that they were largely, if not exclusively used in spearing or shooting fish. Upon this point, Holm said of the Delaware Indians,[95] "they take fish in the same manner (by shooting) : when the waters are high, the fish run up the creeks and return at ebb tide ; so that the Indians can easily shoot them at low water, and drag them ashore."

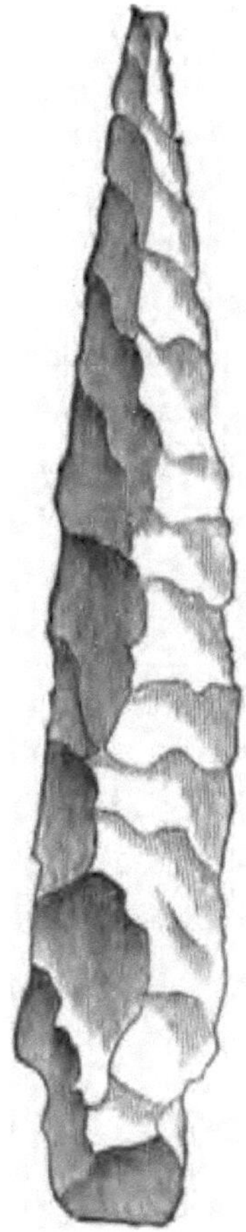

Fig. 247. — New Jersey. ½.

This form of spear is also abundantly met with along the shore of and in the many islands in the Susquehanna river, in the neighborhood of Columbia, Pennsylvania. In September, 1877, in company with the late Prof. S. S. Haldeman, the author visited several points in the river near Chickies, Lancaster county, and found a typical specimen of these slender spearpoints. Subsequently, several specimens were collected by boys in the neighborhood, and the proportion found along the river shore, in comparison to those found on the fields, showed I think, if we may judge of the uses of stone implements from the character of the locality where the majority of them are found, that they belong to a class of spearpoints that were used principally for capturing fish. The fact that a few specimens of bone

[95] Desc. of Province of New Sweden, by T. Campanius Holm, p. 121, Philadelphia, 1834.

fish-spears, such as are now used by the Eskimo, have been found in the Susquehanna valley, does not invalidate the theory, if so we must consider it, that these stone spearpoints were also used in fishing. The Eskimo of late date made stone spears that are ruder than fig. 247. Compare Lubbock's illustration of an Eskimo spear (Prehistoric Times, p. 492, fig. 218) with fig. 247 of this volume.

Loskiel,[96] describing the customs of the Delaware Indians, says "I am now to describe one of the most favorite diversions of the Indians next to hunting, namely, that of *fishing*. Little boys are even frequently seen wading in shallow brooks, shooting fishes with their bows and arrows.

"The Indians always carry hooks and *small harpoons* with them, whenever they are on a hunting party; but at certain seasons of the year they go out purposely to fish, either alone or in parties. They make use of the neat and light canoes made of birch bark, * * * and venture with them into spacious rivers."

While it is evident from the abundance of plummets or weights for fishing lines, found in Essex Co., Massachusetts, that shooting fish was not so generally practised as fishing by lines and nets, there occasionally occur in that vicinity spearpoints of the same general pattern as the preceding; and it is at least probable, that they were used for the same purpose.

Fig. 248 represents an implement of this kind from Salem, Mass. It is a very long, narrow and thin flint blade, which strongly resembles those found in central New Jersey, except that it is thinner, for its width, than the majority of those found in the Delaware valley. As it is without a stemmed base, it is possible that it was intended for use as an awl or knife, or both, rather than a fishing spear.

In a letter from the late Prof. Haldeman, bearing date of Feb. 12, 1878, I am informed of an interesting "find," as follows: "About one-fourth of a mile north of the rock-retreat, discovered by me in 1876, the Chickiswalungo creek enters the Susquehanna. Part of the bank

---

[96] Loskiel, *l. c.*, p. 95.

of the creek has been washed away so as to undermine a buttonwood tree (*Platanus*), nine feet two inches in circumference at four feet from the ground, under the roots of which, and four feet from the surface, my collecting boys showed me a deposit older than the tree, in black mold without gravel, but probably in part due to occasional floods of the Susquehanna. Among the objects are fragments of pottery, parts of one vessel with a row of semiperforated holes below the margin outside; bones of food animals, the round ones split, as usual; one good triangular arrowhead; a chipped knife (limestone of the vicinity); *one like your fish-spears, but not with a worked base;* many chips, etc." This specimen, which I had afterwards the pleasure of examining, varies in no important particular from the specimen from Salem, Mass. Like it, it is of flint or jasper, and more delicately worked than most of the limestone spears found in that neighborhood. The antiquity of the whole "find" is of much interest, from the fact that pottery was found with the implements; as it is evidence of the antiquity of the Indian, and not indicative of any preceding race.

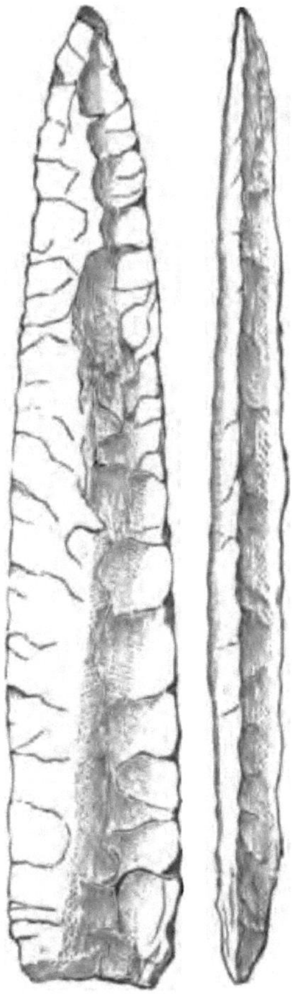

FIG. 248.—Massachusetts. ½.

Though there may be doubt as to the identity of purpose of the implement just described, with that of fig. 247, there can be none at all in the case of fig. 249. It is, indeed, far more gracefully shaped and delicately chipped than the common "fish-spears" of the Delaware valley, and yet it is so well designed for spearing fish, and so little likely to have been used as a weapon, that it is classed among them. Like the preceding, this specimen is very thin, and in this respect it shows an important variation from those made of argillite, or even the later forms of silicious materials, in New Jersey. Large numbers of this pattern occur in the Connecticut valley, but the majority of them have the width more uniform along the greater portion of the length of the blade, which tapers more suddenly to the point. As a rule, spearpoints of equal length do not taper to the point directly from the base, as in this instance, but preserve a uniform width for a third or half their length.

FIG. 249.—Massachusetts. ½.

In some localities west of the Alleghany Mountains, spearpoints of this pattern and size are quite abundant. Many are made of chalcedony and are very beautiful examples of the skill acquired in chipping flint by the Indians of that portion of the country formerly occupied by the Delaware or Lenni Lenapè nation.

Along the northern seaboard of New Jersey, where the mainland is washed by the sea, occasional specimens of these spearpoints, made of white and rose-colored quartz, have been found, which are equally as well made as any specimens from western Pennsylvania, or even from the mound regions of Ohio. Their shape is such as to suggest their use in fishing rather than inland hunting, or as weapons; but strangely enough no specimens have as yet been found in the shell-heaps of the neighborhood, some of which have a large percentage of fish-bones.

Fig. 250 represents a shorter and broader specimen of supposed fish-spear, neatly chipped from a bluish rock, of flint-like appearance.

Though found elsewhere to some extent, the majority occur along the river banks and in the neighborhood of the larger inland lakes and creeks. Many have recently been found on the Pennsylvania shore of the Delaware river, opposite Bordentown, N. J. Here they are mingled with the ever-shifting sands and are brought to light and then buried, by the changes of every tide.

Fish-spears of this pattern are not as common as those with a notched base, though in other respects there is such a general resemblance between them, that it cannot be doubted that they are two forms of one and the same implement.

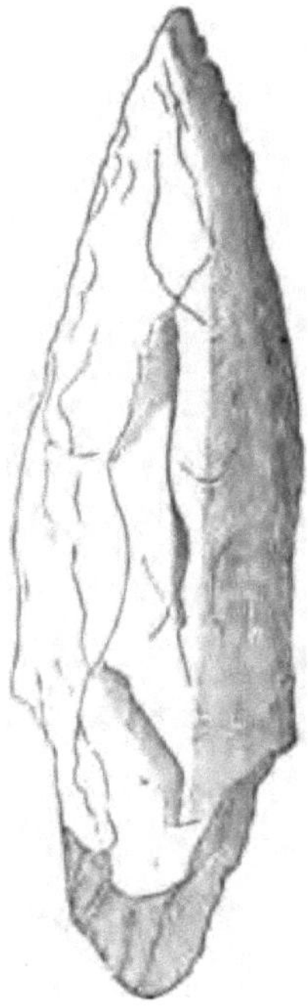
FIG. 250.—New Jersey. ½.

This same pattern of spear is quite common along the coast, and scores have been gathered near Tuckerton, and Barnegat, in Burlington county, and about Beesley's Point, Cape May county, New Jersey, and I am informed that spearpoints of this and allied patterns have occasionally been found in the extensive shell-heaps on Long Island.

One of those curious and interesting "deposits" of chipped stone implements, which are occasionally discovered in various localities throughout the United States, was lately brought to light, in Burlington county, New Jersey; which consisted entirely of spearpoints of this pattern. Unfortunately, they were not counted when first seen, so the exact number is not known, but over four hundred were traced, and the specimens compared, one with another. There was no difference of importance, either in size, shape or finish. They were evidently all made of the same material, a dark, blue-gray flint, and none showed any evidence of having been used, or, indeed, of having been exposed at all to the air. They were remarkably "fresh" in appearance and had evidently been buried very soon after they were made. The locality, where this deposit was found, is on the south bank of Crosswicks Creek, about three miles from its mouth. The proximity of good fishing grounds, the size and shape of the implements themselves, and the fact that such specimens are found, as a rule, near the water, warrant us in considering the spears of this pattern as fishing implements primarily; although their frequent use in other ways is highly probable.

In fig. 251 we have a more elaborately fashioned example of such fish-spears, if indeed, it can be so classed. In general appearance it suggests the idea of drills or perforators; but being very thin, as compared with its width, it is evident that as a drill, it would be of but little value. As it is broken at the base, it is impossible to determine how it was there finished, but it is not probable that there was a duplication of the broad, barb-like projections. Spearpoints of this pattern are very unusual, judged by their scarcity in the various large collections from New Jersey and New England.

In the very large collection of the late Professor Haldeman, of chipped implements from the Susquehanna valley, are several specimens of jasper and limestone spears of this pattern, except that there is but one barb-like projection. These non-symmetrical forms are not uncommon in the class of arrowheads, and many of the so-called stemmed knives have also this peculiarity. If we are justified in considering these implements as spearpoints, it is not evident what advantage there was, in having the one barb. The character of the chipping shows, I think, that it was intentional, and not necessarily that one of the barbs being broken, the fractured margin was smoothed down by re-chipping.

Fig. 252 represents a form of spearpoint which, by its size, approaches closely the boundary between spearpoints and arrowheads. This form is not as characteristic of fishing grounds, as are some of the others, especially figs. 247 and 253, but there is so close a resemblance as to make it probable, that it should be considered as a fish-spear.

While the chipping of these slender jasper points is, in many cases, very carefully done, yet as a rule such implements cannot compare in finish with other forms that were evidently intended for hunting, or as weapons. These latter are much larger and show a higher grade of workmanship than any of the slender fish-spears. Was it because the fish-spears were far more liable to be lost?

Spearpoints of this pattern made of quartz and jasper are much more frequently found in New York and Connecticut, than either in the northern New England states, or in Pennsylvania and New Jersey. In the Susquehanna valley (Pennsylvania) they appear to be very rare.

The preceding examples of these slender spearpoints were all from

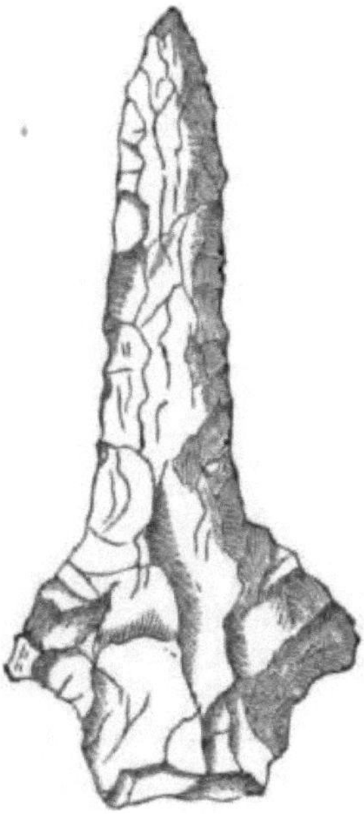

FIG. 251.—New Jersey. 1/1.

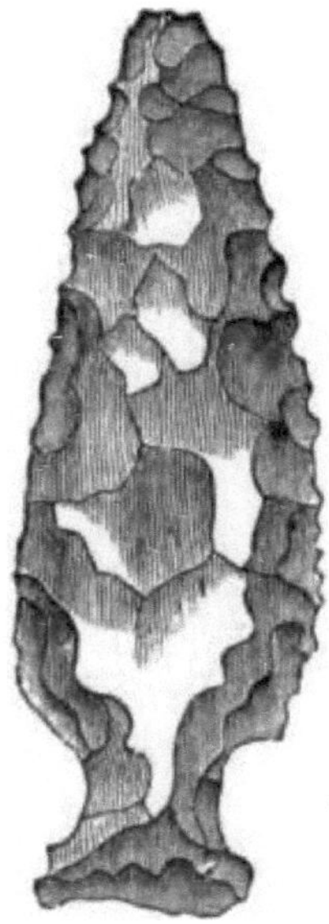

FIG. 252.—New Jersey. 1/1.

the surface, and are found not only in the immediate vicinity of the rivers and smaller streams but to some extent in upland fields, associated with the common forms of stone implements. There is, of course, nothing to indicate that they are of other than Indian manufacture. Associated with them, when found upon the surface, and occurring in scanty numbers in fields, far from any water course, are other examples of these slender spearpoints which do appear to tell another tale. While sometimes found with jasper spears they are more frequently met with alone, and hence, though they may be said to be with them, they are not of them. These spearpoints are all of argillite, of nearly uniform size, and vary but little in the finish of the base.

Fig. 253 represents a good average specimen of hundreds that have been collected from the alluvial deposits, through which various creeks have now worn their channels. In this alluvial mud which has been for centuries, and is still accumulating, many specimens of these argillite spearpoints have been found at various depths, even to five feet, and nowhere do they occur in such abundance as *in* this deposit, which forms the tide-water meadows that skirt the banks of the Delaware river from Trenton to the sea.

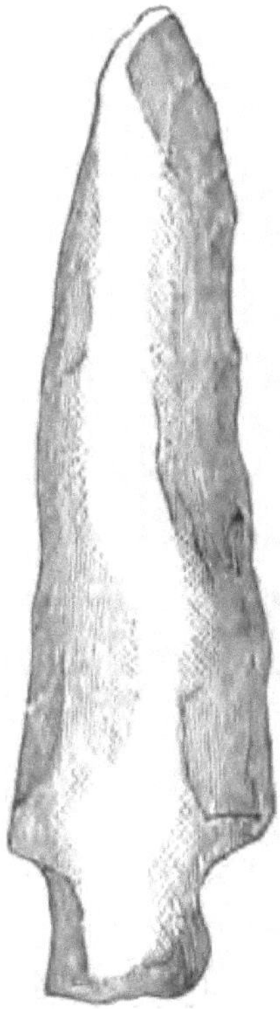

FIG. 253.—New Jersey. ½.

Fig. 254 represents the only variation of importance from the typical form as given in the preceding illustration. This specimen, which strongly resembles an ordinary flint-drill, represents about one per cent. of the fish-spears found in the mud deposits referred to. Occasionally one of this pattern will be found with one or more shallow notches on the edge, near the base, and only on one side. These lateral notches are an indication of the means employed in securing the implement to a staff. Why not notched upon both sides is difficult to imagine; but it is not only in these supposed fish-spears that this peculiarity is found. In quite a large percentage of the jasper and slate hoe-blades—if such they are—this same feature of a notch or notches on one side occurs.

It may be considered that we are without a warrant in assuming that the use of any implement can be determined by the character of the locality where the implement was found. To a certain extent, this is unquestionably true. A bead is none the less an ornament, whether dredged from the river bottom or found in an upland field, and yet how very seldom does any implement or other relic of the Indians occur, except where we should expect to find them. In basing any conclusions upon the characteristic features of the locality, where the implements under consideration are found, it is obviously necessary to determine if there has been any recent general disturbance of the surface. Alterations of the surface caused by the removal of heavy forest growths must be always borne in mind; the possibility of implements being brought from some distance by floods, or even the temporary currents of heavy rainfalls. Occurrences like these can, in nearly every instance, be readily determined, and all objects that have evidently been brought to the surface or the spot where found, through such occurrences, should not enter into account when we come to study a representative series from undisturbed localities. It is clearly evident that, in the vast majority of instances, stone implements are in practically the same position that they were when buried, lost or discarded. A single specimen, or even a hundred might readily mislead one, and give rise to very erroneous impressions as to the character of the locality where such objects were sometimes found, as in the instance given in the second volume of the Reports of the Archæological Museum at Cambridge, Mass., p. 243, where palæolithic implements were found associated with a grooved axe and fragments of pottery. It is only when hundreds have been carefully gathered in person, that it becomes safe to base any conclusions upon locality; but when we have the material in such abundance, as in the case of these argillite spearpoints, and find that over eighty per cent. are from an alluvial deposit skirting the river, it does become highly probable, at least, that they were used in and about the river, or in other words, as a means of capturing fish.

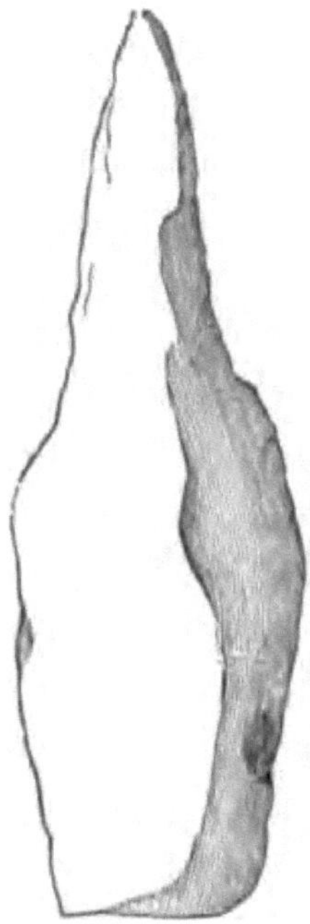

FIG. 254.—New Jersey. ½.

It does not necessarily follow, that because these implements were used mostly as fish-spears, that they were not used in capturing such animals as frequent the banks of our rivers and inland streams. The number of these was really very large, and it can scarcely be doubted but that many were speared or shot with arrows, tipped by these long, slender argillite points.

The beaver (*Castor fiber*), musk-rat (*Fiber zibethicus*), otter (*Lutra canadensis*), mink (*Putorius vison*), among our mammals, and various turtles and water-fowl; all frequented, in vast numbers, the very streams where such large numbers of these argillite spears are found; and while it is true that ordinary arrowheads are frequently found on the shores of these streams, it cannot be maintained, with reason, that the tapering argillite points were used solely for capturing fish, and the few arrowheads that we find were those lost in unsuccessful efforts to capture other animals. On the other hand, nets, lines and traps of various patterns, for securing fish, were so generally used by the Indians, that it may well be doubted, if these argillite spearpoints are really the handiwork of the later Indians. Spears made of other material like the ordinary jasper arrowheads, with which they are associated, are unquestionably of Indian make; but these strongly suggest another and an earlier origin.

The origin of the mud deposit containing these rude fish-spears is a subject requiring fuller treatment than can be here given to it, but its bearing on the subject of the age of the contained implements requires more than a passing notice. This deposit of mud is of a deep blue-black color, stiff in consistency, and almost wholly free from pebbles. It is composed of decomposed vegetable matter and a large percentage of very fine sand. It varies in depth from four to twenty feet, and rests on an old gravel, of an origin antedating the river gravels that contain palæolithic implements. This mud is the geological formation next succeeding the palæolithic implement-bearing gravels. In the implements it contains we have apparently a link between the oldest traces of man, and the recent handiwork of the Indians of historic times.

A careful survey of this mud deposit, made at several distant points, leads to the conclusion that its formation dates from the exposure of the older gravel upon which it rests, through the gradual lessening of the bulk of the river until it occupied only its present channel. How gradually the river retired from the gravel bluffs that formerly marked its boundary, and how rapidly this mud accumulated at the outset, and whether at a more rapid rate then than now, are all points almost beyond solution. The indications are, however, that the present volume and channel of the river have been essentially as they now are, for a very long period; and the character of the deposit is such that its accumulation, if principally from decomposition of vegetable matter, must necessarily be very gradual. Since its accumulation to a depth sufficient to sustain tree growth, forests have grown, decayed and been replaced by a growth of other timber. While so recent in origin that it seems scarcely to warrant the attention of the geologist, its years of growth are nevertheless to be numbered by centuries, and the traces of man found at all depths through it, hint of a distant, shadowy past, that is difficult to realize.

The same objection, it may be, will be urged in this instance, as in all others where the comparative antiquity of man is based upon the depth at which stone implements are found—that all these traces have been left upon the present surface of the ground, and subsequently have gotten, by unexplained means, to the various depths, at which they now occur. It is, indeed, difficult to realize, how some of these argillite spearpoints have finally sunk through a compact peaty mass, until they have reached the very base of the deposit. For those who urge that this sinking process explains the occurrence of implements at great depths, it remains to demonstrate that the people who made these argillite fish-spears either made only these, or were careful to take no other evidences of their handicraft with them when they wandered about these meadows; for certainly nothing else appears to have shared the same fate, of sinking deeply into the mud. In fact, the objection mentioned is met, in this case, as in that of the palæolithic implements, that if these fish-spears are of the same age and origin as the ordinary Indian relics of the surface, then all alike should be found at great depths. This, we know, is not the case. Furthermore, the character of the deposit is not that of a loose mud or quicksand, but more like that of peat. It has a close texture, is tough and unyielding to a degree, and offers decided resistance to the sinking of comparatively light objects deeply into it. This is, of course, lessened when the deposit is subject to tidal overflows, and in the immediate vicinity of springs, which, bubbling through it, have caused a deposit of quicksand. While here, an object sinks instantly out of sight, it is not here that we must judge of the character of the formation as a whole; and over the greater portion of its area, we find no evidence of objects disappearing beneath the surface at a more rapid rate than the accumulation of decomposing vegetable matter would explain. Efforts have been made to determine the rate of progress of this growth of mould, but they are not wholly satisfactory; nevertheless the indications are sufficient to warrant our belief that the rate is so gradual as to invest, with great archæological interest, the characteristic traces of man found in these alluvial deposits.

The relationship of these supposed older spearpoints to those made of jasper and quartzite, is a subject that demands most careful consideration, and an abundance of data, that we scarcely yet possess. The subject will be more fully treated in a subsequent chapter. Here it is only necessary to add that as the origin of the oldest traces of man yet discovered on the northern Atlantic seaboard—whether Eskimo or Indian—is yet, perhaps, an open question; therefore these older and newer spearpoints are considered together, as implements of the same character, though not necessarily the handiwork of the same people.

That class of small chipped implements, known as arrowheads, is of such common occurrence throughout the country, that it seems scarcely necessary to more than draw attention to the several patterns that are found, and make brief mention of their relative abundance.

Obvious as is the purpose of an arrowhead, it is not always easy to determine whether all that appear suitable for heading the shafts of arrows were really so used. Doubtlessly, in the present series of arrowheads of the usual patterns found along our Atlantic coast, there are several that were used as knives; and, *per contra*, in the series of knives there are arrowheads. It is also very probable the same form was frequently used both as a knife and an arrowhead.

Holm[97] writes of the Delawares, "Although the Indians, when the Swedes first came into the country, had no instrument or tools made of iron or any other metal; nevertheless, they could perform every kind of work with their hands with such dexterity and neatness, that the Christians were struck with astonishment. They make their bows with the limb of a tree, of above a man's length, and their bowstrings out of the sinews of animals; they make their arrows out of a reed, a yard and a half long and at one end they fix in a piece of hard wood of about a quarter's length; at the end of which they make a hole to fix in the head of the arrow, which is made of a black flint stone, or of hard bone or horn, or the teeth of large fishes or animals, which they fasten in with fish glue in such a manner, that the water cannot penetrate; at the other end of the arrow, they put feathers."

In the illustration of an arrow with a stone head, fig. 255, found in Peru[98], we have an exemplification of the manner in which arrowshafts were made; the smaller, upper portion, in the Peruvian example, being of hard wood, securely fastened to a reed. In this instance, however, there are no feathers at the base.

Peter Kalm, the Swede, who visited New Jersey in the middle of the last century, has left us an excellent account of the resident tribes, and of their customs, which were at that time slowly changing in consequence of the introduction of iron implements by the Swedish settlers. Of their arrows,

FIG. 255. — Peru. $\frac{1}{4}$.

[97] Holm, *l. c.*, p. 129.

[98] For the use of this cut, the author is indebted to Prof. F. W. Putnam.

he says (Travels in North America, vol. ii, p. 39, London, 1771): "At the end of their arrows they fasten narrow angulated pieces of stone; they made use of them, having no iron to make them sharp again, or a wood of sufficient hardness: these points were commonly flints or quartzes, but sometimes likewise another kind of a stone. Some employed the bones of animals, or the claws of birds and beasts. Some of these ancient harpoons are very blunt, and it seems that the Indians might kill birds and small quadrupeds with them; but whether they could enter deep into the body of a great beast or of a man, by the velocity which they get from the bow, I cannot ascertain; yet some have been found very sharp and well made."

Throughout the entire area of the country treated of in the present work, there does not appear to be any material difference in the character or distribution of the arrowheads found, save that a preponderance of the more delicately chipped forms are found in some localities. A thousand arrowheads from the valley of the Delaware river would not differ materially from a thousand gathered in the Susquehanna valley, and but little, if any, from as many gathered in the Connecticut valley; but as compared with an equal number from Massachusetts, they would show a somewhat larger percentage of the more carefully finished specimens of jasper and chalcedony. In Vermont[99], as elsewhere, arrowheads are "more abundant than any other class of specimens, and all the varieties figured by Col. Foster (Prehistoric Races) are found with others differing from these. Some very singular inequilateral forms occur, like those figured by Professor Haldeman[100] in a recent number of the naturalist, and many others. Indeed, a close examination of any large collection of flint points will show that entirely symmetrical forms were rarely attained; by far the larger part are more or less unequal, both as regards curvature or straightness of the edge and convexity of the surface. One edge is usually more strongly curved than the other, and one surface more

---

[99] Perkins. Amer. Nat., vol. xiii, p. 747.

[100] Haldeman. Amer. Nat., vol. xiii, 292.

convex than the other. Often the blade is not in the same plane with the stem, but seems twisted upon it; due, as I think, less to the intention of the maker than to the fracture of the stone. From these slightly, often almost inperceptibly unequal points, we have every gradation to those which are nearly as unsymmetrical as possible, and of these latter, some are so well chipped that I cannot regard them as 'failures,' but for some unknown reason intentionally of the form we find them. In comparing the specimens from the Champlain valley with those from Georgia, figured by Col. C. C. Jones, jr., I have been struck with the close resemblance between them; there are comparatively few of the objects described in antiquities of the Southern Indians, that cannot be duplicated, often exactly, in Vermont specimens. This resemblance is more noticeable because, among Dr. Abbott's New Jersey specimens, I find many unlike those which we have with us."

Professor Perkins, in the above quotation, refers to such objects from New Jersey, as were described in 1873, and prior to any systematic search for the discarded handiwork of our resident tribes. In the vast amount of material since gathered, there are objects very similar to, or identical with, the collections made by Col. Jones, in Georgia and other southern states; and the differences are now ascertained not to be so marked as they seemed to be, when the brief article on New Jersey archæology was written in 1873.

Whatever may have been the object in making arrowheads of different shapes, it is certain that they vary more than any other form of implement, and in no instance has any particular pattern been found exclusively in one locality. The proportion, however, of the various patterns is quite different in different localities; as, for instance, the leaf-shaped specimens constitute fully six per cent.[101] of all found in New Jersey, although in New England, they are far less common.

While in the immediate vicinity of the seacoast, arrowheads do not

---

[101] Abbott. Nature, vol. vi, p. 515. The author here stated the proportion to be about four and one-half per cent.; but subsequent collecting shows them to be even more abundant.

appear to be as abundant as along our river valleys, they are not altogether wanting even there, but are sometimes found in considerable numbers. It is not improbable that fishing, rather than shooting, largely occupied the time of the coast Indians, and therefore the bow was comparatively little used. Still, it is evident, from the character of the country, that arrowheads once lost in these ever-shifting sands, or in the water, would seldom be brought to light; and thousands may now lie buried in the sand, which, if it ever becomes solid rock, will contain these certain evidences of man's former presence.

Although in no instance has any one pattern of arrowhead been found so characteristic of a given locality, as are the argillite fish-spears of the alluvial deposits along the river, it has frequently been observed by collectors that some particular form occurred in considerable numbers in a locality of very limited area, as a field or other small plat of ground. In my own collecting tours I have frequently noticed this, and can recall now certain fields that appeared to have only leaf-shaped arrowheads, and others where the triangular pattern was alone met with. Even this is noticeable with other forms of chipped implements, and local collectors report fields, or other spots of a few acres, where only scrapers are found. This localizing of certain forms has been so frequently noticed that it cannot be considered as a mere chance occurrence, yet it is scarcely susceptible of any rational explanation.

Unsymmetrical arrowheads, like those described by the late Professor Haldeman, and referred to by Professor Perkins, are of quite common occurrence in New Jersey, but I am not disposed to consider them as really used as points for arrows. Their numbers, the character of the chipping, and the size of the greater proportion of them, render it highly probable that they were knives, and, as such, a small number have been already described. The very want of symmetry that characterizes these specimens renders them absolutely valueless if used as arrowheads; for, unless the tip or point of the head is in a direct line with the shaft, the arrow loses, in a great degree, its penetrative power.

For whatever other purposes arrowheads may have occasionally been used, their main purpose, as their name implies, was for making their arrows more effective as weapons. Considerable controversy has arisen concerning how large a proportion of these implements were used in the manner described, many writers insisting that only the smallest specimens were used as points for arrows, and the great majority were simply knives. It is obviously impossible to determine now the precise character of many intermediate sizes of these objects, but, in some instances, examples have been found under circumstances which at once set aside all doubt, so far, at least, as these specimens are concerned.

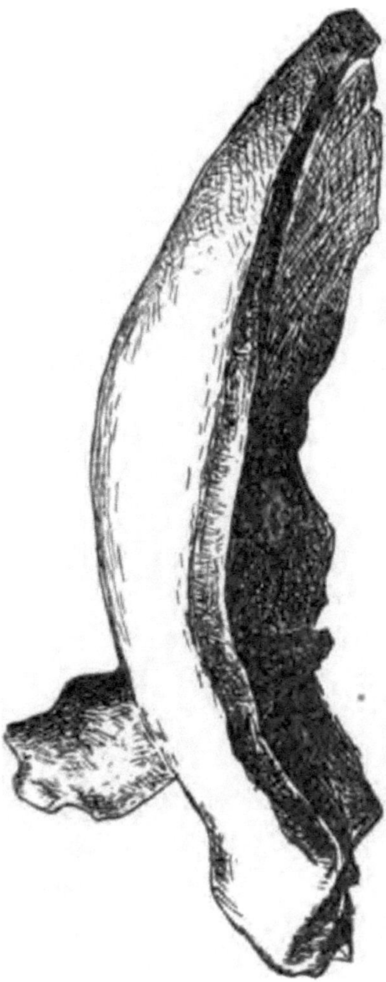
FIG. 256.—New Jersey. ½.

Fig. 256 represents an instance of this kind. We have here a water-worn fragment of a human frontal bone that has been pierced by a quartz arrowhead. The specimen tells its own story. In this fragment of bone, with the arrowhead still sticking in it, we have the use of the bow, the tipping of the arrow's shaft with stone, the custom of war,—all told us in a clear, unmistakable way. Could we but find more examples of the various patterns of stone implements in such telling positions, we should then more clearly realize the character of the daily lives of the native races of America in prehistoric times.

Although it is not practicable to trace any development of the more elaborate patterns of arrowheads from pointed flakes, yet it will be found advantageous to consider first the simpler forms, with the understanding that there is no evidence that any one form of highly finished arrowhead antedates another. It is probable, however, that simple pointed flakes were first used, and that the various modifications and elaboration of details were subsequent to the first arming of an arrow with a sharp fragment of stone.

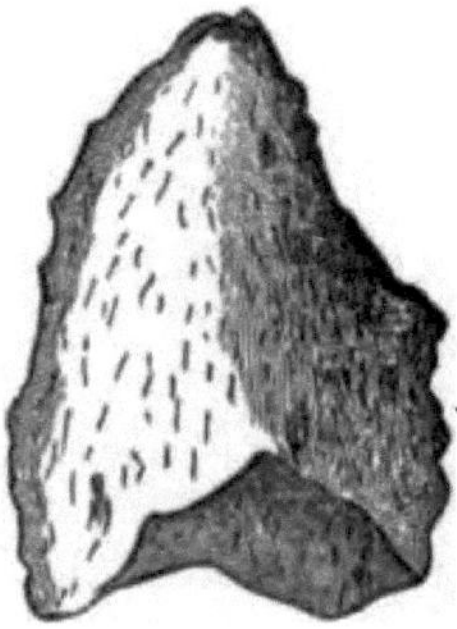

FIG. 257.—New Jersey. ½.

In figs. 257, 258 and 259 are represented flakes of the simplest patterns, which are supposed to have been arrowheads. These, and others altogether similar, have been so frequently found singly, and yet are so evidently artificially chipped, that there need be no hesitation in considering them as being of human origin. The fact of their being found singly is evidence also, that they were implements and not merely refuse chips flaked off in the manufacture of other objects. Some such forms were doubtlessly used to point the first arrows made by man; but these specimens bear no evidence of a greater antiquity than the modern Indian.

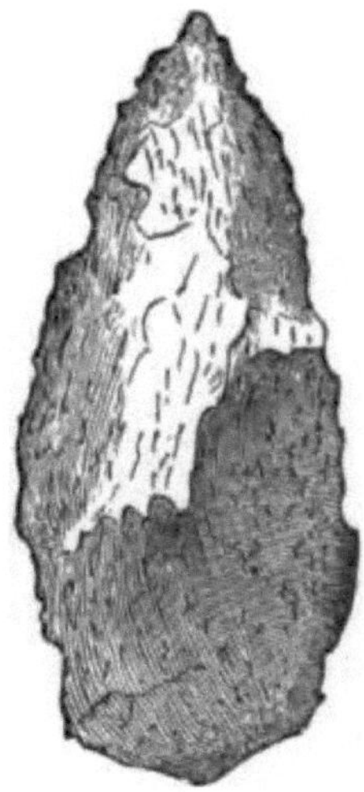

FIG. 258.—New Jersey. ½.

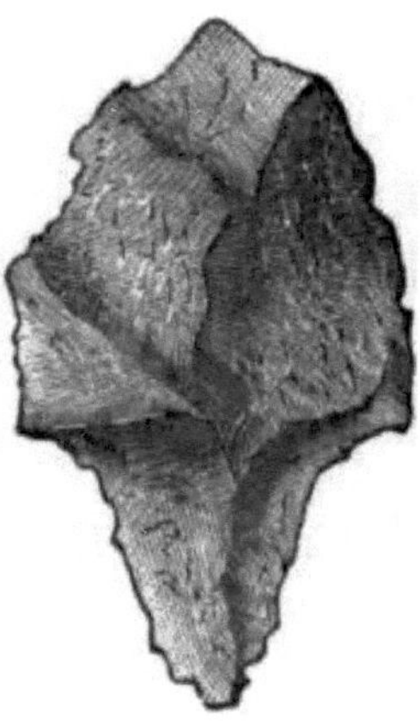

FIG. 259.—New Jersey. ½.

In the rock-retreat discovered at Chickies, Lancaster Co., Pa., by the late Professor Haldeman, in 1876, there was found an unusually large number of rude flakes of the patterns here figured, which had apparently been selected from the mass of chips which had accumulated during the process of making better finished specimens. These "arrowhead-like flakes" were apparently set aside as available for arrows, or were intended to be subsequently chipped into other and better forms. To a certain extent they might have been used as knives, and they recall the remark of the early traveller Kalm, who said of the Delaware Indians, "instead of *knives* they were satisfied with little sharp pieces of flint or quartz, or else some other hard kind of stone, or with a sharp shell, or with a piece of a bone which they had sharpened."

In fig. 260 we have a more specialized form, and one that is exceedingly common. That it should be classed as an arrowhead is reasonable, even though it is by no means as desirable a pattern as many others. When made of jasper or quartz, arrowheads of this form are generally much more carefully worked, and present smoother surfaces and more evenly chipped edges.

FIG. 260. — New Jersey. ½.

Arrowheads of this shape, made of slate and other comparatively soft stone, and a few of argillite, are also found in New Jersey. In some localities they are very abundant, and even outnumber the jasper specimens. In consequence of the characteristic rude finish of them all, and the degree of weathering of their surfaces, which is to some extent an indication of antiquity, these ruder specimens of stemmed arrowheads, like fig. 260, give rise to the impression that they are, as a class, really older than the same forms made of silicious mineral. This, however, may not be the case, and it is as yet uncertain, whether the Indian made a step in advance by adopting a material for the manufacture of his weapons more difficult to work than slate or argillite, but affording far better results to the skilled workman.

I believe the evidence to be complete, that the race preceding the Indians used argillite invariably for all their implements, but the few facts that seem to indicate progress on the part of the Indian, during his occupancy of the Atlantic coast, are, of themselves, insufficient to warrant our basing any conclusions upon them.

Fig. 261 represents a second form of stemmed arrowhead, which is of better workmanship, and foreshadows the elaborate patterns that were in common use, after the art of working in flint had been carried to perfection. If the one object, in making arrowheads, was to secure that shape of point which would convert the completed arrow into the most effective weapon, then the simple, triangular specimens, with moderately concave bases, would meet the purpose; and in fact, we do find this simple and effective form more abundant than any other, but so many intricate forms are also found that it will always be a source of wonder, and a subject of endless controversy, why such simple objects should have been fashioned in so many curious shapes.

Fig. 262 represents a neatly chipped arrowhead, usually of jasper,

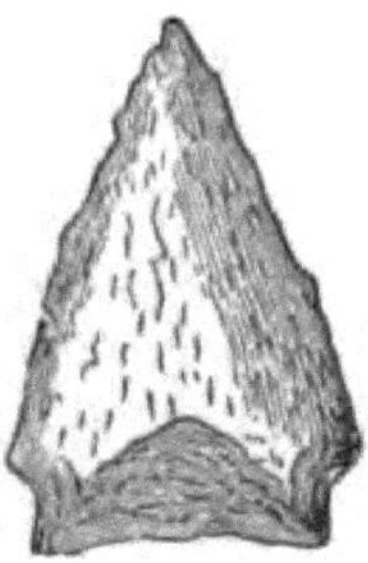

FIG. 261.— New Jersey. 1/1.

FIG. 262.—New Jersey 1/1.

FIG. 263.— New Jersey. 1/1.

which may be taken as a typical example of that form, known as the arrowhead with a "notched base." A large proportion of the arrowheads found along the northern Atlantic seaboard are of this form, though it is not in such abundance as to be considered peculiar to this portion of the continent. Arrowheads of this pattern vary greatly in size, and when much longer than fig. 262 are not distinguishable from knives, and, possibly, spearpoints.

Fig. 263 represents a longer, but narrower example of this form which is of equally common occurrence. Throughout New England, arrowheads identical with this are very common, and it may be considered as one of the characteristic forms north of the Connecticut valley.

Figs. 264 and 265 represent beautiful specimens of the smallest size of arrowheads, both of this pattern with the notched base. It is an unusual circumstance to find arrowheads of this size along the Atlantic seaboard, or indeed anywhere east of the Alleghany Mts. In Ohio and Indiana, they seem to be more abundant. As a rule, the smallest arrowheads found along the coast are triangular in shape. Of several thousand specimens of all shapes, there are less than a dozen of this pattern, that are as small as fig. 264, and no other specimen has been collected by the author, as small as fig. 265. As with the very small scrapers, these diminutive arrowheads are chipped with great nicety, and are among the finest examples of this kind of work that have been found on the Atlantic seaboard.

FIG. 264.—New Jersey. $\frac{1}{1}$.

Their size is apparently an objection to their ordinary use as arrowheads, and so few have been found, that it is not likely they were made for ordinary hunting or warlike purposes. Still, we find on comparing them with modern stone and glass arrowheads, and especially with the beautiful obsidian specimens found in the southwestern territories and the agate examples found in Oregon, that they are about of the same size. Arrowheads of this size constitute but a very small proportion of the whole number found; probably not more than one in a thousand. That they are really more abundant may be true, and their apparent rarity due to the fact that their diminutive size renders them more liable to be lost in sandy soil, and to be broken, if lying in loose gravel. Unless of some bright color, such small arrowheads as fig. 265 are very inconspicuous objects.

FIG. 265.—New Jersey. $\frac{1}{1}$.

Fig. 266 represents a remarkably true and carefully worked example of an arrowhead with a notch at the base. Specimens as carefully chipped as this are now seldom found, except in a fragmentary condition. From the extreme care with which the edges have been worked, it is possible that this implement was used as a knife. It is made of light brown or yellowish-brown jasper, which occurs in great abundance as bowlders and pebbles, in the Delaware river. Of this material, all the finest examples of flint work, found in New Jersey and in New England, are made.

Fig. 267 represents a somewhat similar specimen, except that it is much narrower. Unlike the generality of these worked flint implements, this specimen is a thin flake of chalcedony, without surface chippings and only altered along the edges, so as to give it a symmetrical outline. This handsome specimen, also, may have

FIG. 266.—New Jersey. 1/1.

FIG. 267.—New Jersey. 1/1.

FIG. 268.—New Jersey. 1/1.

been used as a knife, but it is as probable that it is properly classed as an arrowhead.

Fig. 268 represents a modified form of this same pattern, in which the sides are parallel for one-half their length, and then slope evenly, by straight margins to an acute point. Like fig. 266, this specimen is chipped from a jasper pebble, and is of equally artistic workmanship.

Arrowheads of this pattern are not abundant in New Jersey, and are very rarely found in New England. In western New York, they are occasionally found, and are of more common occurrence in Ohio and Indiana. The late Professor Haldeman found but three specimens during several years careful collecting in the valley of the Susquehanna.

All the arrowheads of this pattern, that I have seen, appear to be of the same grade of workmanship, and of the one material. That they possess any advantage from the peculiar finish of the edges seems very questionable, but that they were arrowheads cannot be doubted.

In figs. 269 to 273, inclusive, is represented a second form of arrowheads, that is almost as common as the preceding. Instead of the notched base, they have a straight stem, and vary among themselves indefinitely, as to the relative length and breadth of both the stem and the blade. Fig. 269 may be taken as a type of this pattern, as the great majority of these arrowheads have the edges slightly convex. While perhaps a majority of them are carefully chipped, there is a very large proportion that are rudely made. Apparently, the workman ceased to bestow any care in finishing the implement when he had once obtained a sharp point.

FIG. 269. — New Jersey. ½.

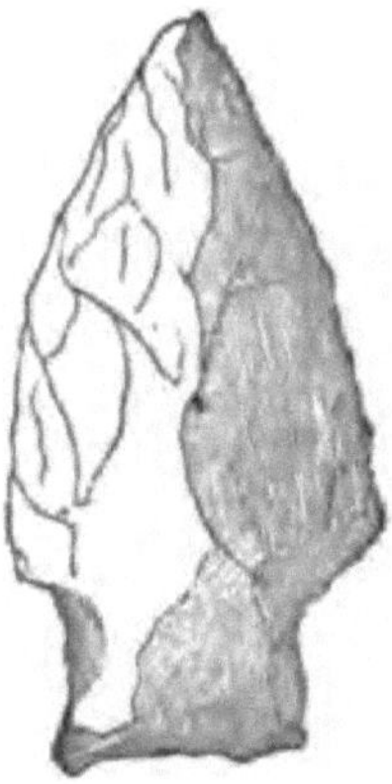

FIG. 270. — New Jersey. ½.

In every large series of stemmed arrowheads, there will be a small percentage as pronounced in pattern as figs. 271 and 272, but generally they more nearly approach figs. 270 and 273.

Fig. 272 is a rare form of stemmed arrowhead, that is of common occurrence in Europe. In the Clement collection of Swiss Lake stone implements in the Archæological Museum at Cambridge, Mass., there are many specimens that are of this size and shape. Others are contained in the Rose collection of Danish antiquities, belonging to the same museum.

FIG. 271. — New Jersey. ½.

Fig. 273 represents about the minimum size of these stemmed arrowheads. Implements of this size appear to be in larger proportion in some localities than in others, but they are not in such numbers as to justify the belief that they were ever in very general use.

In the very large series of arrowheads from the valley of the Susquehanna, contained in the collection of the late Professor Haldeman, are hundreds of specimens of this pattern, varying in outline to about the extent illustrated in the series of specimens here figured. As shown by the material gathered in the neighborhood of Columbia, Pa., it would appear that these stemmed specimens were more abundant than any other form, not excepting those with a notched base.

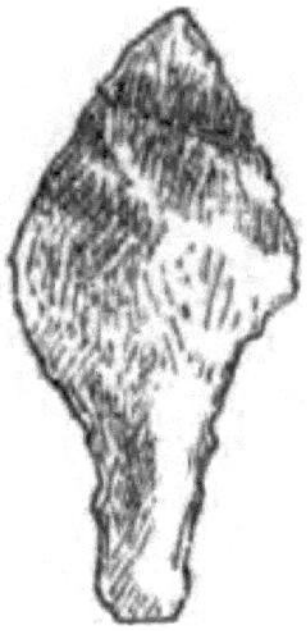

FIG. 272. — New Jersey. ½.

The various patterns of varieties of stemmed arrowheads, that are here figured, all occur in abundance in New England and New York; yet in but few limited localities are they found in such great abundance as in the valleys of the Delaware and Susquehanna rivers.

Fig. 274 represents a splendidly chipped implement which may be out of place in association with arrowheads, though it cannot certainly be classed either with the spearpoints or knives.

While an unusual pattern in New Jersey and Pennsylvania, and of even greater rarity throughout New England, specimens of this shape are recorded from most of the northern Atlantic coast states. In New York they have been found in the valleys of the Mohawk and the Hudson, and on the shore of several of the inland lakes. In Ohio, they are much more common. Of all that have been examined the workmanship is excellent, and only the choicest materials have been chosen in making them. The very deep lateral notches seem to indicate that a very secure hold to the shaft or handle was required, and that the implement, thus fastened, was intended for serious work, such as would put a great strain upon it. So seldom are these large arrowheads, or knives, met with along the seaboard, as compared with Ohio and the southwestern states, that it is probable that all that have been found in

FIG. 273. — New Jersey. ½.

New England and New Jersey were brought from the interior states, as was probably the case with many of the large spearpoints of finer finish. Some doubt, however, is thrown upon this suggestion by the fact that there have been found a small number of delicately chipped arrowheads of small size, of precisely this pattern. These arrowheads were made of yellow jasper, and found in a field near Princeton, New Jersey, associated with dozens of other patterns.

I have also had an opportunity of seeing in a very complete local collection of stone implements, made in the neighborhood of Lake Hopatcong, Morris Co., New Jersey, three specimens of these implements of the same size as fig. 274, and all made of a beautiful green jasper. The workmanship was excellent. They were found together near the shore of the lake, and in the immediate vicinity were found seven copper beads and a fine chalcedony spearpoint. It is not improbable that all these objects had been washed from a grave.

FIG. 274.—New Jersey. 1/1.

This form of arrowhead or knife does not occur in Europe, so far as I can learn. Evans, in his elaborate volume on the "Stone Implements of Great Britain," figures nothing that bears any resemblance to this form. All the larger spearpoints and arrowheads of England appear to be of the stemmed, or stemmed and barbed patterns.

Figs. 275 and 276 represent two excellent examples of a pattern of arrowhead or small spearpoint, which is very frequently found in New Jersey. They vary indefinitely in size, from the larger of the two figured to others that are considerably smaller than fig. 275. This form is one of a few that appears to be every way desirable, and yet it is not as abundant as some others. When made of jasper, thin, and with a good edge, it would be difficult to design a better weapon than fig.

276. The stem is of such shape and size as to render a secure attachment to the shaft easily practicable, and with a moderate exercise of force, a most ugly wound could be inflicted with this implement.

In fig. 276 we have an implement that bears considerable resemblance to the specimen described in the preceding paragraph (fig. 274). By merely widening the base of the stem of fig. 276, we convert it into the pattern of fig. 274. Thus, while possibly both are too large to be used as arrowheads, they were probably used in the same manner, either as spears or knives.

Implements of this pattern, as large as fig. 276, are not frequently found in New England, even in the Connecticut valley where so large

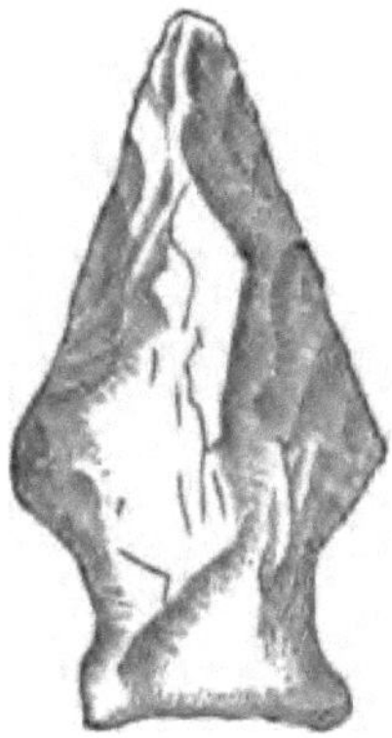

FIG. 275. — New Jersey. ½.

FIG. 276. — New Jersey. ½.

a variety of forms have been obtained. In western New York they are more abundant than in the Hudson river valley; while in Pennsylvania, although recorded in few numbers in some localities, they were poorly represented in the collection from the Susquehanna valley, made by the late Professor Haldeman. From a field of a few acres, near Trenton, New Jersey, I gathered twenty specimens of this form in the summer of 1878, and six more in the ensuing year. Other specimens were also found which I have failed to secure. In every case these implements were made of jasper, either red or yellow, and are among the best specimens of flint chipping in my collections from New Jersey, now in the Archæological Museum at Cambridge, Mass.

FIG. 277.—New Jersey. ½.

Barbed arrowheads, like figs. 277 and 278, are not uncommon wherever arrowheads of any pattern are found. Barbs can be added to almost all the types, and seem to be the result of fancy on the part of the maker. The examination of broken specimens, among the refuse of chips of an arrowmaker's "shop," does not indicate that they were ever very abundant; and nowhere do we find a preponderance of this pattern. When the barbs were related to the stem, as in fig. 277, it is difficult to see of what practical use they were, as they could not add to the effectiveness of the weapon. On the other hand, when they were associated with straight, stem-like bases, as in fig. 278, it is evident that when once they had been shot or thrust into a body, they could not readily be removed,and hence in war, they would be more deadly than such as could be promptly withdrawn from the wound.

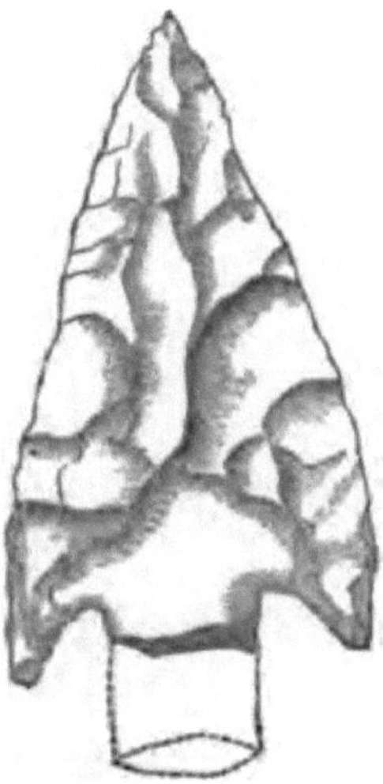

FIG. 278.—New Jersey. ½.

Arrowheads, that are distinctly barbed, do not appear to be of more common occurrence in the western and southern states, than along the Atlantic seaboard. On the Pacific coast, while not unknown, they are rare. Of the series collected in the small islands off the coast of California, by the officers of the U. S. Geog. Survey, West of the 100th Meridian, there is but a single specimen. (Vol. vii, Archæology, pl. iii, fig. 1.)

Of a series of nearly thirty-three hundred arrowheads, collected in a single county in New Jersey, but fifty-six are distinctly barbed, and probably this is as large a proportion as occurs anywhere along the seaboard. Of thousands of specimens from the Connecticut valley, the percentage of those that are distinctly barbed is not as large as in the New Jersey series.

In Europe, the barbed arrowhead is a prevailing type, almost numbering one-half the specimens found in many localities. Of a series of seventy-two arrowheads from Great Britain, figured by Mr. Evans, twenty-eight are distinctly barbed. Some of them have the barbs of remarkable length, and terminating in blunt ends. This latter feature is almost unknown among American flint arrowheads, and never, I believe, occurs in the various forms common to the Atlantic seaboard.

That form of arrowhead which is a modification of the barbed, and known generally as the triple-notched arrowhead, is one of quite com-

Fig. 279. — New Jersey. 1/1.

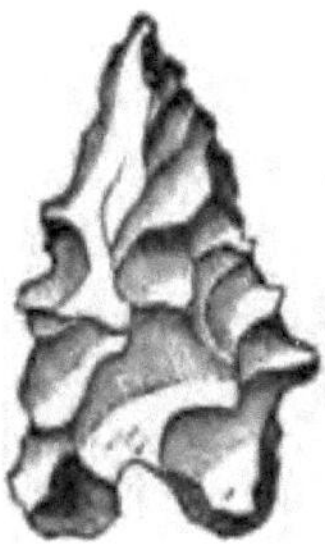

Fig. 280. — New Jersey. 1/1.

Fig. 281. — New Jersey. 1/1.

mon occurrence in New Jersey, Pennsylvania and Delaware, and also in Connecticut, but is unusual in Massachusetts and more northern localities. As seen in the accompanying illustrations, the specimens of this pattern vary considerably in form and size.

Figs. 279 and 280 represent the more common sizes; fig. 279 the minimum, whilst the curiously designed specimen, fig. 281, seems too blunt to be classed as an arrowhead, and ought rather to be classed as a knife; though in this case it is very difficult to imagine any object in the prominent side barbs.

The great majority of these triple-notched arrowheads are of quartz and jasper, and, as a rule, are well made. Two of the five illustrations, here given, are distinctly serrated, but it must not be supposed that such a large proportion is usual; nevertheless the majority of the arrowheads with distinctly serrated edges are of this pattern.

FIG. 282.—New Jersey. ½.

In a large series of arrowheads collected in the state of Delaware, and kindly forwarded to me for examination by the owner, Mr. T. J. Bennett of Dover, Delaware, there was a large number of specimens of this pattern, all of which were made of jasper. Many had the edges distinctly serrated.

The late Professor Haldeman found this pattern to be quite well represented in the extensive series of arrowheads collected near Chickies, Lancaster Co., Pa., and noticed the large proportion of serrated specimens among them. The majority of these specimens, however, were similar in size and finish, to fig. 282.

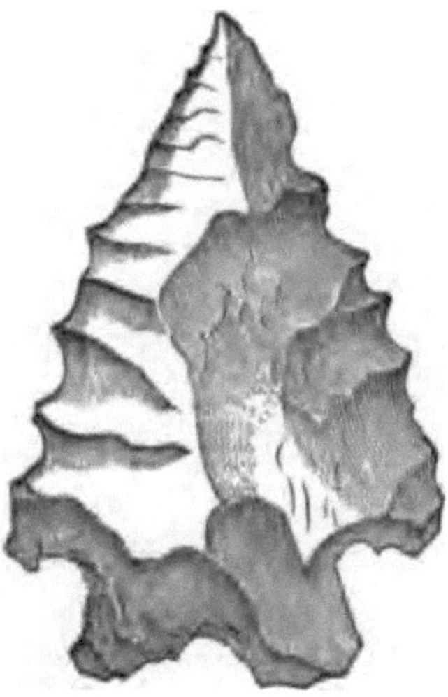

FIG. 283.—New Jersey. ½.

Fig. 283, of the series here given, is apparently too large to be classed as an arrowhead. The serrated edges, however, seem to suggest this use, rather than that of a knife; and it is scarcely large enough to be classed as a spearpoint.

Triangular arrowheads are of every imaginable shade of width and length, and in workmanship they vary from the rudest to the most exquisitely chipped specimens. As is the case with so many other kinds of implements, there will often arise a doubt as to whether all these specimens, varying as they do in form and size, were used for the same purpose. It seems unquestionable that figs. 284 and 285 were arrowheads. Their size, shape and, it may be added, their abundance, all indicate it; but this may be doubted when we come to consider such an example as fig. 288. This, however, is not too large for an arrowhead; and, until it can be shown that specimens of this size were otherwise utilized, it seems safe to class it as such, though one of the largest size.

When long and slender, like figs. 284 and 287, these implements are certainly admirably adapted for tips to the shafts of arrows, and being so largely made of black flint or chert, they recall forcibly the words of Holm that "the head of the arrow * * * is made of a black flint stone."

Of the triangular arrowheads found in New Jersey, but very few have been found which have serrated edges, and all of these were of the smallest size of this pattern, several being even smaller than fig. 286.

FIG. 284.—New Jersey. $\frac{1}{1}$.

FIG. 285.—New Jersey. $\frac{1}{1}$.

FIG. 286.—New Jersey. $\frac{1}{1}$.

There is a prevalent opinion concerning arrowheads of this pattern, that they were used almost exclusively for "war arrows," and were usually poisoned. Upon what grounds this opinion is based I have not been able to learn, except that it has been suggested that these points or heads were placed loosely in a slit on the end of the shaft, and so would be retained in the body of the person shot, however carefully the shaft might be withdrawn. How far this may be true is difficult to determine, but it is highly improbable that any one form of these implements should have been designed or set aside for any particular purpose. In New Jersey there is no known locality where these are found unassociated with other forms, and if "war" points, they should in some localities be a marked feature of ancient battle-fields, several of which are supposed to be known to the local historians, and even to the collectors of Indian relics.

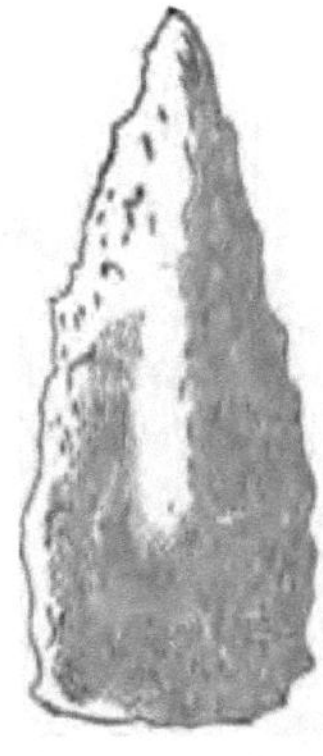

FIG. 287.—New Jersey. ½.

In figs. 289 to 293 inclusive, we have examples of short, broad triangular arrowheads, such as occur in even greater numbers than the preceding. When of this size and shape, they are often carelessly chipped, the absence of a sharp point being very noticeable. Their bluntness is not caused by the loss of these points, but was evidently produced intentionally. Occasionally these triangular arrowheads have two notches near the ends of the base, thus making the implement one of the barbed pattern (see fig. 309). Whenever these notches occur, they are invariably associated with a style of chipping much superior to that upon the specimens here figured. The peculiar needle-like point on fig. 292 is not an accident, but was deliberately chipped out. Of a large series of these broad triangular specimens from New Jersey, there were many with this peculiarity; and the

FIG. 288.—New Jersey. ½.

FIG. 289.—New Jersey. ½.

FIG. 290.—New Jersey. ½.

FIG. 291.—New Jersey. ½.

same was noted as occurring on a series of the same pattern from the Connecticut valley.

Of a series of thirty-three hundred arrowheads from Mercer county,

New Jersey, to which reference has been made, fourteen hundred and twenty-eight belong to the triangular pattern. This shows how very common it was, and proves clearly that, among all the various patterns of arrowheads, this was emphatically the favorite form.

FIG. 292.—New Jersey. ½.

The proportion of triangular to other shapes of arrowheads is less in New England, than in the middle states, judging from the collections preserved in the public museums; and in Pennsylvania they are also less common than in New Jersey, so far as the collections of the late Professor Haldeman, of the Susquehanna valley forms, and of my own, of the Delaware valley patterns, enable one to make a comparison.

Considered as a class, these triangular arrowheads will not compare favorably with those of the same pattern found in Ohio and Indiana. From various localities in these states, are derived most marvellous specimens of skilful chipping of flint, which in some cases is only equalled by the finest specimens of Danish arrowheads of the same pattern.

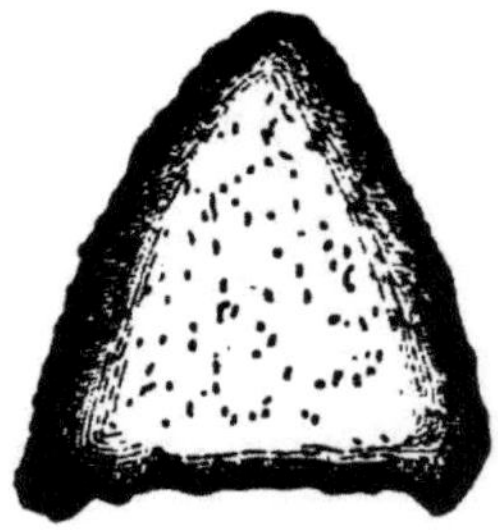

FIG. 293.—New Jersey. ½.

Figs. 294 and 295 represent two irregularly shaped triangular arrowheads which differ from the preceding, in having concave bases. This feature is much more marked in some specimens than in those figured; in some cases, being not a curved indentation, but rather like an inverted V.

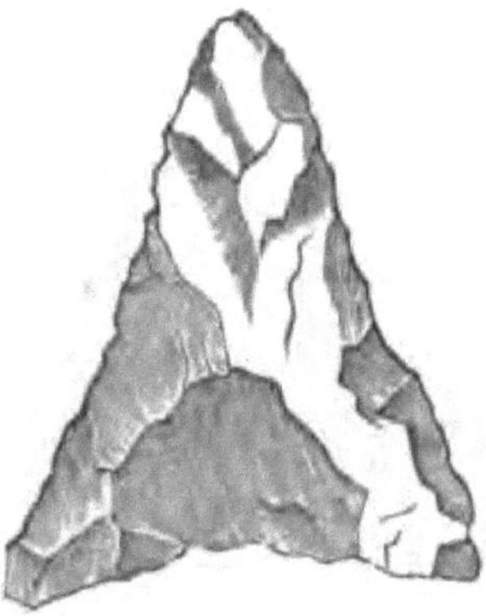

FIG. 294.—New Jersey. ½.

In fig. 295 we have a modified form of the triangular pattern, which is so rude, that at first it might be considered an unfinished implement; but so many are found of exactly this shape that it is probably an intentional and not a chance form.

Of the ordinary form of triangular arrowheads, with a concave base, such as fig. 294, there appears to be a varying number as compared with those that have a straight base, as they occur in different localities. In central New Jersey, they constitute but a small proportion, in Pennsylvania they outnumber the others, while in New England they are about equally divided.

FIG. 295.—New Jersey. ½.

While the whole subject is merely a matter of conjecture, I am led to believe, from the circumstances under which many of these relics are found, and also their numbers in many localities, which bear evidence of great antiquity, that the triangular pattern is the original form of arrowhead, if there ever was such a thing. Attention has already been called to various forms of flakes that were probably used for tipping the shafts of arrows. Flakes of themselves would soon be found to be inconvenient from the fact of their irregularity of outline, and an attempt to reduce them to a symmetrical shape would soon be made, and the first efforts in this direction were apparently to make a shapely triangular point for the arrowshaft. The oldest jasper arrowheads I have found in New Jersey were simply flakes, with such secondary chipping as was required to make them triangular; and the oldest specimens, apparently, of the series collected by the late Professor Haldeman, in the valley of the Susquehanna, Pa., gave that archæologist the same impression, as stated to the writer.

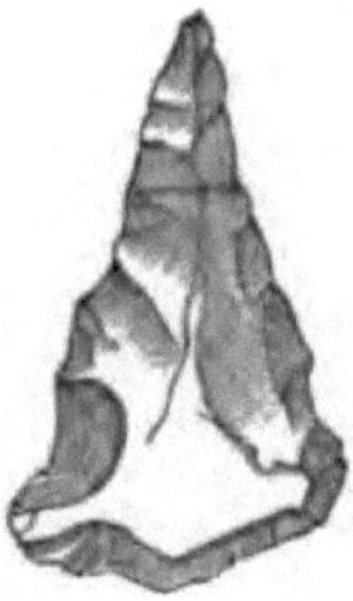

FIG. 296.—New Jersey. ½.

A well-marked and very handsome pattern of arrowhead is the long triangular with convex base. In this form they are well known as the leaf-shaped arrowheads, and constitute a considerable proportion of the whole number found.

Fig. 296 represents a poorly finished specimen, which is here introduced as a "connecting link" between the triangular specimens, and such as are distinctly curved at the base.

Fig. 297 more nearly approaches the typical examples of this form of arrowhead, given in the succeeding figures.

Figs. 298 to 300, inclusive, represent good average specimens of this implement. A small proportion of the whole number found are somewhat larger than the specimens here figured, and occasionally we meet with one which is considerably smaller than fig. 299, but these larger and smaller specimens are so few in number, as scarcely to affect the statement that in size, as well as finish, those figured fairly represent the whole series.

Whether made of quartz or jasper, arrowheads of this pattern are,

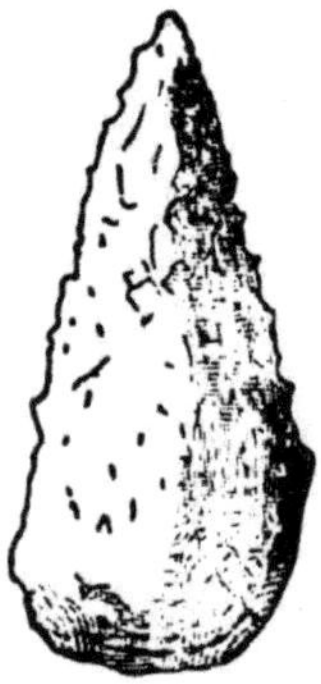

FIG. 297. — New Jersey. ½.

FIG. 298. — New Jersey. ½.

FIG. 299. — New Jersey. ½.

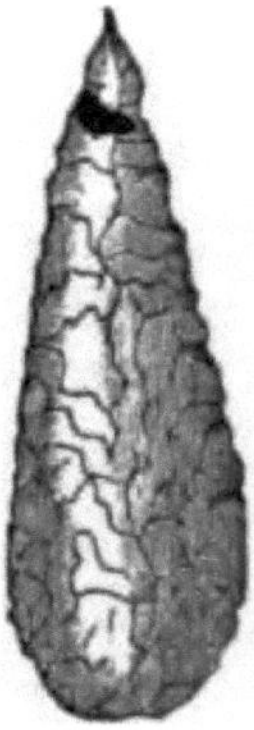

FIG. 300. — New Jersey. ½.

as a rule, very carefully chipped, and many are still very sharply pointed. Few specimens of flint chipping are more delicate than the white quartz example, fig. 300; which still retains its original point, and sharp cutting edges.

Though this form of arrowhead differs very materially in numbers, in the various river valleys, throughout the New England and middle states, yet it cannot be considered a rare form. In some limited localities it reaches fully to seven per cent. of all found.

The leaf-shaped arrowhead occurs throughout the southern and western states, and was a favorite pattern with the natives of the islands off the coast of California (U. S. Geol. Survey, west of 100th Meridian, vol. vii, Archæology, pl. iii), and is found also along the Pacific coast, northward to British America.

In Europe it is a well known form. Mr. Evans figures a large series of English specimens, none of which differ materially from the American forms. He includes, however, with them, those that are pointed at each end. It is doubtful if these were really used as arrowheads. The same pattern is found in the Atlantic coast states, and abundantly in Ohio and the southwest. I am disposed to consider them as knives.

Lozenge-shaped arrowheads, as they are generally called, are simply

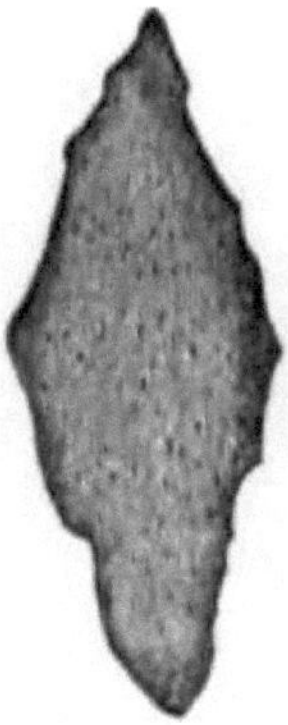

FIG. 301.—New Jersey. ½.

FIG. 302.—New Jersey. ½.

FIG. 303.—New Jersey. ½.

FIG. 304.—New Jersey. ½.

modifications of the leaf-shaped pattern, in which the bases are angular instead of being convex in outline. Fig. 301 represents an example that approaches more nearly to the stemmed arrowheads; and fig. 302 varies but little from a typical leaf-shaped specimen. In figs. 303 to 305 inclusive we have typical examples of the true lozenge-shaped specimens. Between these forms there are intermediate patterns, and a closely connected series can readily be made, which extends from specimens as pronounced as fig. 304 to the leaf-shaped arrowheads of the foregoing page.

Arrowheads of this pattern are largely confined to certain localities, and seem to be wholly wanting over very extended districts. In the Susquehanna valley, but very few specimens were found by the late Professor Haldeman, and those that were collected were generally of such large size as to suggest the knife rather than the arrowhead. In the Connecticut valley, this form is nowhere common, and less than a score of specimens, from northern New England, are in the collections of the Museum of Archæology, at Cambridge, Mass. Of a series of about fifty specimens from New Jersey, all are from one limited locality, in Gloucester county, except half a dozen of the more indefinite pattern, like fig. 302.

FIG. 305.—New Jersey. ½.

It is not improbable that the great majority of these lozenge-shaped implements were really used as knives, and not as arrowheads. In many respects they resemble what are known to be chipped knives, found in distant localities, as in California, and the shape of the base is such as to suggest that they were inserted into a broad wooden handle, instead of a slender reed. In this case, as in all others, there must ever be a doubt as to the purpose for which these chipped flints were made; and the fact borne in mind, that they were not confined to any one use. So far as they proved available as knives, doubtlessly they were used as such, and the same may be said of their use as arrowheads.

Arrowheads with distinctly serrated edges are not abundant either in New England or the middle states. Slight serrations, such as occur on some roughly chipped specimens, are rather frequent, but with this exception, they are comparatively rare.

Attention has already been called to another pattern of arrowhead, which is frequently serrated along the sides, and which constitutes the majority of the serrated examples of these implements. Other forms are also met with, as the notched-based, stemmed and barbed arrowheads.

Fig. 306 represents an example of an arrowhead, with finely serrated edges, of the so-called notched-based pattern. Although now rather indistinct, the serrations can readily be traced and show that they were intentionally produced, and have no connection with the chipping of the surfaces.

Fig. 307 represents an ordinary stemmed arrowhead with serrated sides. Examples with this feature so marked are very rarely met with along the Atlantic coast. I have seen none from New England that were as distinctly serrated, and but three specimens from the Susquehanna valley.

Fig. 308, with barbs as well as serrated edges, is a pattern that occurs

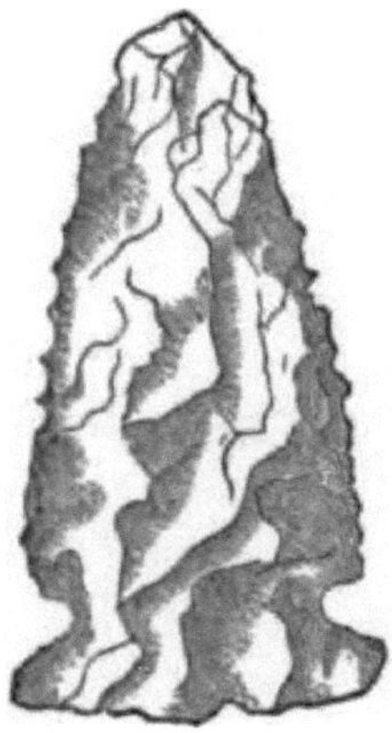

FIG. 306.—New Jersey. $\frac{1}{1}$.

FIG. 307.—New Jersey. $\frac{1}{1}$.

FIG. 308.—Indiana. $\frac{1}{1}$.

but rarely anywhere. In New Jersey, but few examples have been found, and these are generally broken. In New England, even in the Connecticut valley, they are rare. The specimen here figured is from Indiana.

Of American arrowheads, with serrated edges, it may be said that they do not form a class of themselves, or that they were made for any particular purpose. The art of chipping flint, so as to produce these tooth-like projections, was not sufficiently well understood to enable the Indian to make much use of this feature, and no examples of implements with such evenly wrought serrations and ripple-marked surfaces, as are seen on arrowheads, saws and daggers of flint, found in Denmark, have been found in this country, or at least, on the Atlantic coast of the continent.

Fig. 309 represents a form of arrowhead, that combines the characteristic features of several other patterns. It is, in fact, a triangular arrowhead, with barb-like angles at the base. This form, also, occurs frequently with serrated edges.

FIG. 309.—New Jersey. 1/1.

While it might properly have been described under one or more of the several divisions of arrowheads mentioned in the preceding pages, it is here placed by itself that its peculiarities, which are very marked, might the more readily be recognized.

Arrowheads of this pattern are quite abundant in many limited localities in New Jersey and Pennsylvania, and are remarkable for the delicacy of the chipping, the uniformity of size, and the fact that all are made either of quartz or jasper, the former mineral being that which was generally used.

CHAPTER XX.

FLINT DAGGERS.

THE typical chipped flint dagger, which is such a characteristic form of implement in Denmark and throughout northern Europe, is of comparatively rare occurrence in the United States, and particularly so, along the northern Atlantic seaboard. Of the various patterns of weapons described by the early writers, not one is mentioned as being in common use among the American native tribes that even suggest the dagger. Long, slender spearpoints are supposed to have been used in this manner, but there is no reason why the term "dagger," as now understood, should be applied to them.

The class of objects more particularly referred to in the present chapter may be described collectively as carefully chipped flint implements of great length in comparison to their breadth, acutely pointed, occasionally with the base defined by hilt-like projections, and the blade quadrangular or oval in section.

Daggers, like the typical European examples figured by Sir John Lubbock,[102] have been found in the Ohio valley. In the museum of the Academy of Science, at Salem, Mass., there is a beautiful specimen found in the bed of Crosswicks Creek, Burlington Co., New Jersey, and fragments of others have been found near Easton, Pa.

Col. C. C. Jones, jr.,[103] has given a figure and description of a beautiful specimen, seven and one-half inches in length. In this specimen the blade is separated from the handle by barb-like projections, and the end of the handle terminates in similar barb-like projections.

Fig. 310 represents a dull brown jasper implement that is in every

---

[102] Lubbock. Pre-Historic Times, 2nd ed., p. 97, figs. 115, 116 and 117. London, 1869.

[103] Jones, *l. c.*, p. 267, pl. vii, fig. 3.

particular the facsimile of dozens in the magnificent Rose collection of Danish implements, belonging to the museum of Archæology at Cambridge, Mass. This specimen measures nearly seven inches in length, and nowhere exceeds an inch in width. The base or handle is as acutely chipped on the sides as the blade proper, and is too sharp to have been held in the naked hand. A handle of some kind was necessary, though its shape or the material of which it was made is not known.

FIG. 310.—New Jersey. 1/1.

The magnificent chert implements of a dagger-like character, found on the Pacific coast, were attached to handles by means of asphaltum, or had handles made of this material. Most of these implements, however, are quite thin and flat, and evidently were not put to any severe use. Associated with them, however, are others that are thick, and distinctly oval in section, which also were hafted with, or by means of asphaltum; and it is by no means improbable that some gum or resin was used by the Atlantic coast tribes, in a similar manner. Attention has already been called to the fact that the Delaware Indians used fish glue, to secure the stone points to the shafts of the arrows.

Fig. 310 is but little shorter than the larger examples of the chert implements found in California, figured in the volume on Archæology, of the Government Survey of Capt. Wheeler,[104] and possibly may have been used

[104] U. S. Geographical Survey west of 100th Meridian, vol. vii, pl. 1, figs. 5, 6 and 7.

in ceremonial observances only, as Mr. Powers[105] has stated, was the case with the Pacific coast implements. According to this author, among the various objects worn and paraded in various ceremonial dances, are the flakes or knives of obsidian or jasper. "I have seen several which were fifteen inches or more in length and about two and one-half inches wide in the widest part. Pieces as large as these are carried aloft in the hand in the dance, wrapped with skin or cloth to prevent the rough edges from lacerating the hand, but *the smaller ones are mounted on wooden handles and glued fast.*"

Considering the rarity of such jasper implements as fig. 310 and others, that more nearly correspond with those of the largest size mentioned by Mr. Powers, may it not be that, on the Atlantic coast also, they were produced only upon ceremonial occasions, and did not serve any of the purposes that have been ascribed to them?

In the museum of the Academy of Science at Salem, Mass., is a beautiful example of a chipped jasper implement, similar to those described by Mr. Powers. It is made of jasper, and measures eleven and one-fourth inches in length by two and seven-eighths inches in width.

This specimen was presented by Jos. Story, esq., in 1824, and is labelled as coming from New Jersey; associated with it is a specimen of another class, identical with the western spades or shovels. It is of the same mineral, also from New Jersey, and presented by Mr. Story in 1824.

As specimens of such large size are very rare in the eastern or middle states it is more than probable that they were the peculiar property of "chiefs" and possibly were used on state occasions as a badge of office, rather than on the field of battle. There is too much work on such an implement to risk its being broken in a fight.

Messrs. Squier and Davis[106] figure a "flint" similar in size and

---

[105] Powers. Tribes of California: Contributors to North American Ethnology, vol. iii, p. 79. Washington, 1877.

[106] Squier and Davis. Anc. Mon. Miss. Valley, p. 211, fig. 99 (No. 3).

mineral, but varying from it in being pointed at each end. After mentioning the use to which the stemmed examples were put, they add, "There are others, however, the manner of using which is not so obvious. No. 3 is an example. It measures eleven inches in length by two and a half in greatest breadth. It has been suggested that it was fastened at right angles to a handle and used as a sort of battle-axe."

CHAPTER XXI.

GROOVED STONE CLUB-HEADS.

A grooved globular or oval pebble is an object so readily utilized in many different ways that it is well nigh impossible to determine to what use any particular specimen was put. Furthermore, these stone implements are common to a large portion of the globe, and the use to which the natives of one country put them is not necessarily the same as that of a different people.

Grooved pebbles found in Denmark are referred to by Nilsson[107] as "plummets" and considered as of such use only. The suggestion that they were "sling-shots" he regards as a mere supposition. Evans[108] suggests that, in England, at least, they were used as "sinkers" for nets and lines, inasmuch as they generally are not battered at the ends, and so show no indication of use as hammers. Col. C. C. Jones, jr.,[109] does not refer to this form, when occurring in the southern states, as being distinct from the more abundant notched pebbles, but considers them alike as sinkers for nets and lines.

While it is very probable that the above suggested use of grooved globular pebbles is correct, so far, at least, as it applies to the localities, to which these authors refer, there are some reasons for considering that possibly they were also occasionally, if not habitually, put to a far different use.

From the fact that many of the grooved globular pebbles are carefully worked to a very symmetrical form, and the groove finished with as much care as it is on many of the common axes; and for the

---

107 Nilsson. Stone Age in Scandinavia, p. 25. London, 1868.

108 Evans. Ancient Stone Implements of Great Britain, p. 211, London, 1873.

109 Jones. Antiquities of the Southern Indians, p. 338. New York, 1873.

additional reason, that they are found in fewer numbers and associated with ordinary forms of weapons and domestic implements, rather than in the vicinity of rivers or the larger creeks, it is believed that many, at least, of the larger examples, were used, not as sinkers, or plummets, but as club-heads, and therefore should be classed as weapons. There is no historical evidence to confirm this opinion, so far as relates to the Delaware Indians, but it is known that clubs of various patterns were, and indeed still are, a favorite weapon of the native races of America.

The Iroquois, according to Morgan,[110] had war-clubs of two patterns; one, the *Gä-jë-wä*, with a globular head; the other, *Ga-ne-u'-ga-o-dus-ha*, armed with a point of deer's horn. Of these, he remarks: "before the tomahawk came into use among the Iroquois, their principal weapons were the bow, the stone tomahawk and war-club. The *Gä-jë-wä* was a heavy weapon usually made of ironwood, with a large ball of knot at the head. It was usually about two feet in length, and the ball five or six inches in diameter."

Besides these, Mr. Morgan adds, what is of more importance in this instance, "oval stones, with grooves around their greatest circumference, were also secured in the heads of war-clubs, and thus made dangerous weapons." It is recorded, also, of the Massachusetts Indians,[111] that "they made use in their wars, of the 'balista' (a war-club with a stone head covered with hide) which instrument is represented several times, agreeably to Chingwauk's interpretation, on Dighton Rock." West of the Mississippi river, a form of weapon was long ago described by the explorers, Lewis and Clarke,[112] as a poggamoggon, which, although not necessarily a grooved stone, is further evidence that globular stones were used as club-heads.

Fig. 311 represents an excellent example of these larger grooved pebbles, which are supposed to have had other uses than as net-sinkers. In this example, the groove is accurately finished, and quite smooth, and the ends of the pebble are free from every indication of hard

---

[110] League of the Iroquois, p. 362. New York, 1850.

[111] Schoolcraft, *l. c.*, pt. 1, p. 284, pl. 15, figs. 1 and 2.

[112] Exp. up the Missouri, vol. i, page 415. Philadel., 1814.

usage of any kind. The stone itself is very hard and heavy, and might be used indefinitely for cracking skulls without exhibiting any wear. Occasionally, equally well-wrought implements of this pattern are found, that have two grooves crossing each other at right angles. The object of this is not clear. No traces of a handle, such as that in which it is supposed these grooved stones were mounted, have been found in any of the Indian graves in New Jersey; but it is of interest to note that these stones frequently occur in these graves, and there is no instance of an ordinary net-weight having been found among the contents of any ancient burial. As the large mauls were necessarily used with a handle, and these smaller grooved pebbles are simply mauls in miniature, it is quite reasonable to suppose, that when they are highly finished, and have a certain smoothness indicative of a leather covering, they were used as weapons.

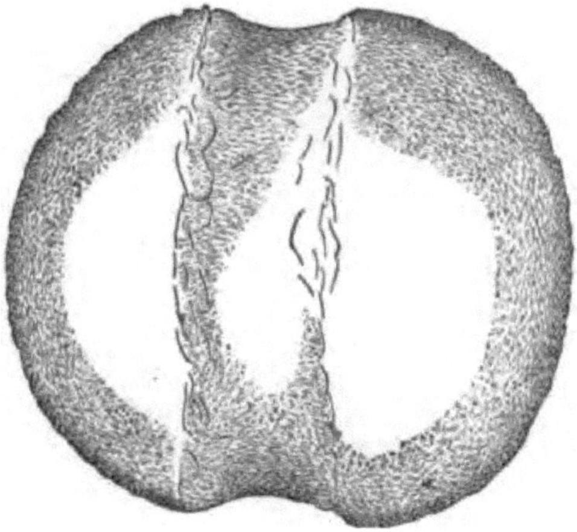

FIG. 311.—New Jersey. ½.

Fig 312 represents a second example of these larger grooved pebbles, which though not actually polished, is exceedingly smooth and free from all inequalities of surface. The groove, in this specimen, is unusually narrow and deep, and very smooth; and, if originally pecked, has since been intentionally ground or worn away by the friction of a leather cord. This specimen, whatever it may prove to have been, would be more intelligible were it not for the deep cuplike depressions on the sides, so situated as to break the continuity of the groove. These cups, or hollows, are larger than the finger pits of common hammer stones, and, like the groove, are carefully ground; and, except that they are oval, would appear to have been drilled. Putting aside the presence of these lateral depressions, it is probable, that this specimen, like the foregoing, was mounted in a flexible handle, and used as a club.

Other uses, however, may have been found for these globular stones; for instance, we find in Schoolcraft's work an illustration of a war-club with such a small round stone inserted in a notch in the end of the club, giving the completed implement the appearance of an ordinary wrench with a small object held in its jaws. Such a specimen, as fig. 311 might have been used in this manner, also.

Mr. George C. Musters[113] has given us a most interesting account of

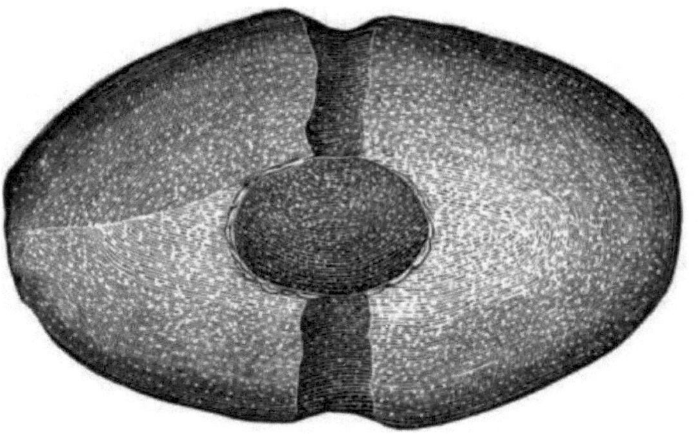

FIG. 312.—New Jersey. 1/1.

the weapons and hunting implements of the Tehuelche Indians, and in his description of the "bolas" now in use by these savages, refers to ancient bolas, which seem to be identical with the grooved pebbles, or "net-weights," that we have described. Mr. Musters writes: "Ancient bolas (globular stones) are not unfrequently met with. These are highly valued by the Indians, and differ from those in present use *by having grooves cut around them*, and by their larger size and greater weight." There is no reason for believing that the Patagonian bolas were ever in use among the North American Indians, but it is not at all improba-

[113] At home with the Patagonians, p. 166. London, 1872.

ble that a weapon should have been in common use among the native populations of the Atlantic seaboard, of which these same globular pebbles formed a part. The long cords of the bolas might have given place in this country to a flexible handle, and a more terrible blow might have been inflicted with this implement thus mounted, than could have been given with wooden war-clubs of greater size and weight.

As in the case of many other kinds of stone implements we find these grooved pebbles grading into other and totally different objects. When materially larger than the specimens figured, these objects become mauls; and in every series of a few hundreds of these, it will be found, that in many the groove is not continuous, and so they approach the notched pebbles. This is specially marked, when the pebbles are somewhat flattened. These facts, however, do not bear upon the questions of probable uses of typical forms; for in no instance yet, has it been possible to draw a dividing line between one class of objects, and that which most nearly resembles it. Perhaps, of the many puzzling forms of stone implements yet found, these small grooved pebbles are the most difficult to decipher. Occasionally, one has been found, less than an inch in diameter.

CHAPTER XXII.

PIPES.

NOTHING of all the handiwork, in stone or clay, possesses so great an interest, and recalls so vividly the early history of America, as the tobacco-pipe. Whether as the merest fragment of a pipe of clay, or as a carved and polished stone specimen, they bring back to us the image of the dusky warrior, gorgeous in feathers and vermilion, bearing with the dignity of a king his ever-present pipe.

The sorrowful history of the long series of broken treaties between the American Indian and the European settler is made up of promises and ceremonial smoking; but the clouds of the fragrant smoke and the white man's recollection of his promises were equally evanescent.

When, happily, we find an occasional pipe in our rambles, or, disturbing the ashes of some forgotten hero, we rifle the grave of this, the chiefest of his possessions, we are apt to think of it only as a mere "medium through which the narcotic influences of tobacco were imparted."

Smoking pipes, however, have other and more interesting significances, and are inseparably connected with the whole social system of the American native races. To know the whole history of tobacco, of the custom of smoking, and of the origin of the pipe, would be to solve many of the most interesting problems of American ethnology.

To bring together the little that has been recorded by the earliest European travellers concerning pipes and the custom of tobacco smoking is not within the scope of the present volume. It must suffice here, simply to record the fact that the later, if not the earlier natives of the Atlantic coast were habitual smokers, as the vast numbers of fragmentary pipes attest; and that they who used them also made them is the testimony of those who first came in contact with the Indian.

Holm[114] says of the Delaware Indians, "they make tobacco pipes out of reeds about a man's length; the bowl is made of horn, and to contain a great quantity of tobacco; they generally present these pipes to their friends. They made them, otherwise, of red, yellow and blue clay, of which there is a great quantity in the country; also of white, grey, green, brown, black and blue stones, which are so soft that they can be cut with a knife; of these they make their pipes a yard and a half long or longer."

While, in the above quotation, there is a somewhat vague description of the stone pipes, it is evident that the stem has been included with the bowl, as though but one object, instead of two. In the valley of the Delaware, to-day, pipes of steatite of various colors, and of silurian slate, the "black and blue stone;" and yet others of clay, "red, yellow and blue," are to be gathered if we carefully search for them.

Peter Kalm, the Swede, who made many careful notes of the antiquities of the Delaware Indians, so long ago as 1749, refers, as follows (Travels in North America, vol. II, p. 42), to their smoking-pipes. He says "the old *tobacco-pipes* of the *Indians* are * * * * made of clay, or pot-stone, or serpentine stone. The first sort are shaped like our tobacco-pipes, though much coarser and not so well made. The tube is thick and short, hardly an inch long, but sometimes as long as a finger; their color comes nearest to that of our tobacco-pipes which have been long used. Their tobacco-pipes of pot-stone are made of the same stone as their kettles. Some of them are pretty well made, though they had neither iron nor steel. But besides these kinds of tobacco-pipes, we find another sort of pipes, which are made with great ingenuity, of a very fine, red pot-stone, or a kind of serpentine marble. They are very scarce, and seldom made use of by any other than the *Indian sachems*, or elders. The fine red stone, of which

---

[114] Holm. History of New Sweden, p. 130. Philadelphia, 1834.

these pipes are made, is likewise very scarce, and is found only in the country of those *Indians* who * * * * live on the other (western) side of the river Mississippi. The *Indians* themselves commonly value a pipe of this kind as much as a piece of silver of the same size, and sometimes they make it still dearer. Of the same kind of stone commonly consists their *pipe of peace*, which the *French* call *calumet de paix*, and which they make use of in their treaties of peace, and alliances." The same writer, when in the neighborhood of Quebec, has also recorded that "in some places, hereabouts, they find among the states, a stratum about four inches thick of a gray, compact, but pretty soft limestone, of which the *Indians* for many centuries have made, and the *French* at present still make, tobacco-pipes."

Smoking-pipes, as found in New England and in New Jersey, do not show any marked peculiarities from which to infer the occurrence of certain patterns in one locality and not in the other. So far as can be ascertained, there is a larger proportion of certain forms found in New England than in New Jersey, and *vice versa*, but nothing further.

Pipes made of Catlinite, to which Kalm refers, as "red pot-stone," are of very rare occurrence in New England, and even more so in New Jersey or Pennsylvania. In western New York, occasional specimens have been found.

Of the whole number of pipes found along our Atlantic seaboard, and preserved in the museums, there is no one specimen or series of one pattern that can be considered as a typical form. The simplest patterns have frequently as much polish and symmetry, as the most elaborately carved and otherwise embellished specimens. Taken as a whole, pipes suggest very strongly that they were made by those who proposed to use them, and the differences exhibited are those which always occur in the respective handiwork of painstaking and careless people.

While, for convenience of description, the various examples of pipes have been rudely classified, it is to be understood that the order of the descriptions is not based upon relative abundance, neither has it any ethnological significance. If, indeed, every Indian was his own pipe maker, this must necessarily be the case. It is certainly true of some of the southern and western patterns of pipes, that they are characteristic of the localities where they occur; but the same objects when found on the Atlantic seaboard are, collectively, a heterogeneous mass; the outcome of transient fancy, and not of settled custom.

If we are to grade the culture-development of the eastern seacoast Indians, by the excellence and artistic finish of their smoking-pipes alone, they will rank lower than the southern and some of the western tribes; but if their handiwork in stone, as a whole, is considered, they will compare favorably with any and all others.

Fig. 313 represents a beautiful example of a common Atlantic coast pattern of smoking-pipe. As the illustration plainly shows, these pipes are simply cylindrical bowls placed at right angles upon flat stems, or bases. Inasmuch as these pipes are cut from one piece of stone of a kind that does not occur as variously shaped pebbles, so that those that were somewhat pipe-shaped might be chosen, it is apparent how great must have been the amount of labor, to work to its present shape, a pipe like 313. Taking into consideration the fact that the mineral of which they are made is usually steatite, they nevertheless show a great amount of patience, and a certain degree of artistic skill.

Fig. 313 was found in an Indian grave at Revere Beach, Essex Co., Mass. Many of the finest examples of this pattern of pipe have been recovered from graves in this vicinity. In the museum of the Academy of Science at Salem, Mass., are three handsome specimens of this pattern. They were found by Professor Putnam,[115] in graves, near Beverly, Mass. One of these pipes has several holes bored along the margin of the base, probably for the attachment of ornaments. Associated with these pipes, which were taken from three graves, were various objects, as arrowheads, celts, and three large pendants. There was also, in one of the graves, a thin slab of smooth sandstone with a

---

[115] Putnam. Bulletin of the Essex Institute, vol. iii, p. 123. Salem, Mass.

few scratches so arranged as to represent a ladder such as a child might draw. Finally, such an amount of ochre had been placed in

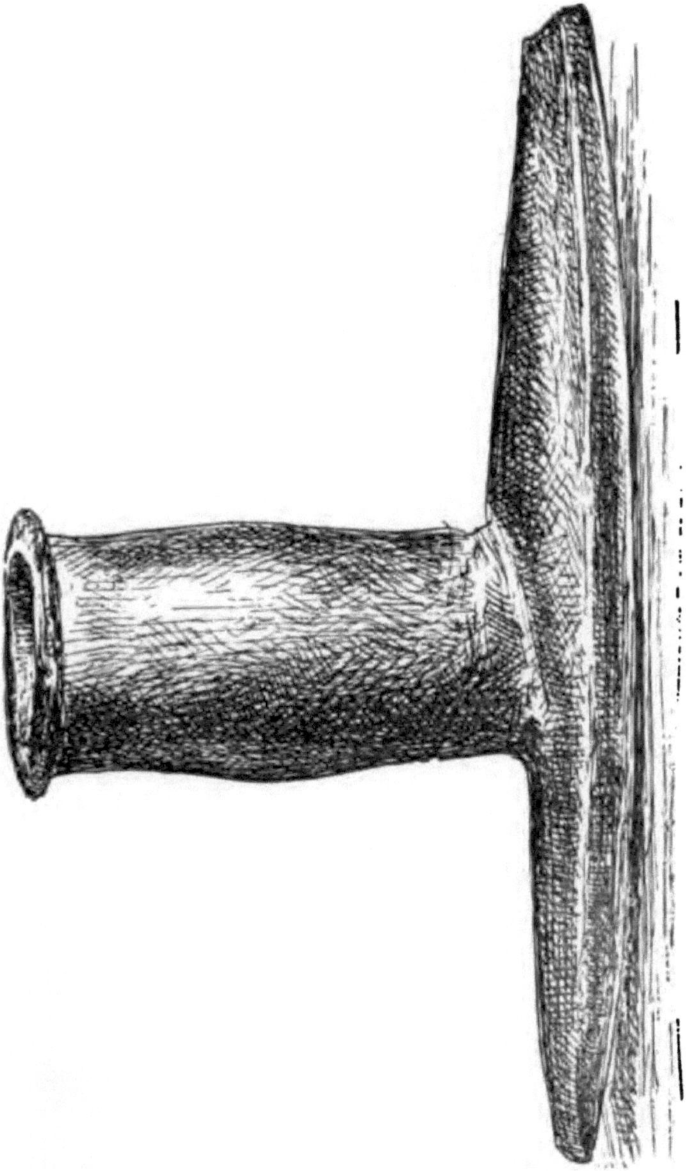

FIG. 313. — Massachusetts. 1/1.

these graves that the entire contents were more or less colored.

Of this pattern is a pipe figured by Morgan,[116] of which he speaks

---

116 Morgan. League of the Iroquois, p. 356.

as "doubtless a relic of the moundbuilders" which, having found its way into the hands of a Seneca, was finally buried by his side, in the valley of the Genessee. The author further remarks, "in material and finish, it is unlike, and superior to the pipes of the Iroquois." The fact of being made of black marble would seem to indicate that such a pipe was not of Algonkin origin, though in shape and finish, there is nothing to distinguish it from the pipes of the Algonkin neighbors of the Iroquois.

Fig. 314 represents a very handsome specimen of pipe, made of compact steatite, of a green-black color, which differs from the preceding in several particulars, although about the same size. This

FIG. 314.—Virginia. ½.

specimen is of a pattern seldom found in the New England and middle states, except of a much smaller size and made of clay. (See fig. 318). In fig. 314, as in the preceding, there is no attempt at ornamentation by incised lines or other methods, the surface being simply polished. The walls of the bowl are of uniform thickness, and there are still visible on the interior, the marks of the tool used in excavating it. The stem, which is a flattened oval in section, tapering gradually from the bowl to its termination, has been carefully drilled, and the perforation is perfectly straight.

This specimen was found in Isle of Wight Co., Virginia.

Fig. 315 represents an interesting example of a steatite pipe, found many years ago, in the village of Princeton, New Jersey. It is remarkable for the size of the stem as compared with that of the bowl. Whether the broad and deeply incised lines on the side of the stem are of Indian origin is not known, but they do not look as if they had been very recently cut. If intended merely as ornamental lines, the result can hardly be considered satisfactory. The raised figure, with a deep dorsal notch, is common to all of these pipes. Whether placed there simply as an ornament, or as an aid in securing the long reed stems, which Holm says were of a man's length, is certainly an open question.

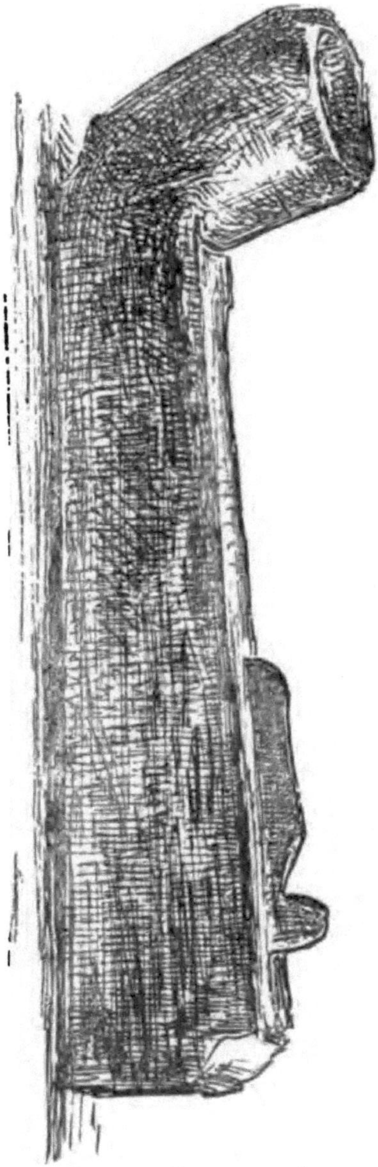

FIG. 315.—New Jersey. $\frac{1}{1}$.

Pipes of this pattern are not as frequently found in New Jersey as are those with the thin flat stem, like fig. 313, while in New England they are scarcely known.

Fig. 316 represents a second example of this pattern of pipe, but with a more generous bowl, and a far less inconvenient stem. The finish and design in this specimen are alike creditable to the maker, and we have in it all the requirements for a comfortable smoke.

Like the preceding, fig. 316 is made of a dark greenish, compact serpentine. The bowl has been made by boring with a stone drill of about two-thirds its width. This cavity is of uniform width and corresponds in direction with the inner perpendicular margin of the bowl, and so is at right angles with the perforation of the stem. In this instance the stem ornament, though notched, is not sufficiently so, to afford any aid in fastening a cord from the long reed stem to this portion of the pipe. Indeed, the whole appearance of this ridge-like ornament is such, that it cannot be considered as having any other than an ornamental purpose.

Fig. 317 represents a beautifully sculptured pipe made of a very dense, unyielding stone, of a jet black color. This pipe was found near Lewes, Delaware. While this specimen exceeds, in finish and

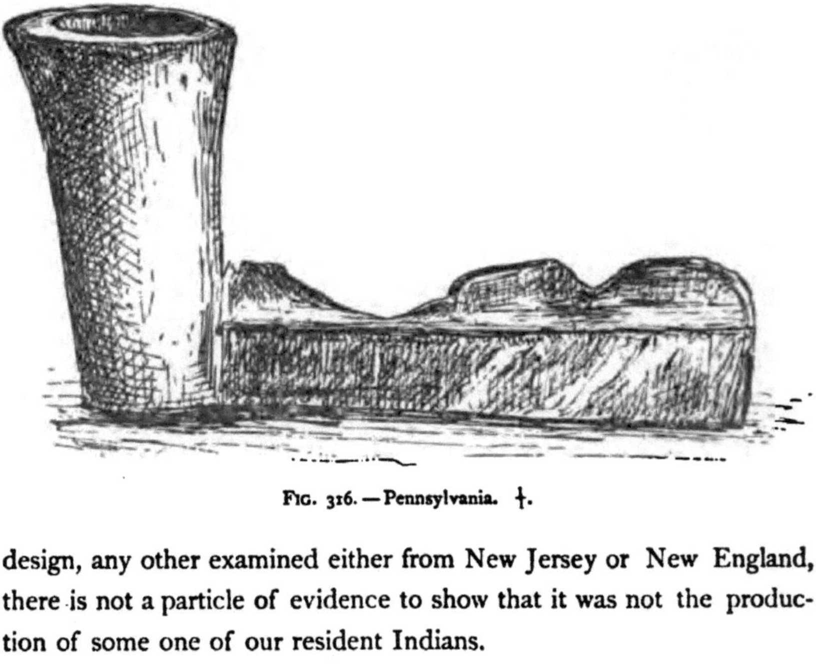

FIG. 316.—Pennsylvania. ½.

design, any other examined either from New Jersey or New England, there is not a particle of evidence to show that it was not the production of some one of our resident Indians.

Pipes of this pattern are of common occurrence in the extreme southwest, but are found less frequently as we come north of the Carolinas. That this specimen may have been brought from a distance is possible, but the character of the work upon it does not necessarily imply that such was the case.

If we assume that in many, if not in most instances, every Indian made his own pipe, there is nothing strange in the fact that an elaborate specimen should occur occasionally among those of a much plainer pattern. A well made stone pipe would last an Indian a lifetime, and we can well imagine that spare hours during the greater part of that lifetime might be spent in its ornamentation, even long after it was so far finished as to be ready for use.

Lifelike and artistic as is the turtle carved upon the stem of fig. 317, it must be remembered that without this ornament, the pipe is of the same pattern as the preceding, and this fact, with that of the common occurrence of representations of the turtle, in other ways, renders the

FIG. 317.—Delaware. ½.

combination as we here find it, of the pipe and the turtle, a very natural one, among the productions of the Delaware Indians.

The pipe, represented by fig. 318, is in the Archæological Museum at Cambridge. At the Boston meeting of the American Association for the Advancement of Science, Mr. F. W. Putnam made it the subject of a communication, of which he has kindly furnished me with the following abstract:

"The pipe is made of the dark slate from which so many pipes and objects are carved by the tribes of the northwestern coast of America, and it unquestionably originally came from that region. The style of carving, representing the peculiar human figure combined with the head and wings of a bird, is so characteristic of the northwestern carvings as to leave no doubt as to its origin. The question that will be asked is, how did it happen to be found on the coast of Massachusetts? Of course the natural reply to this question will be that it was brought by some sailor from the Pacific, and long ago lost. This may have been the case, but to me it seems as probable that it may have been brought *across* the continent long before our ships sailed to and from the Pacific coasts. Did we know only of a few instances of the discovery of objects far away from their proper natural positions

FIG. 318. — Massachusetts. ½.

we might rest contented with the 'sailor theory,' but too many similar facts are now known to leave us satisfied with the theory that such carved pipes, stones, shells, and other objects were brought from distant parts in modern times, to be thrown away, or lost, and accidentally buried. The contents of the burial mounds and of old Indian graves, all over the country, furnish us with unquestionable data of the early exchange of objects from distant places. The fact that beads and disks made from the large marine shells of the southern coasts have been found in mounds far in the interior of the country, and that the shells of the marine *Busycon* of our southern coasts have been found in numbers in the mounds in the vicinity of the great lakes, as well as pretty generally in old burial places over the whole region from Michigan to Florida, not to mention many other instances of a similar character, is sufficient to show that trade, or intertribal exchange of some kind, took place long before the white man pushed the red from the Atlantic coast. From the large number of facts that we now have, I therefore fail to see why it is not decidedly probable that this pipe was obtained from the northwestern coast by intertribal communication.

The following quotations are of interest in this connection as they show that pipes were a common article of exchange among certain recent tribes:

'Common sort of pipes are made of a kind of *ruddle*, dug by the Indians living to the west of the Mississippi, on the Marble river, who sometimes bring them to these countries for sale. Pipes made of red marble used only by chiefs and captains, this sort of marble being rare and found only on the Mississippi.' (Loskiel, p. 51.)

'Sometimes they make such great pipes both of wood and stone, that they are too foot long, with men or beasts carved . . . but these commonly come from the Mangawwoop, or the man eaters, three or four hundred miles from us.' (Roger Williams, *Key to the Indian Language*, p. 55.)

The Indians 'barter pipes, *et cet.*, for raw skins.' (Lawson, Carolina, p. 207.)

'Black marble pipes are made with great patience and labor, by one person only, throughout the whole nation.—He lives in Natchez, and being the only man that knows where the stone can be found, monopolizes the business entirely, and sell his common pipes at half the price of a blanket.' (Schoolcraft, vol. v, p. 692.)

That Indians made at least occasional visits to other nations in far distant portions of the continent is evident from the statement by DuPratz, who mentions that '*Moncacht-apé* among the *Yazous*, a nation about forty leagues north from the *Natches*,' in his extended wanderings, reached the Pacific coast, as is evident from the following quotation from his narrative, as given to DuPratz. 'After this expedition,' he says, 'I thought of nothing but proceeding on my journey, and with that design I let the red men return home, and joined myself to those who inhabited more westward on the coast, with whom I travelled along the shore of the *Great Water*, which bends directly betwixt the north and the sun-setting.' *Moncacht-apé* concludes by saying 'it was five years before I returned to my relations among the *Yazous*.' (DuPratz, History of Louisiana. Vol. ii, p. 128. London, 1763.)

The pipe was found by Mr. Walter B. Cobb, in 1853, between two and three feet underground, on his father's place in North Carver, a small town about four miles from Plymouth, and was secured for the Museum at Cambridge through Mr. A. M. Harrison of Plymouth, who has no doubt about the pipe having been dug up as stated. In confirmation of the theory that this pipe was brought across the continent by some kind of Indian interchange, I call your attention to the photograph of a similar northwestern-coast pipe, now in the Natural History Society of Amesbury, Massachusetts, which was found in digging a town road through a gravel hill in New Hampshire, several years ago; and also to the drawing of another pipe, in general character the same as the one from Plymouth, which was found in Canterbury, New Hampshire, and now owned by Dr. E. E. Gans of Boscawen, New Hampshire. It seems hardly probable that these three pipes, of unquestionable northwestern origin, should have been brought from the Pacific coast by sailors and lost in the states of Massachusetts and New Hampshire."

Fig. 319 represents a very large and rudely finished pipe bowl, made of sandstone. While comparatively smooth, there is no appearance that it has ever been polished or in a more finished condition, than at present. It was found near Trenton, N. J., and is a facsimile of large numbers both of stone and clay, found in the southwestern states. Pipes of this pattern are seldom found in New England, and are of rare occurrence in New Jersey. An allied form made of steatite, but with both the bowl and stem square instead of cylindrical, is common in Georgia, and occasionally is found in New England. Pipes made of baked clay, of the shape and size of fig. 317 on the other hand, are very frequently found in the mounds of Tennessee and Arkansas,[117] while no specimens of pipes of this pattern made of clay are known to have been found in New Jersey or northward.

Fig. 320 is a plain, but neatly finished pipe bowl, made of the Silurian striped slate, that was so highly regarded by the Indians and moundbuilders, for all ornamental stone work. Pipes of this pattern,

FIG. 319.—New Jersey. $\frac{1}{1}$.

and particularly of this material, are of somewhat rare occurrence in New Jersey or New England.

A small pipe bowl of this shape, but made of compact, fine-grained sandstone, and of about one-half the size of fig. 320, is contained in the series of pipes in the collections of the Archæological museum at Cambridge, Mass. (P. M. No. 15698). Like the specimen figured, it has no ornamentation upon it.

Dr. Chas. Rau gives a figure of a somewhat similar specimen, but

[117] Putnam. Eleventh Annual Report, Peabody Museum, p. 347, fig. 35.

which has a squared base. (Smithsonian Contributions, to Knowledge, No. 287.) He refers to pipes of this pattern, as "reminding one of the corn-cob pipes in use among the farmers of this country."

Of the series of pipes preserved in the museum at Cambridge, Mass., all made of this material are from the mounds of Ohio and other western states. The finding of this specimen in New Jersey is the less surprising, however, from the fact that the majority of ornamental carvings are of this same mineral, thus showing that the supply was sufficient for making pipes, had the Indians chosen to do so.

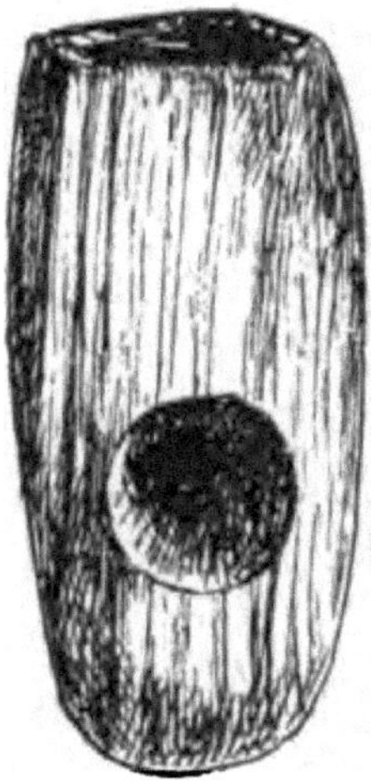

FIG. 320.—New Jersey. ½.

A few of the most remarkable specimens of pipe sculpture from Ohio were made of this material; and as some of these have been derived from mounds, it is of interest to know, that the same mineral should have been so generally used for such purposes, by both peoples; a fact that does not, of itself, indicate any racial relationship, and care should be exercised not to class, inferentially, objects of striped slate found on the surface, with similar specimens taken from the mounds. Identity of form, and of material, does not prove an identity of origin. The tendency to ascribe to the superior (?) skill of the moundbuilders all the creditable stone implements found on the surface, in moundbuilder regions, has led to much unfortunate confusion.

A pipe somewhat similar to fig. 320, found in Vermont, has a rude attempt at ornamentation, in a projection at the rim, of what is supposed to represent the beak of a bird. This pipe is over two and a half inches long and rather more than one inch in its longest diameter, the cross section being oval. Pipes of other patterns made of stone "are rarely found," in Vermont; but those that have been preserved "are well made and polished, and while as compared with the elaborately carved specimens from the mounds, our Vermont specimens appear very plain, yet they are not of inferior workmanship though in simpler form."[118]

Fig. 321 represents an example of a pipe, differing from the preceding, in the position of the hole for the insertion of the stem. In the former the stem was at right angles with the bowl; in this pipe, it was so inserted, at the end of the bowl, as to bring the two nearly, or quite in a line.

This perfect specimen, like the majority of pipes of this shape, is made of soapstone, and bears no trace of ornamentation. It is an elongated oval bowl two and a half inches in length, and a little more than one inch in diameter at the mouth and five-eighths of an inch in diameter at the base. The front of the bowl is somewhat convex in outline; the opposite outline is more nearly straight. A little above the middle of the front of the bowl commences a projection a quarter of an inch in width and a little less than an inch in length.

FIG. 321.—New Jersey. ½.

The pipe used by the Shoshones at a ceremonial smoking and speech-making, in honor of Capts. Lewis and Clarke, is described as "made of a dense, but almost transparent, green stone, very highly polished, about two and a half inches in length, and of an oval form, 'the bowl being in the same line' with the stem." A small piece of burnt clay was placed at the bottom of the bowl, to separate the tobacco from the end of the stem. "This was an irregular round figure, not fitting the tube perfectly close, in order that the smoke might pass with facility."[119]

---

[118] Perkins, *l. c.*, p. 740.

[119] Stevens. Flint Chips, p. 516; quoting Lewis and Clarke.

The above description very closely applies to fig. 321, which, from the position of the stem-hole must have been used in a similar manner. It closely resembles the long tubular pipes so characteristic of the islands off the coast of California. While many of the latter are three and four times as long as this specimen, others are of about the same length, and so almost, if not entirely, lose their tubular character, and are simply elongated pipe-bowls. Tubular pipes of the character of the largest examples found on the Pacific coast are, however, not wanting on the Atlantic seaboard.

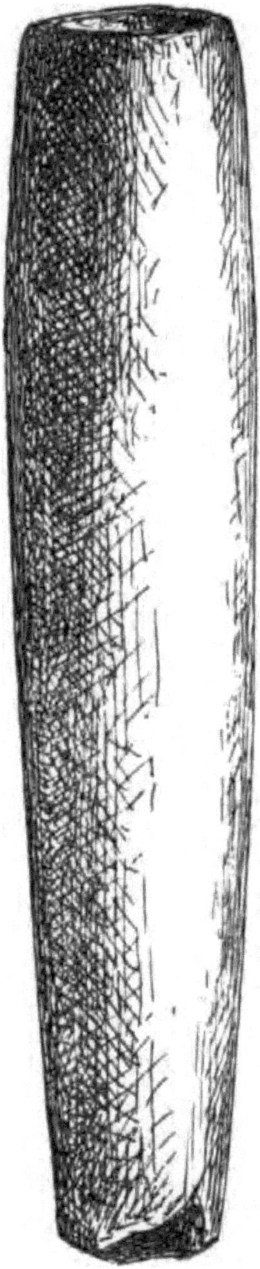

FIG. 322.— Massachusetts. ½.

Fig. 322 represents a very characteristic specimen of these pipes, found near Lawrence, Mass. In no particular does it differ from the remarkable series of smoking pipes that have been found in California. Fortunately, these have the bone mouthpieces still in them, and hence there can be no doubt as to their use. While there have been many stone tubes discovered in various localities, which from their size, the large and uniform diameter of the bore, and other features, should probably not be classed as smoking pipes,[120] it is equally evident, that those of this shape were so used, if we may be guided by what is known of similar specimens found in other localities.

[120] See chapter xvi, with accompanying plate (xxi), of Jones' Antiquities of the Southern Indians, for a detailed account of Stone Tubes.

It is certainly most probable, that the remarkable series of clay-slate tubes, found at Swanton, Vermont, and described in detail by Professor Perkins, were used as smoking pipes, for the reason that in all of them the bore is not of uniform size. These tubes are described as "all of similar form, being cylindrical; the perforation, at one end about half an inch in diameter, *enlarges to nearly an inch in diameter at the other end.* They are of smooth, hard stone, of a drab color in some specimens, brown in others. They are very nicely formed and finished, the surface being smooth and almost polished. The small end of the bore was stopped somewhat imperfectly by a stone plug ground into shape." (See description of Shoshone pipe given by Lewis and Clarke, quoted on page 329.) "The length of the tubes varies from seven to thirteen inches. Similar tubes have been found on one of the islands in Lake Champlain, and near Burlington, Vermont."[121]

These tubes have also been found in New York, in ancient graves, and associated with them were implements of the same general character, as those found with similar specimens in Vermont. Mr. S. L. Frey[122] has given us an excellent description of the results of relic-hunting in the Mohawk valley; and in the examination of certain graves, he found a stone tube, "four and a quarter inches long; the perforation has at one end a diameter of one-quarter of an inch, gradually enlarging until it reaches at the other end, a diameter of three-quarters of an inch." With this tube, which was unquestionably used as a smoking pipe, were a "sea shell, somewhat modified for a drinking vessel, its longest diameter being four inches, a beaver's tooth, several bone awls, three arrowheads, a number of flint flakes, pieces of a tortoise shell, some fragments of deer-horn implements, a bone gouge and a large wing bone of a bird." A subsequent examination of the graves resulted in finding a second tube, apparently of the same material, but different in shape and length. This second specimen "is eight and one-half inches long, and one inch in diameter, having

121 Perkins, *l. c.*, p. 734.

122 Frey. American Naturalist, vol. xiii, p. 637.

a bore of five-eighths of an inch at one end and two-eighths of an inch at the other. It is smoothly made but has no polish at present, being covered with an earthy coat, and in patches with a thick concrete of lime and sand. With this tube were found, lying side by side, three hornstone implements, of large size, and pointed at each end." A stone tube, of striped slate, is recorded as having been found at Trenton, N. J.,[123] six inches in length by one and one-half inches in greatest diameter, near one end, thus making the implement slope abruptly. This specimen, except in material, is identical with the smoking pipes found in southern California; and does not differ materially from the New York or Vermont specimens.

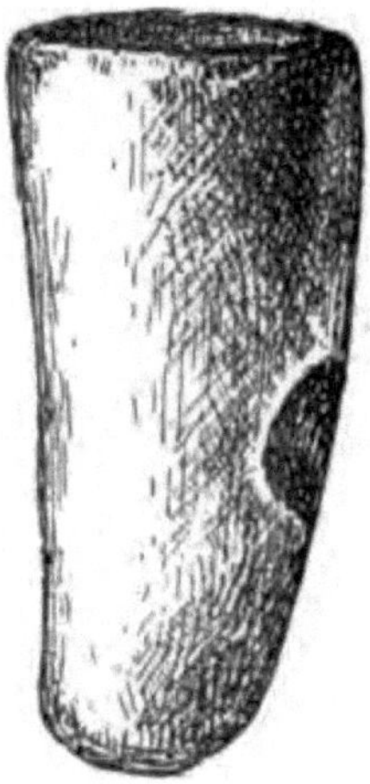

FIG. 323. — Pennsylvania. ½.

A series of clay-slate tubes were recently found in a grave near Bridgeport, Gloucester Co., New Jersey, which are in size, color and character of perforation much like those found in Vermont. Others, of steatite, said to have been found in other Indian graves in the same neighborhood, are, one and all, unquestionable frauds. Several of the latter have found their way into various private cabinets. The genuine clay-slate examples are at present in the collection of Wm S. Vaux, esq., of Philadelphia, Pa.

Fig. 323 represents an admirable example of steatite pipe, which is of unusual interest in that it is a connecting link, as it were, between the tubular pipes, such as the preceding and the plain bowls, with a side stem-hole, as fig. 320. This specimen, which is made of light, mottled, gray steatite, is very highly polished, and symmetrical in outline. The cavity is quite large (the sides being very thin), and terminates in a narrow stem-hole extending to the lower end of the specimen. There is also a lateral stem-hole of greater diameter, opening into the bowl just above the commencement of the other.

[123]Abbott. Nature, vol. xiv, p. 154, fig. 1. London, 1876.

Smoking pipes of this pattern are seldom met with, only two examples having been found, to the author's knowledge, in the localities treated of in the present volume.

This specimen was found near Bainbridge, Lancaster Co., Penna., by Mr. F. G. Galbraith, and is now in the collection of the late Professor Haldeman.

The occurrence of pipes made of metal is of great interest, from the fact that while copper pipes are recorded as having been in use by the Manhattan Indians, no specimens are now known to be in collections. In a foot-note to page 45 of Smithsonian Contributions to Knowledge, No. 284, Dr. Rau remarks that "the navigators who first visited the Atlantic coast of North America noticed copper pipes among the natives; as for instance, Robert Juet, who served under Hudson as mate in the Half Moon. Such pipes must be very rare. There are none in the Smithsonian collection."

In the museum at Cambridge, in a large series of wooden pipes from Alaska, are several which have the bowls formed of sheet copper (P. M. No. 1858), but whether any of these are of an antiquity antedating European contact is very doubtful. Occasionally, plain clay pipes have been found in graves, which were partially encased in broad copper bands; and it is very probable that pipes thus ornamented were seen by the early voyagers from Europe, who mistook them for pipes made wholly of metal.

At Cambridge, there is preserved a rare and most interesting specimen of a pipe (P. M. No. 14172), which bears a great resemblance to the plain clay pipe, fig. 324. This specimen is made of sheet lead carefully rolled, so that the symmetry of the bowl and stem is preserved. The specimen was found at Revere, Massachusetts. The appearance of this specimen is such as to suggest that it is of Indian manufacture, although made of course after association with the Europeans.

The occurrence of what have been termed "compound calumets," or "council-pipes," has been reported in the American Antiquarian, vol. i, p. 113. One such is described as made "of a hard, light gray, almost white steatite. It is cylindrical or tapering in form, and nearly six inches in height. About two inches from the base * * * extends a horizontal groove in which have been pierced four equidistant stem-holes, which extend obliquely downwards to the base of the bowl. * * * * The size of the specimen, and the existence of four orifices for the insertion of stems, prove conclusively that it was not an ordinary pipe, but was in all probability smoked in some formal ceremony."

If there were any historical evidence that such pipes ever had been used, there might then be no reason for looking with suspicion upon the above described specimen, and even if it be genuine, it is scarcely necessary to conclude that the four stem-holes were simultaneously used. Three of them may have been plugged with some ornamental

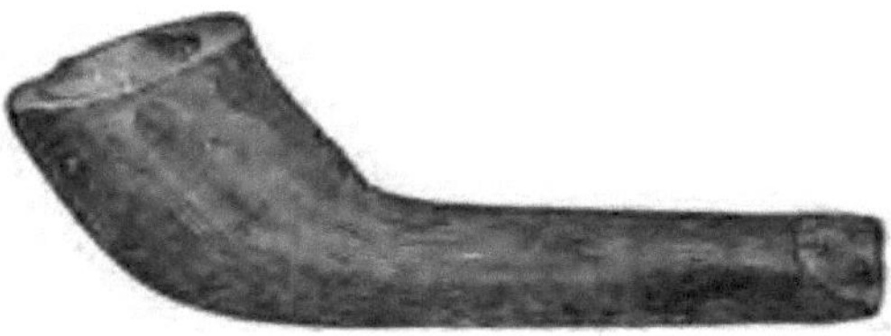

FIG. 324.—New Jersey. ½.

carving, just as certain of the moundbuilder pipes were decorated with pearls and polished pebbles; and the California pipes were inlaid with mother-of-pearl. Had this one specimen only been obtained, it might be classed with the remarkable pipe sculptures that occur so sparingly along the Atlantic seaboard; but the fact that from the same general locality others of various patterns have also *mysteriously* turned up, as the result of explorations of certain parties, it is very evident, that as a class, these "compound calumets" are frauds; and were possibly copied from the specimen here mentioned, which was the first discovered, and which, if genuine, in spite of the existence of four "stem-holes," need not necessarily be a so-called "council pipe."

Fig. 324 represents a pattern of small clay pipe, of which fragments are found in great abundance but which, as a perfect or nearly perfect example, is but seldom seen. The excellence of workmanship exhibited in these clay pipes is very uniform, and leads to the belief that they were moulded by other than the native potters. There is much difference of opinion on this point. When, however, a large series is brought together, and the character of the ornamentation carefully examined, it is found that there exist many grades of excellence in this respect, though neither is beyond the capabilities of the redman. Furthermore, both Holm and Kalm distinctly refer to pipes of clay, made by the Indians. Fig. 324 is an example of these pipes which has no ornamentation. They are quite rare, in comparison with those which have the bowl covered with carefully stamped lines, dots and other depressions.

The clay, of which these small pipes are made, is of a much finer quality than that used in the ordinary earthenware, already described. Except occasional traces of very fine white sand, such as occurs naturally in many of the veins of clay, there appears to be no foreign substance. When burnt, these pipes are of a yellow-gray, or a brick red color, generally the former; but there appears to be no difference in the quality of the pipes, whatever the color when burned. The great variety of tints in the clay, blue, black, and red, all burn to a yellow, when now used for drain-pipe and terra-cotta ware.

Figs. 325 to 329, inclusive, represent fragments of clay smoking pipes and stems, and show the various patterns of decorations used in ornamenting them. In all cases, this work, whether mere lines, or combinations of lines and dots, is very superior to that on any of the pottery found in the same locality, so much so, in fact, that were it not that there are different grades of merit in the work, it might be ascribed to European origin.

Attention has already been called to stone pipe-bowls, as occasionally occurring, though quite rare in comparison with those that have a stem complete in itself, or one that can be used by the aid of a supplementary stem of wood or reed.

Even more rare, it is believed, are clay pipe-bowls, without stems or mouth-pieces of the same material, such as figure 330.

This beautiful specimen, which is really artistically designed, varies not only in the shape of the bowl, and especially in the base, but is of harder clay than those that have been figured. An interesting feature of this specimen is the small hole passing through the base, below the large hole intended for the stem or mouth-piece. Ornate as this pipe-bowl still is, by reason of its graceful shape, it was probably not sufficiently so in the eyes of its primitive owner. By means of a cord

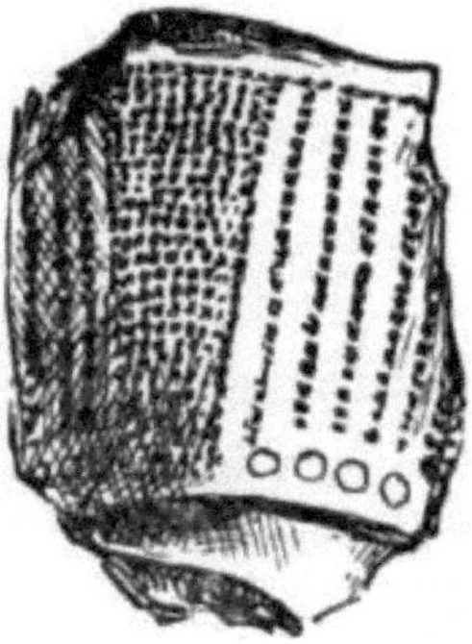

FIG. 325.—New Jersey. ½.

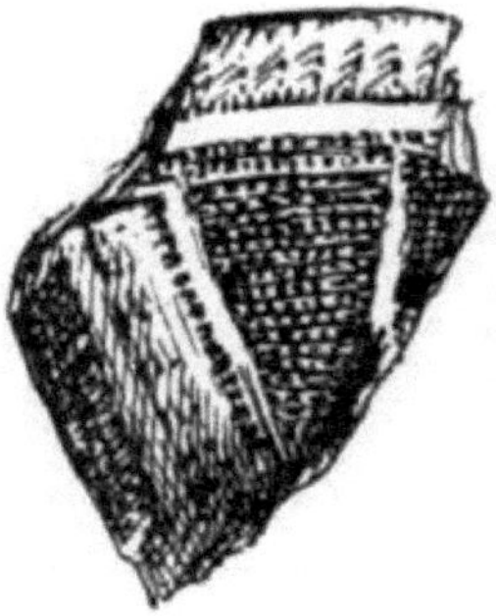

FIG. 326.—New Jersey. ½.

FIG. 327.—New Jersey. ½.

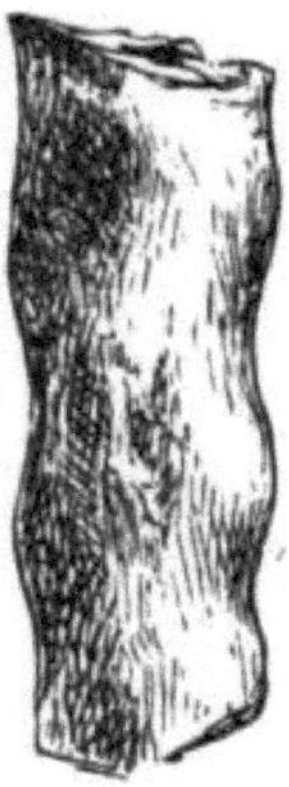

FIG. 328.—New Jersey. ½.

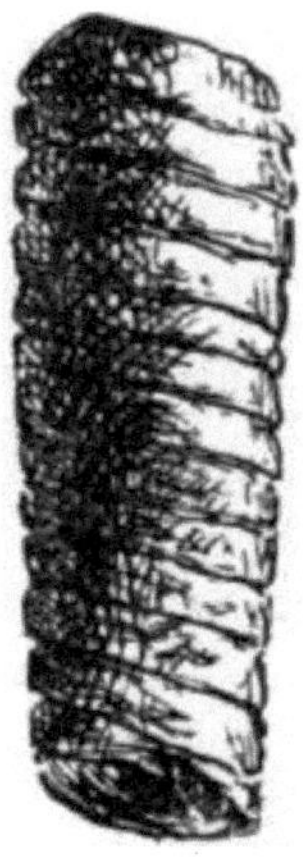

FIG. 329.—New Jersey. ½.

passed through this small opening, feathers, beads and brilliant trinkets of any kind, were readily suspended; and this, it is believed, is the object of this small perforation.

These holes are not unusual, when the stems of *stone* pipes are thin, flat and projecting, as in fig. 313. They are of frequent occurrence, also, in the more modern Catlinite pipes.

In the collection of North American smoking pipes in the museum at Cambridge, are examples of this pattern, made of steatite and slate. One specimen (No. 15697) from Lafayette, Indiana, is strikingly similar to fig. 330. The pattern is one not characteristic of any locality, but does not occur abundantly anywhere.

FIG. 331.—Penna. ½.

Dr. Chas. Rau, in Smithsonian Contributions to Knowledge, No. 287, figures a pipe of this shape, made of argillaceous stone and found in Ohio. He believes that pipes of this character may not be very old; adding that "the type occurs among the pipes carved by modern Indians."

Fig. 330 was found in Lancaster Co., Pa., and is now in the cabinet of the late Prof. S. S. Haldeman, to whom I am indebted for the opportunity of describing and figuring this specimen.

FIG. 330.—Penna. ½.

Throughout New Jersey and New England generally, there is not found that variety of forms in smoking pipes, which is comparatively common in many portions of New York. Mr. Frey[124] found among the many relics of the Indians, once living in the Mohawk valley, many fine examples of clay pipes of intricate and ornate designs. He refers to the Mohawk Indians as "showing great ingenuity in making pipes, the bowls of which are frequently in the form of a bird or mammal, and these always being very true to nature."

Fig. 331 represents an interesting earthen fragment, probably broken from a pipe such as referred to by Mr. Frey. Its color, consistency and size all suggest that it was a portion of the bowl of a small clay pipe, such as is represented on page 334, although no perfect pipes, having a like

[124] Frey, *l. c.*, Amer. Naturalist, vol. xii, p. 781, figs. 8 to 11 inclusive.

ornament have been found, or are known to the author, from New Jersey or the New England states, notwithstanding they are so abundant in New York.

In the Magazine of American History, for Sept., 1878, there is given a figure of an Iroquois pipe with a raccoon's head projecting from one side of the bowl, in such a position as to face the smoker, when the pipe was used. Mr. Stone refers to this specimen as a "rough and uncouth Iroquois pipe in the shape of a raccoon's head which was found at Lake George;" and draws some unwarranted conclusions from it, when compared with a somewhat better finished specimen of an Iroquois pipe of essentially the same type.

Mr. Schoolcraft, in his History of the Indian Tribes, Pt. II, p. 90, and plate 47, briefly describes and figures specimens of Iroquois pipes, one of which has a bear's head, of the same character of finish as the specimen of animal head represented in figure 331. Referring to the objects on plate 47 of his work, Schoolcraft remarks, "the articles grouped in plate 47, from Ellisburgh, Jefferson county, New York, exhibit the same ready tact in moulding images of the human face and the distinctive heads of animals on the plastic basis of clay pipes, which is found extensively in that area." In Vermont, also, "a few fragments of pipes have been found made of terra-cotta." One is a perfect specimen resembling a trumpet, and similar to some figured by Schoolcraft.[125]

Whether the first pipes were merely shapeless lumps of clay, hardened by much use, and thus suggested the workable stones as a more desirable material for their manufacture, is perhaps uncertain; but there is evidence that the two forms of stone and clay were used at one and the same time, and that those of clay continued in use, after the others had ceased to be manufactured in large numbers. While it is very rare, that we come across a rude and apparently very old clay pipe—one that might possibly antedate the earliest stone pipes—there is probably nothing so modern or more common, in the whole range

125 Schoolcraft. Hist., etc., of Indian Tribes, Part 1, pls. 8 and 10.

of Indian handiwork, as it is still found in the Atlantic coast states. Many were probably made after the introduction of clay pipes, of somewhat similar patterns, by the Europeans.

Clay pipes of the plainer patterns are not as abundant in Massachusetts and Connecticut, as in New York and New Jersey; but they are sufficiently well represented in the contents of ancient graves, and in "surface finds," to be accounted as not uncommon.

Clay tubes are occasionally met with, which in many cases were undoubtedly used as pipes, but others are of such a small calibre, that their purpose is problematical.

Among the many objects of interest gathered from mounds in southeastern Ohio, by Prof. E. B. Andrews,[126] is a clay tube, which is described as a cylinder of yellow clay, but slightly baked, if at all. One end is closed, except a small circular opening. Mr. F. W. Putnam[127] has remarked in a foot-note to Professor Andrews' report that "these tubes of stone, *clay*, and copper discovered by Professor Andrews approach so near to the long tube-like pipes made of stone, and still used by the Utes, that I can hardly refrain from classing them with pipes. The principal difference consists in these tubes having what would be the mouth-piece made by the termination of the pipe itself; while in the stone tubes, that are unquestionably pipes, the mouthpiece is probably made by inserting a hollow bone or reed. These tube-like pipes have been found in numbers in the old burial-places of California, and there has recently been one received at the museum, which was collected in Massachusetts. * * * * In Squier and Davis' 'Ancient Monuments of the Mississippi valley' several of these *stone* tubes are described, one of them identical with figure 5 (*the clay tube*) figured in this article, and the authors of that work also suggest that these tubes may be pipes."

Two specimens of elaborately ornamented clay tubes,[128] found in the same grave in New Jersey, were about seven inches in length, and

---

[126] Andrews. Tenth Report of Peabody Museum, p. 63, fig. 5. Cambridge, Mass.

[127] Putnam. Tenth Report of Peabody Museum, p. 63, fig. 5. Cambridge, Mass.

[128] Abbott. Nature, vol. xiv, p. 154. London, 1876.

of unusual interest, as they have so far proved unique. In each case, there was a projecting ornament which, prior to their inhumation, apparently, had been partially broken off, so that its precise character could not be ascertained. Unlike most tubes, the diameter of the bone was uniform and very small, and their use therefore, as smoking pipes, was impracticable.

A smaller and plainer pattern of clay tube found frequently upon the surface is represented in fig. 332. This specimen has much the appearance of the stems of such plain clay pipes as have been described; but it is not improbable that many such specimens as fig. 332

FIG. 332.—New Jersey. ½.

were made as they now are, and were used in various ways. Many probably are simply elongated beads, and, if originally pipe-stems, have had the fractured end carefully smoothed, until now no trace of a broken surface can be detected. These tubes have been considered also as "whistles" (Amer. Naturalist, vol. ix, fig. 150) and certainly can be very readily used as such.

The specimen here figured was found in an Indian grave, associated with the usual types of implements occurring in ancient burials. This would seem to indicate that whether a utilized pipe-stem or an implement *de novo*, it had some special use.

CHAPTER XXIII.

DISCOIDAL STONES.

THE name "Chungkè Stones," given to the peculiar biconcave stone disks found in the southern and western states, and also applied to the more abundant plain stone disks which occur in scanty numbers, in the middle states, is certainly calculated to mislead unless it be admitted that the plain disks, and those that are biconcave and occasionally perforated, are one and the same implement.

Col. C. C. Jones, jr.,[129] in his most admirable account of these stones, classes them together; and with those of most elaborate finish, and biconcave, he associates such plain stone disks as have been found in New Jersey and northward of that state.

But two examples of these plain discoidal stones have been examined, that were found in New Jersey. Others, however, are known to have been found, and passed into private collections. These are all perfectly plain circular stones, with straight sides and level margins. Of two examples from Gloucester Co., New Jersey, one is quite smoothly polished and accurate in outline; the other has the margin somewhat sloping, and so will not remain in an upright position, when placed upon a level surface. Both these specimens are made of compact sandstone, and are quite heavy.

Fig. 333 represents a specimen of these discoidal stones found in New Jersey, which is not distinguishable from hundreds of similar implements from the southern states. It is the same "hard, black, close-grained stone, capable of receiving a fine polish" which "formed the favorite material, especially along the coast," as described by Col. C. C. Jones, jr., in his work on the southern Indians.

[129] Jones. Antiquities of the Southern Indians, p. 348, pl. xx. New York, 1873.

A remarkably fine specimen, of the same character as fig. 333, is a polished example, made of trap rock, in the collection of the late Professor Haldeman. In a letter from Professor Haldeman, bearing date of May 29, 1880, he there states, "I have, through the kindness of the family, lately got 'the Wittmer disk' of black stone; two surfaces flat and polished, edge convex—better finished than anything allied from this locality." This specimen, which measures about four inches in diameter, by nearly two inches in thickness, was found "about four miles below Columbia, Lancaster Co., Pa." It was presented to Professor Haldeman by Mr. Jacob H. Wittmer, the inheritor of the farm "upon which it was found by his grandmother when a little girl,

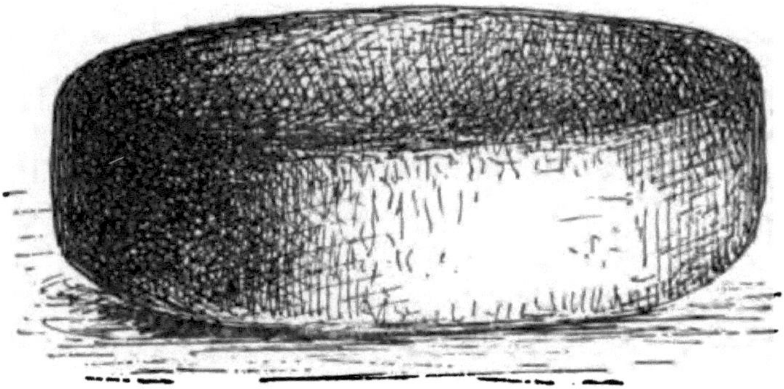

FIG. 333.—New Jersey. 1/1.

about the year 1765. The locality is well known for arrowheads, European beads, etc."

This disk or chungkè stone will compare favorably, in every respect, with the best specimens of this pattern of these objects, from the southern states.

Are we to consider these discoidal stones, found in New Jersey, as chungkè stones? From the several accounts given of the game in which these stones were used, there does not appear to be any reason why a plain disk would not have served the purpose as readily as a biconcave one; and as there is also a regular gradation from some that are even convex to those that are deeply concave and even perforated, it is not improbable that all were used in playing the one game, or possibly, some modifications of it. If such be the case, and the presence of such specimens as fig. 333 indicates that the Indians of the Delaware valley were chungkè players, then we ought also to find the concave stones, that are considered as the typical form of this implement; even if they are not found, then this fact can scarcely be considered as a proof that the plain disks were not gaming stones. Did they not occur in the southern states, associated with the biconcave specimens, it might more reasonably be supposed that they had some other purpose. In this connection also, it is well to recall the fact that the Delaware valley was occupied at one time by bands of Shawnee Indians, who very probably played this game during their northern sojourn, and may have introduced it among their Lenapè neighbors. As the early travellers and missionaries among the resident Delawares or Lenni Lenapè make no reference to the game, it is not likely that it was ever so prominent a pastime among them, as it was among the southern tribes.

CHAPTER XXIV.

INSCRIBED STONES.

WHEN the class of ceremonial objects come to be described, one specimen will be found the most interesting, perhaps, of all, viz.: a carved stone commemorating a treaty between two tribes (chap. XXV, p. 363). Attention will also be frequently called to the series of marginal notches on gorgets and other forms of ornamental stones, which are believed to have a "meaning," and are not simply one of the few methods adopted by the Indian, to ornament such stones as they wore about their persons, or paraded on ceremonial occasions. Stones, however, other than those of a ceremonial, or ornamental character, with marks, dots, lines, figures either singly or in combinations, are of very rare occurrence. Pictured rocks such as are found in various rivers, and along their banks, are not included in this category. But one engraved stone has been found in central New Jersey, so far as known, among the thousands of ordinary stone implements that have been gathered. This specimen bears some slight resemblance to the picture writings, as described by Schoolcraft, Catlin and others.

As the result of his investigations, Catlin[130] remarks: "I have been unable to find anything like a *system* of hieroglyphic writing amongst them; yet, their *picture writings*, on the rocks and on their robes, approach somewhat towards it." The engraved stone figured on page 347, is supposed to be much the same as a "bark letter," of one of which Sir John Lubbock[131] gives an illustrated account, quoting in part from Schoolcraft. This letter has a large number of figures of men and certain animals on it, and thus has the appearance of being

[130] Catlin. North American Indians, vol. 2, p. 246, 4th edition, 8vo. London, 1844.

a more advanced production in this method of communicating ideas. This, perhaps, is more apparent than real, for it is not improbable that the straight lines in the engraved tablet here figured represent rivers or creeks, and the arrow the direction taken by some Indians, who desired to inform a party, following in their track, of the route they had taken. This is in accordance with translations of bark-letters as given by Schoolcraft, and which are doubtlessly correct.

That a few slight scratches upon stone or wood, as made by an Indian might have to him, or to another, considerable meaning, and its significance be readily construed by one of these people, under certain circumstances, is readily seen by the account given by Kalm, in his North American Travels. This writer says of the Delaware Indians, "their good natural parts are proved by the following account, which many people have given me as a true one. When they send their ambassadors to the *English* colonies, in order to settle things of consequence with the governor, they sit down on the ground as soon as they come to his audience, and hear with great attention the governor's demands which they are to make an answer to. His demands are sometimes many. Yet they have only a stick in their hand and make their marks on it with a knife without writing anything else down. But when they return the next day to give in their resolutions, they answer all the governor's articles in the same order, in which he delivered them, without leaving one out, or changing the order, and give such accurate answers, as if they had an account of them at full length in writing."

While in this case, these notches in a stick were mere aids to memory, it shows that the fundamental idea of expressing thoughts or recording facts, by means of signs, was not novel to them, and renders the suggested explanation of the marks upon the inscribed stone, here described, as plausible.

Fig. 334 represents this inscribed stone. It is a nearly oval slab of micaceous slate, about an inch thick, seven inches in length and four

---

[231] Lubbock. Origin of Civilization, 2nd ed., p. 43, fig. 9. London, 1870.

and three-fourths inches in greatest width. The edges have been rudely bevelled, and the specimen brought to its present shape before the figures and lines were inscribed upon it. They consist of a series of well-defined lines, one extending the entire length of the slab, and dividing it into two nearly equal parts. There are also three others

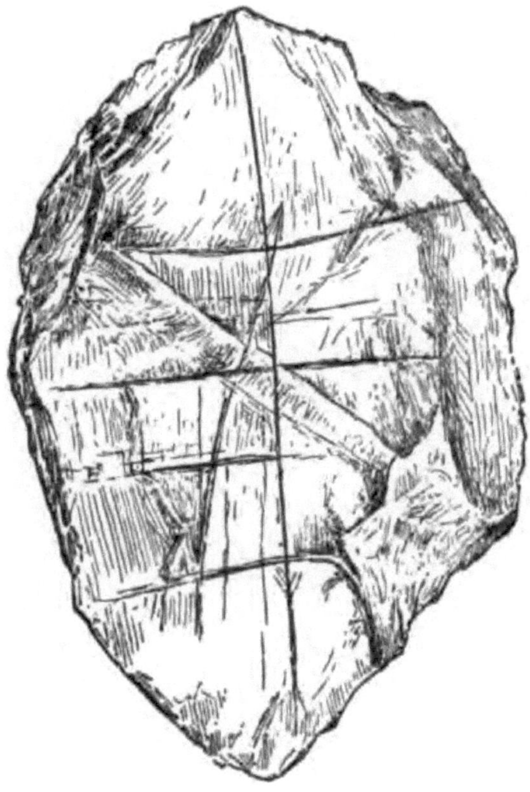

FIG. 334.—New Jersey. ½.

that cross this one at right angles, and a fourth short one, with "split" ends, on the left-hand side, below the centre of the slab.

The most noticeable feature of the inscribed side of the stone is the well-defined arrow, extending obliquely across the stone from right to left. Without this the stone would certainly be wholly unintelligible, but the arrow seems to explain the specimen, or rather furnishes a basis for conjecture.

As the specimen exhibits no attempt at ornamentation whatever, and considering that it was found in a swamp but lately reclaimed, and on the margin of a well-known Indian trail, it can scarcely be doubted that the markings upon it were placed there to convey information to those for whom it was intended,—in a word, that it is a "birch-bark letter" written upon stone, and a very primitive attempt at picture writing.

CHAPTER XXV.

CEREMONIAL OBJECTS.

UNDER this vague title has been classed a group of perforated stone objects, very artistically designed, which vary greatly in size and pattern, and yet have so much in common, that we appear to be justified in considering them as intended for practically the one purpose.

However uncertain we may be as to that purpose, there is sufficient evidence in the objects themselves to show that they were not weapons, and it is equally improbable that they were ever used in any such manner, as would warrant our calling them "implements," and hence the use of such names in connection with these objects, as "tomahawk," "amazonian axe," and "hatchet," is to be avoided as altogether misleading.

It must not be understood that, in giving the above name to the objects here described, they were the only ceremonial objects in use among the native tribes of our Atlantic coast. Many of the pendants, gorgets and perforated plates of mica, were very probably worn or carried only on particular occasions, and were not all ordinary personal ornaments.

Perhaps one of the most interesting features of these objects is their very general distribution, throughout the country, east of the Mississippi river. They may, indeed, be found far westward of that river, but the known localities, in which they are of comparatively common occurrence, are all on the eastern half of the continent. In the mounds of the Ohio valley, it is said, some of these objects occur occasionally. Many of them are exceedingly beautiful in design and finish. They are therefore one of those peculiar forms, neither weapon nor implement, which are common to both the moundbuilders and the Indians. In the case of simple weapons, or primitive agricultural or domestic implements, the same form might readily occur independently, but that a purely ceremonial object of the same pattern should be designed is not so probable; and the fact of the presence of these perforated stone objects in mounds is either evidence of a close relationship between the two people, or else it must be admitted that these objects, when found along our Atlantic coast, were either brought there by the later Indians, or were copied by them from originals seen in the interior. If it were necessary to admit that the Indians, who were in possession of this country, at the date of its discovery, were of more recent origin than the moundbuilders, then the supposition that they had adopted a moundbuilder's implement might hold good; but as yet, there is not one jot or tittle of evidence that proves that the native races of the northern Atlantic seaboard were not as old as the moundbuilders. The latter seem to be the older, simply because the traces of antiquity on the seaboard have been overlooked, or strangely disrègarded, because so uninviting, when compared with the rich harvest of strange objects that rewards the explorers of the western mounds.

Throughout all the river valleys, east of the Alleghany mountains, these perforated ceremonial objects are found in about equal abundance. In every locality, where there is a navigable stream, there was an Indian village. Often there were several, as at the mouth of each of the smaller creeks, and wherever these villages stood, we may confidently expect to find fragments, at least, of these pretty objects.

In New England, they are probably not so frequently found, though they are by no means uncommon. In Maine, besides those of common form, one has been found of remarkable character.[132] Professor Perkins[133] has described them from the Champlain valley; and the Archæological museum at Cambridge, Mass., contains several from Massachusetts and Connecticut. Throughout New York, they are of

[132] Putnam. Bulletin of the Essex Institute, vol. iii, p. 92. Salem, Mass.

[133] Perkins. Amer. Naturalist, vol. v, p. 12, figs. 3 and 4.

common occurrence. The collection of stone implements from New Jersey, in the Museum at Cambridge, contains forty-seven specimens; many, of course, in a fragmentary condition. Of these, some are so far unfinished that the perforation has not been begun, or is only partially complete; thus demonstrating the interesting fact, that these objects were otherwise finished, before the perforation was begun.

Dr. Chas. H. Stubbs of Wakefield, Lancaster Co., Penn., has been fortunate enough to find a very fine series of these implements, mostly made from the Potsdam slates. They are in the various stages of manufacture, and show that the slate was first coarsely chipped, then pecked or more delicately chipped until the outline was secured; after which they were carefully polished, and finally perforated. This was done not only with a hollow reed, but sometimes with a solid stone drill. It would seem from their unusual abundance in some portions of the Susquehanna river valley, that many were made there for barter with other tribes or communities, as was the case with some forms of chipped implements, as the arrowheads.

Of the series of specimens from New Jersey, the greater number are made of steatite, and of the striped Silurian slate, so much used for all ornamental objects. Not all, however, are of such easily worked material. Marble, diorite, compact serpentine, quartz and jasper are all represented in the series referred to. One small specimen is beautifully worked from a yellow jasper pebble, and has been drilled with a reed or other hollow drill, with sand and water. Being broken in the line of the perforation, the striæ are very plainly seen. While many of these objects are of beautifully colored stone, others, of equally fine workmanship, are made from the dullest tinted sandstone pebbles. When color was so greatly prized in every article of personal adornment, it seems strange that so large a number of these objects should have been made of plainly tinted stones, when other minerals, equally desirable in other respects, and of bright colors, were always to be had in any quantity.

Fig. 335 represents a symmetrically designed example of these ceremonial objects, made of steatite of a yellow-brown color. It may be taken as a fairly representative specimen, so far as the shape is concerned. The perforation in this, as in the majority of steatite specimens, has been made with a hollow drill, worked in but one direction. When, however, material so hard as jasper was perforated, the drill was used from each side. In many cases, the boring from the two directions was not correct, and did not meet, so that the hole through the object was more or less crooked.

The design of such an object as fig. 335 is very clear. The perforation could only have been intended for the insertion of a handle, and the stone thus mounted must have been carried as a truncheon, as the

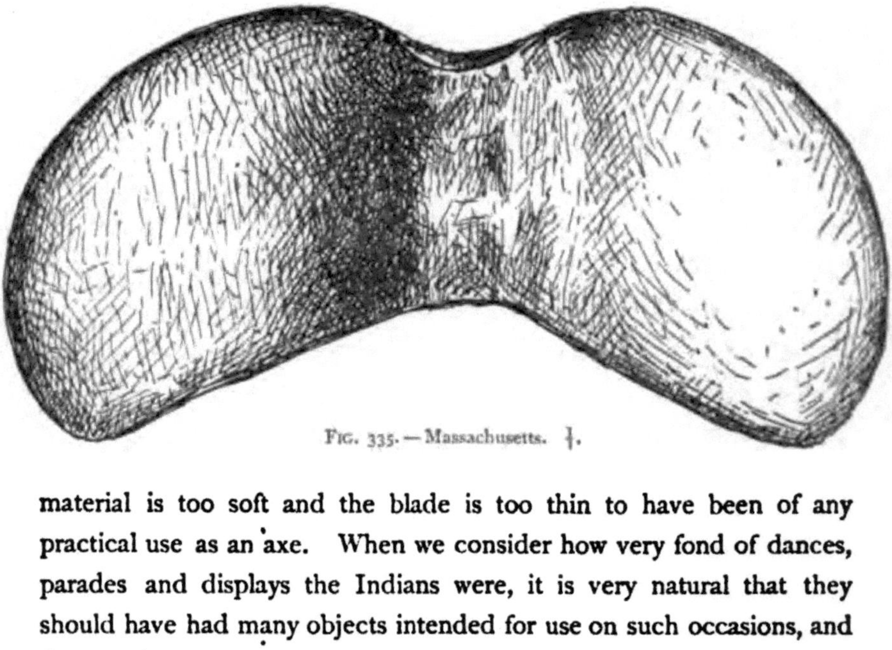

FIG. 335.—Massachusetts. ½.

material is too soft and the blade is too thin to have been of any practical use as an axe. When we consider how very fond of dances, parades and displays the Indians were, it is very natural that they should have had many objects intended for use on such occasions, and for no other purpose.

Fig. 336 represents a beautiful example, of a somewhat different shape from the preceding. It is made of a uniform, compact sandstone, and has been carefully polished until every trace of unevenness has been obliterated. The upper and lower edges are flattened; and the ends, which vary in outline, are so narrowed that it almost amounts to a cutting edge. The perforation is very accurate.

While occasionally one of these long and narrow specimens is even of greater length than fig. 336, the great bulk of them are shorter; and hence it may be considered as of about the maximum size, measured from tip to tip. Of many that I have measured, but four were longer than this specimen.

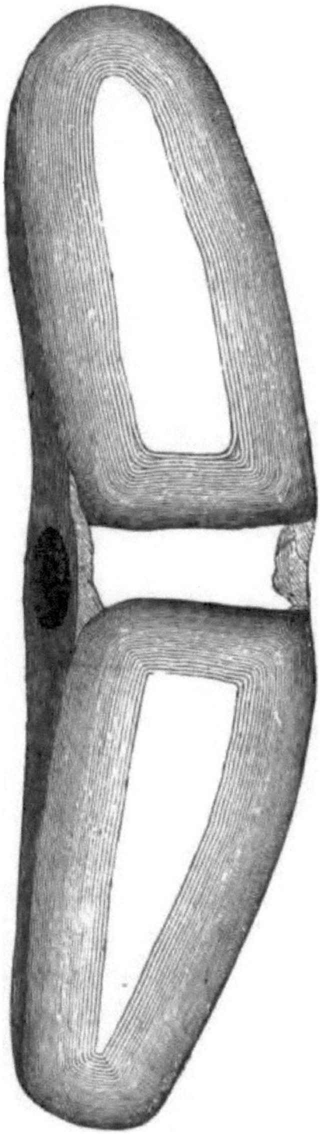

FIG. 336.—New Jersey. ½.

Figs. 337 and 337*a* represent another example of this form of perforated stone. It is shorter and broader than the preceding, but it is well made, and drilled with that smoothness and beauty which are marked features of fig. 336. The outline drawing of a sectional view of the specimen shows that the perforation is somewhat oval, instead of perfectly circular, and the diameter of the drilling is a little less at the apex than at the base. The drilling of this hole must, therefore, have been done with something different from a section or a number of sections of reed of identical diameter. This specimen has probably been drilled by the application of sand and water, in connection with a solid drill, as a pointed wooden stick, but the perforation begun below has been continued but half the distance and recommenced on the other side as is usual in such cases.

Fig. 338 represents a very gracefully designed example, made "of a greenish sandstone, and as smooth as the material allows."[134] It is eight and

[134] Perkins. Amer. Naturalist, vol. v., p. 15.

one-fourth inches long. The points of this specimen are smoother than the general surface, and are supposed therefore to show that it had been put to some use. This is, considering the size and shape of the object, improbable, if not impossible.

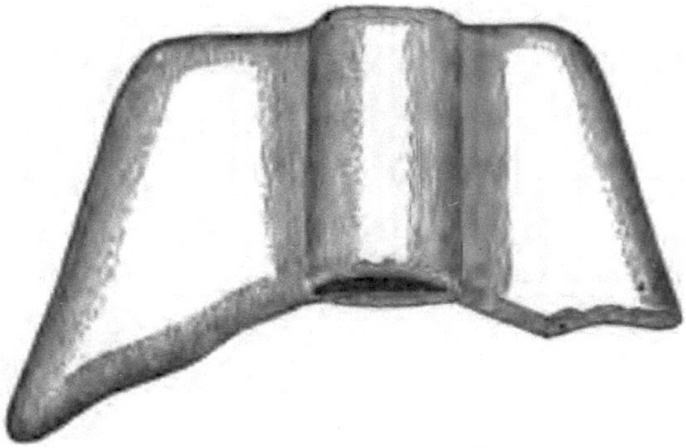

FIG. 337.—New Jersey. $\frac{1}{1}$.

Of the specimens of this character found in Ohio, very many are of this pattern; others have the wing-like projections cylindrical instead of flattened. A duplicated pattern of these implements consists in two such

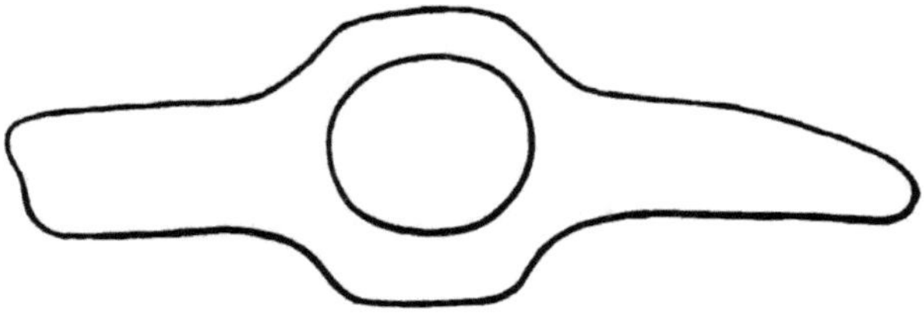

FIG. 337*a*.

specimens as fig. 338 placed together, by their convex faces; so that the perforations of the two shall be continuous. Examples of this pattern are figured in Mac Lean's[135] work on the Archæology of Ohio. No similar specimens have been found, I believe, in New Jersey, or in any of the New England states.

FIG. 338.—Vermont. $\frac{1}{4}$.

Fig. 339 represents a specimen of common shape, yet noticeably different in not being perforated for a handle, but simply grooved upon

---

135 Mac Lean. The Moundbuilders, p. 170, figure 45. Cincinnati, Ohio, 1879.

one side; the groove, too, being narrow and very shallow. The specimen is of hornstone, and was first pecked into its present shape, and subsequently polished over the whole surface of one side and one-half of the surface of the other side. The groove is polished over its entire surface. The margins are all blunt, and, although quite regular in outline, have not been carefully finished.

While the appearance of fig. 339 suggests that it might have been split, and that the groove was the result of an accident, it is evident that such is not the case, as a number of these articles, grooved in the same manner, have been found, and in them we have a simple form of

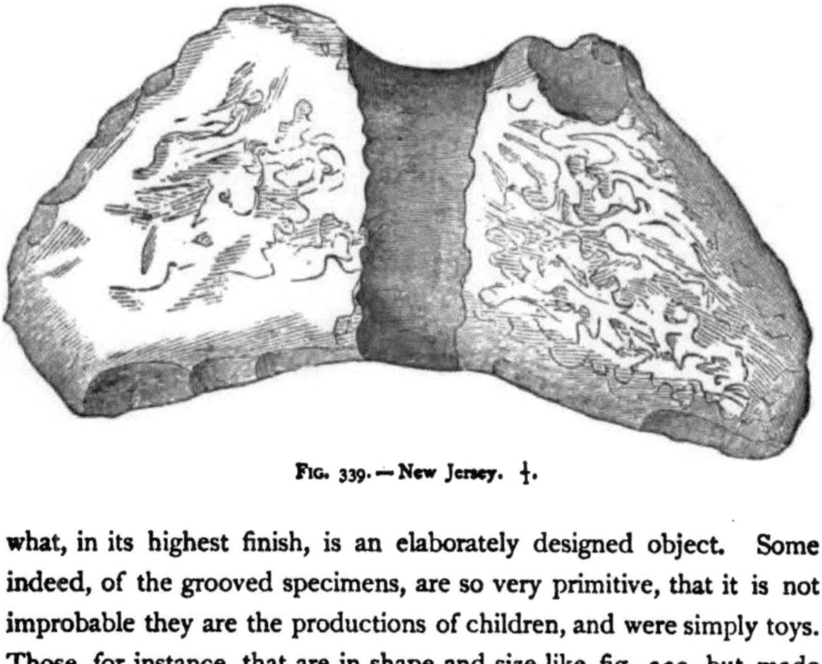

FIG. 339.—New Jersey. ½.

what, in its highest finish, is an elaborately designed object. Some indeed, of the grooved specimens, are so very primitive, that it is not improbable they are the productions of children, and were simply toys. Those, for instance, that are in shape and size like fig. 339, but made of a soft chalky slate, might readily be fashioned by any child.

Fig. 340 represents an example of what is believed to be the same object as those described on the preceding pages. It is of very different shape, however, and may have had a different "meaning," if there was any special significance in any of them. This specimen is four inches in length, and nearly as broad at the top, as shown in the illustration. Specimens of this pattern are frequently found throughout New England, and are of common occurrence in New Jersey. While generally made of Silurian striped slate, or steatite, occasionally more compact mineral was used. The labor of finishing such an object, when made of diorite or jasper, must have been enormous. Of this pattern, the two largest specimens measured respectively, seven and six and one-half inches in length, by six and five inches in breadth. The larger specimen was finished to the polishing, but the perforation has not been begun. In the smaller the perforation was complete. The winglike portions of these broad ceremonial objects vary considerably in width, and in some examples, they are so narrow, that the object looks like a tube with parallel ridges on each side, extending its whole length.

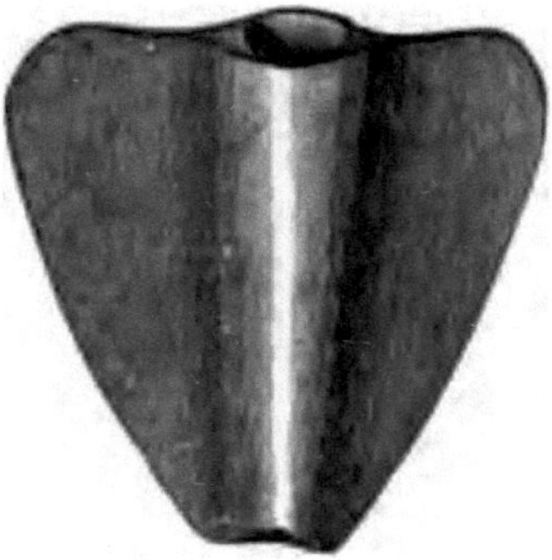

FIG. 340.—Vermont. ½.

A second example of these large and broad ceremonial stones found at Monkton, Vermont, is represented of actual size, in fig. 341.

Its length is four and one-fourth inches; the width at the larger end three and one-half inches, at the smaller, two and one-eighth inches. The greatest thickness is at the larger end and measures one inch, and is one-eighth less at the opposite end.

The perforation, which is complete, is five-eighths of an inch in diameter at one end, and somewhat smaller at the other, as shown in the supplementary figs. 341 *a–b*. It is nearly circular, and retains the marks of the drill. One side of the specimen is nearly plain, with a slight rounding at the edges only. The opposite side slopes to the edge from a line corresponding to that of the perforation. This feature of one flat and one curved or bulging surface is common to the great majority of the specimens of this pattern.

Fig. 341 is smoothly polished, though some of the tool marks made in shaping the object still remain. The material from which it is made is a soft, green, argillaceous slate. One portion of the stone is of a light, the other of a dark, green shade.

For the use of the cuts representing this object, I am indebted to the kindness of Mr. J. M. Currier, of Castleton, Vermont.

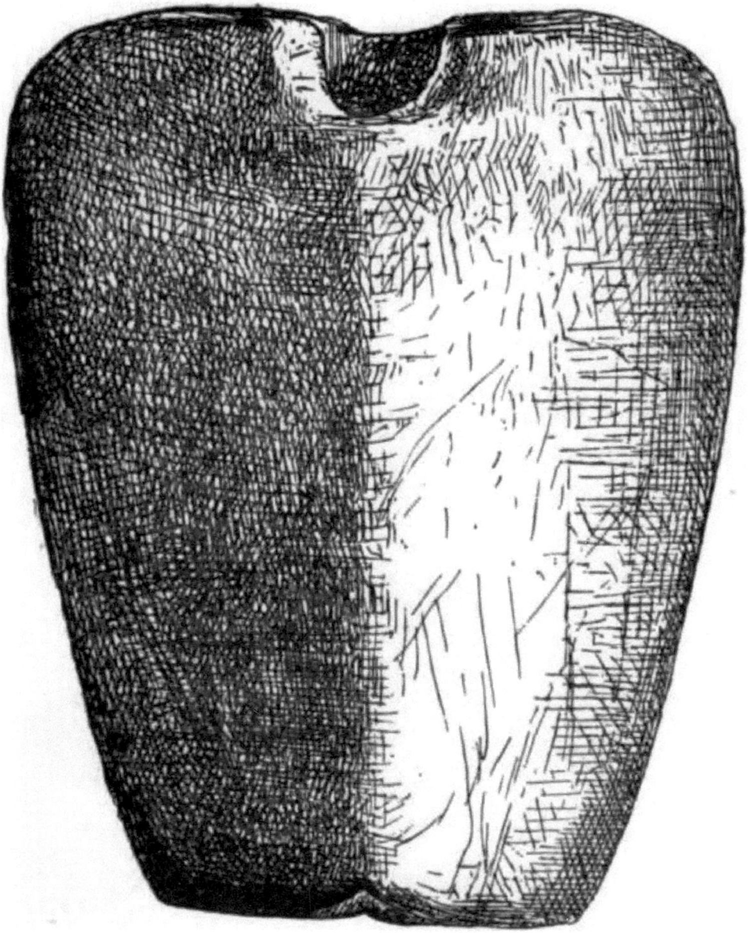

FIG. 341. — Vermont. ½.

Fig. 342 represents a flattened oval piece of striped slate, which, on account of the perforation, is classed with the specimen just described. Of itself, it is not a particularly attractive object, and it seems reasonable to suppose that such decorations were attached to it, or to the handle that is supposed to have passed through the perforation; as the maker's fancy might suggest. It is a little curious that no mention is made of these objects by any of the writers who visited this country, in the seventeenth century. Such prominent and common objects, as they once were, judged by the numbers now found, could scarcely

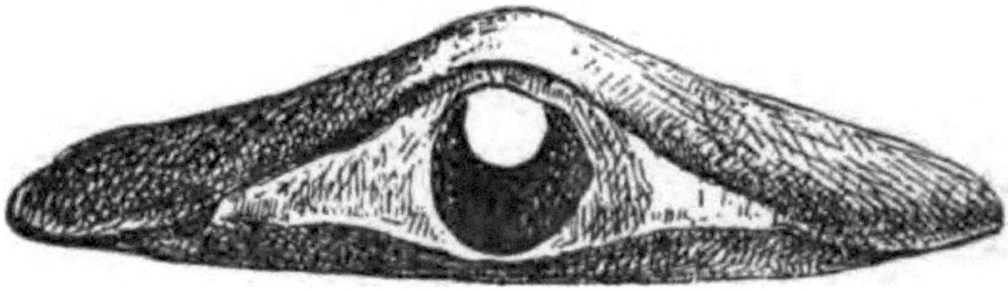

FIG. 341*a*.

have been overlooked had they been then in use. Why they should have been discarded, prior to European contact and the introduction of gaudy metallic baubles, is a mystery. Referring to the numbers of these "ornamental axes," Col. C. C. Jones, jr.,[136] remarks, "it may be that the American war-chief suspended from his belt one of these

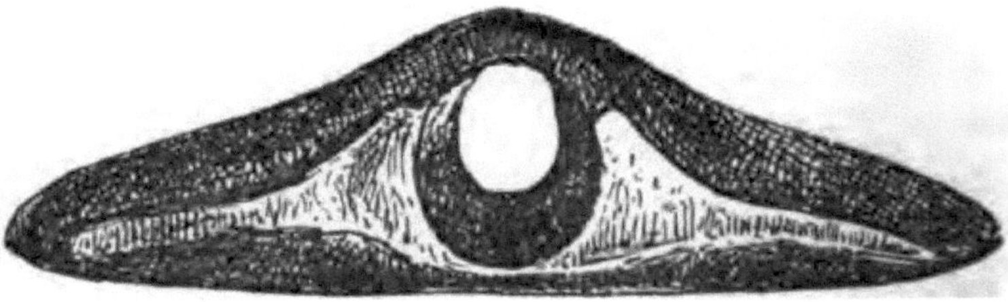

FIG. 341*b*.

delicate implements, and regarded it with emotions near akin to those which possessed the breast of the Scandinavian warrior as he cherished and displayed his *victory-stone*." This possibly explains, in a few words, the entire purport of these interesting objects.

Fig. 343 represents a similar specimen, found near Lawrence, Mass. The peculiarity of this specimen is in the hollow on one side, as shown

[136] Jones. Antiquities of the Southern Indians, p. 284. New York, 1873.

in the illustration. On the opposite side, there is a corresponding ridge, but not equal in height to the depth of the depression. As is seldom the case, when the specimens are small, the perforation, which has been made from each side, is not straight, the two ends not joining accurately.

FIG. 342.—New Jersey. ½.

Fig. 344 represents a very interesting form, of totally different shape from any of the preceding examples. This specimen is an oval or ovoid polished piece of the striped or Silurian slate, so commonly used for making ornamental objects. The two halves, if we divide it through the centre of the perforation which extends from the top to the bottom, will be found identical in every particular of shape and dimension. The base of the specimen is somewhat more flattened than the top, and appears to have been in contact with another stone, as it is worn off smoothly, but with a variable width. This worn surface is of a lighter tint than the other portions of the specimen. The perforation is a little less in diameter than that of fig. 336, but it is of equal beauty of workmanship. The diameter is the same throughout, the perforation being accurately circular, and showing the rings which indicate

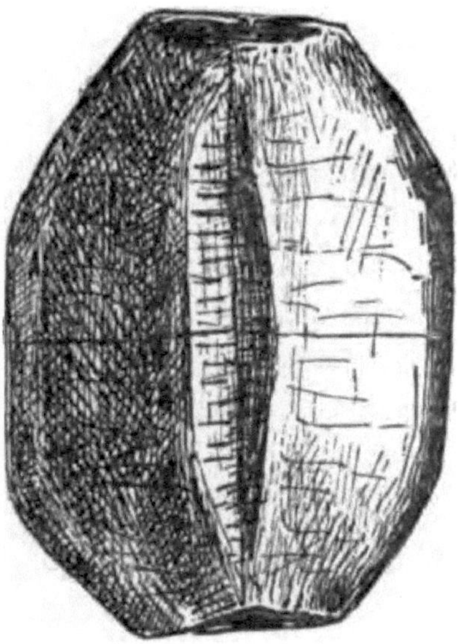

FIG. 343.—Massachusetts. ½.

drilling with a hollow tube. For a short distance from the base, extending upward along the sides of the perforation, the "rings" are not distinguishable, except by the closest scrutiny, and appear to have been worn away by the rubbing of whatever passed through the hole as a handle. From point to point, this specimen measures two and five-eighths inches, and across the middle one inch and a half. The diameter of the perforation is just one-half an inch, or one-third of the total width of the specimen itself.

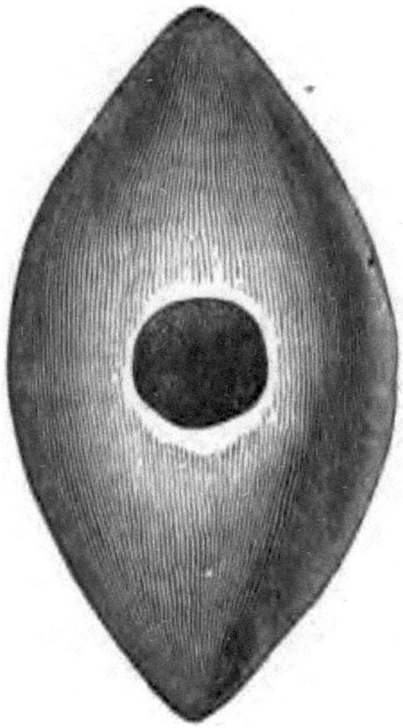
FIG. 344.—New Jersey. 1/1.

Fig. 345 represents one of the most elaborately carved and otherwise interesting specimens of perforated ceremonial objects, that have yet been discovered. It was found[137] "at the point where Lake Wankewan ('Measly Pond') originally emptied into Lake Winnipiseogee, New Hampshire, at a depth of about two feet, in the sandy drift at the head of the lake, where the ground apparently had not been disturbed for centuries."

"The stone is of an oval form, smoothly finished upon the surface, and of as perfect contour, as if turned in a lathe. Its dimensions are three and seven-eighths inches in length, and two and three-eighths inches in thickness. The material is a silicious sandstone of a greenish clay-drab color and of fine grain. The sculptures are mostly in bas-relief, upon a ground sunk below the surface of the stone, and of a higher grade of art than usual in Indian workmanship. It is difficult to conceive that such work could be done without the aid of metal tools. A hole was drilled through the longest diameter which tapered uniformly three-eighths of an inch at the larger end, to one-eighth at

[137] Tapley. American Naturalist, vol. vi, p. 696, figs. 139—142 inclusive.

the smaller, the use of which was probably the same as in the class of stones known as 'gorgets,' to which we should refer it. Around the aperture at each end was a border of points like a star, as will be seen

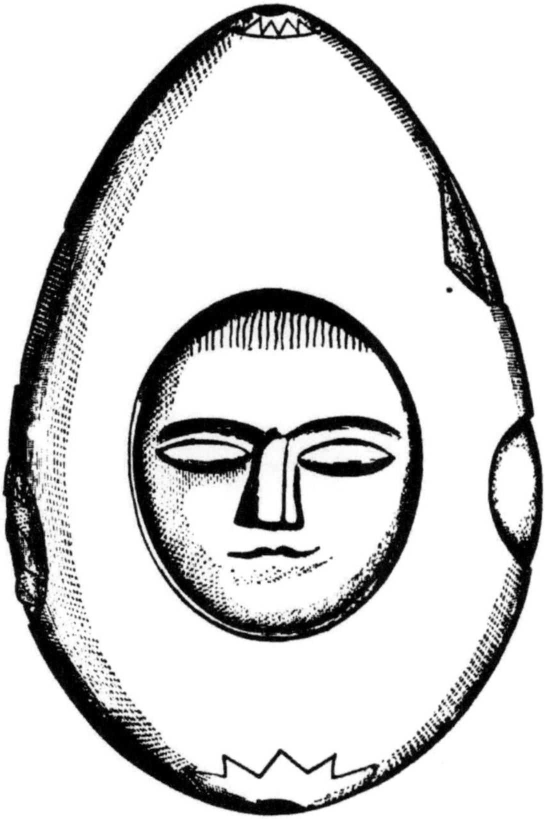

FIG. 345.—New Hampshire. ½.

by reference to the first of the four illustrations representing the specimen.

"Fig. 345 is intended to give an idea of the form of the stone, the figures at the sides being the profiles of figs. 346 and 348. The Indian 'mask' has the characteristic outline and projecting mouth seen in other specimens of Indian art. The wavy lines on the forehead are supposed to indicate the hair. The finish of the whole is quite elaborate.

"In figs. 346, 347 and 348, the dotted line is intended to indicate the position of the pictures on the stone. The lines of the 'wigwam' are regularly drawn and the surface is 'pricked up' or roughened. The

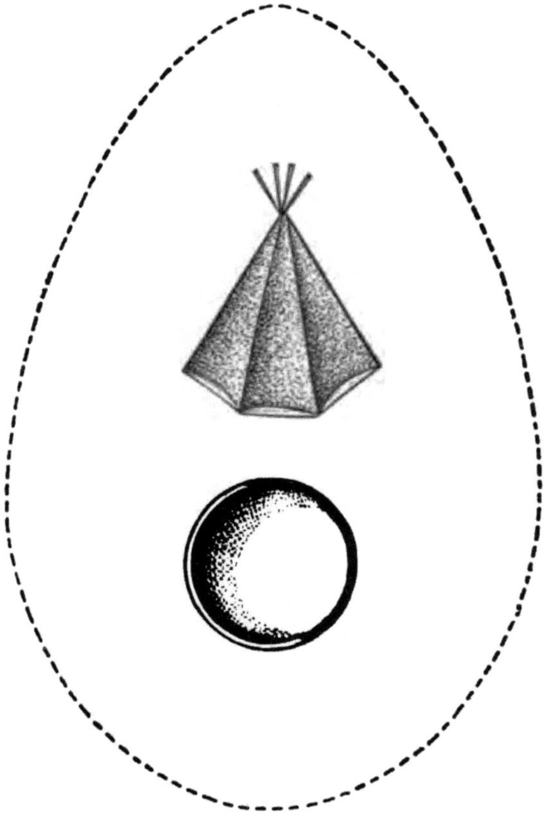

FIG. 346.—New Hampshire. ½.

circle below is perfectly rounded and supposed to represent the full moon, although every one has the privilege of forming his own theories in regard to the significance of the symbols.

"Fig. 347 has a delineation of four 'arrows' inverted. Underneath this is a 'new moon,' and the two round dots that may represent stars. Below this are two 'arrows,' crossed, and a convolute or coil which may be a 'serpent.'

"Fig. 348 shows an 'ear of corn,' nicely cut, and in a depressed circle are three figures, the central one representing a 'deer's leg,' and the others of doubtful interpretation.

"As in illustration of the surmises of those who are interested in

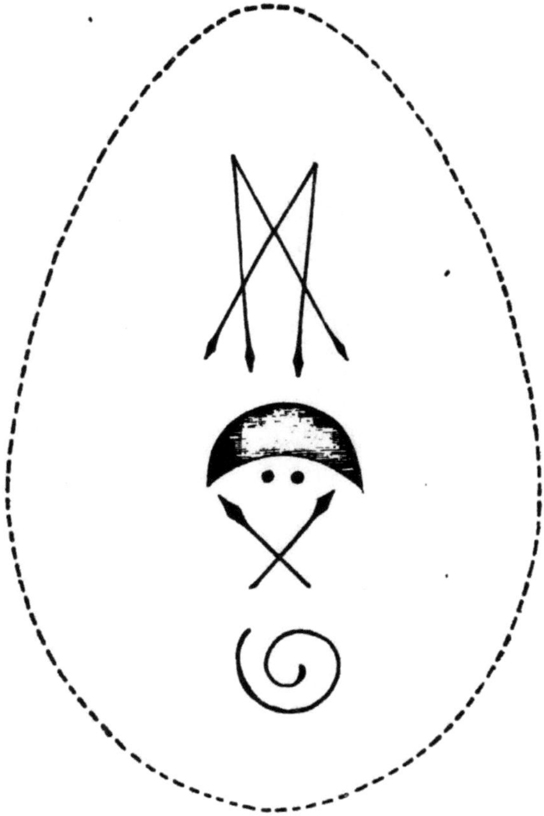

FIG. 347.—New Hampshire. ½.

deciphering such inscriptions we give the following, which is certainly ingenious and even plausible.

"It is suggested that the stone commemorates a treaty between two tribes. The reversed arrows in figure 347 symbolize place; the moon and stars the date; the crossed arrows a union of the two forces for aggressive or defensive purposes, etc. The wigwam might indicate the place where the treaty was consummated, and the corn and other emblems the feast by which it was commemorated."

Prof. F. W. Putnam,[138] in calling attention to this peculiar relic of the New England Indians, has remarked that it appears "to be far more elaborate than anything he had known as the work of the earlier

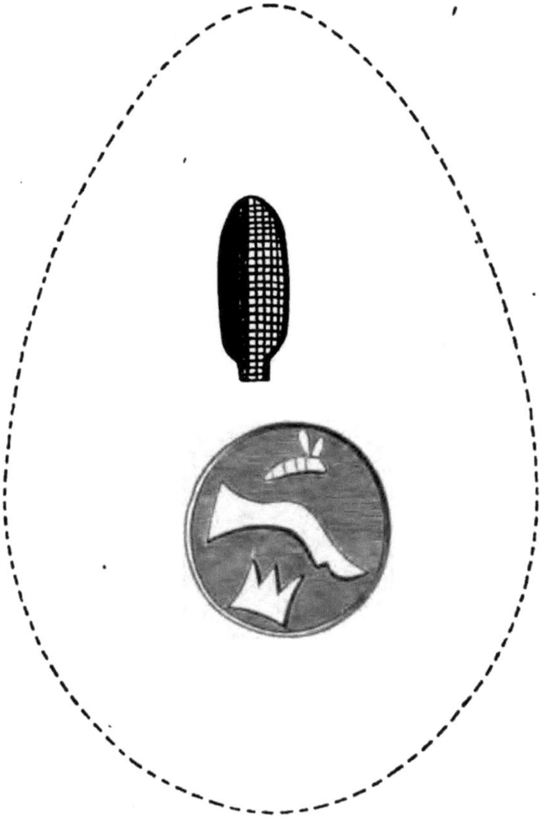

FIG. 348.—New Hampshire. ½.

inhabitants of New England. On this stone we have the characteristic Indian face, similar to the few others that have been found in New England, with an attempt at an artistic result in the finish of the stone and the other figures carved upon it, that would certainly lead us to infer that its maker, if an Indian, was of a far higher caste as an artist, than the distorted and childlike outlines of animals and men, ordinarily

---

138 Putnam. Bulletin of Essex Institute, vol. iv, p. 92. Salem, Mass., 1872.

cut or painted by them, have heretofore impressed us as possible; and were it not for the fact, that the face is so similar to undoubted Indian representations of the human face which we have from New England, he would be inclined to think that it might have been the work of some other race. The position in which the stone was found marked it quite an ancient piece of workmanship; and, from its shape, and the fact of its having a hole through the centre, he believed it would be classed with the singular perforated stones called gorgets, found throughout the country, always more or less elaborately finished, and were supposed to have been worn on the breast as an ornament or badge of office."

The artistic merit of the various carvings on stone, executed by the Indians, is so variable, that it is unsafe, from this cause, to infer that any production may have some other than an Indian origin. That they had the ability to invent tasteful designs and to execute them creditably, is fully shown by some specimens of pipe sculpture that have been preserved. If we compare the pipe with a turtle carved upon the stem, fig. 317, with the caricatures of human faces, rudely cut upon flat oval pebbles, described in this chapter, it is almost impossible to imagine that the three specimens have the same origin. As an artistic production, the carved face, from New York, described on a succeeding page, equals that of the "gorget" from Lake Winnipiseogee; and the curious carved "bird stones," common to our Atlantic coast states, are, as a class, even of greater merit.

Fig. 349 represents another example of this form of carved ornamental stone, made of green steatite, which is of much simpler design, but more nearly approaches the New Hampshire specimen, than either of the three carved faces from New Jersey, illustrations of which are given. As will be seen by reference to the illustration, the perforation extends through the long diameter of the stone, and is of large size, thus suggesting its close relationship with the ceremonial objects of various patterns described in this chapter. Like the preceding very elaborate specimen from Lake Winnipiseogee, it has decorative markings upon it other than the face figured in the accompanying illustration. These consist of shallow depressions which may possibly be parts of an unfinished attempt to represent the human face.

The fact that one surface only of the thin slabs, that are perforated for suspension, is ornamented in any manner, apparently indicates that they were used solely as personal ornaments, and rested upon the clothing of the wearer. This would, of course, prevent the under side of any pendant from being exposed, and render useless any marks put upon it; but in this and the preceding examples, as in all the objects which are here considered of ceremonial import, and not personal ornaments, the perforation is of greater diameter, and extends lengthwise through the specimens, so that a handle or staff could be used as a means of carrying them about in a conspicuous manner; and thus exhibit all sides equally well.

FIG. 349.—Massachusetts. $\frac{1}{1}$.

If it be objected that we have, in fig. 349, too small an object to be used in the manner suggested, then in this specimen, we have an interesting example of a particular form of carved stones, which may be differently classed, in accordance with their size. In this case, fig. 349 should be considered as an ornament.

Fig. 349 was found by Rev. B. F. DeCosta, at Wellfleet, Cape Cod, Mass., and by him kindly presented to the museum of Archæology at Cambridge, Mass.

Common as are these perforated ceremonial stones, they nevertheless were not so readily made as arrowheads, or the thin sandstone disks which so often did duty as ornaments. When broken, they were not always discarded, but were often utilized as ornaments.

Fig. 350 represents an ornament, as it is now supposed to be, which was originally a ceremonial object, of an unusual pattern, made of Silurian striped slate. The specimen has been broken in the line of the perforation, and the fractured edges have subsequently been ground down, until they are as smooth and well finished as any other part. Having no perforation, whereby it could be suspended, it is not clear, how it was worn or used.

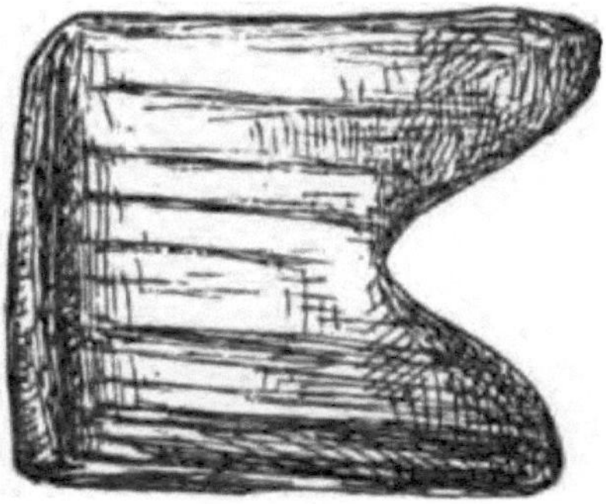

FIG. 350.—New Jersey. ½.

Fig. 351 represents another of these fragments, which has been subsequently utilized.

The broken edges of this specimen, which has been fractured along the perforation, have been carefully smoothed down, and through the middle of one of them a small hole has been drilled. This has been drilled from both sides, showing that the specimen was broken after the larger perforation had been completed.

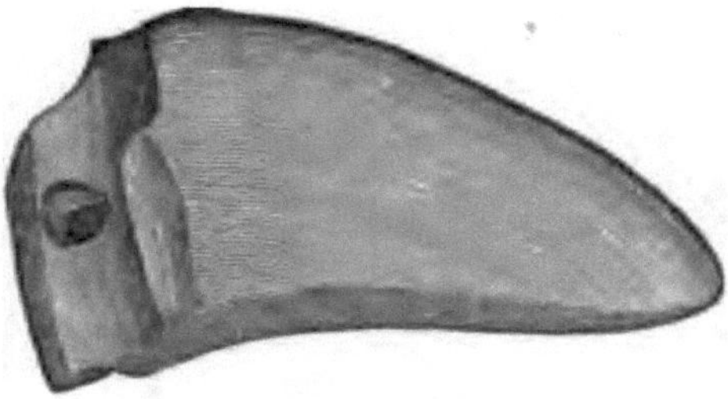

FIG. 351.—New Jersey. ½.

Such an object as fig. 351 might have been used, as were fossil sharks' teeth, shells and pebbles, as a pendant, or one of several, on the same necklace; and therefore might properly be referred to under the head of ornaments.

CHAPTER XXVI.

BIRD-SHAPED STONES.

UNDER this title, which is not accurately descriptive, though of common acceptance, it is proposed to consider a class of carved stone objects, which are of comparatively frequent occurrence over the entire area of the United States, east of the Mississippi river. Their significance has been discussed more than that of any other form of stone implement or ornament made by the American Indians.

Schoolcraft[139] has designated this form of relic as a handle for a knife, the blade of which was obsidian or jasper. One of these "knife-handles" is figured, found on Cunningham's island, lake Erie, New York, which is considered to be "apparently a sacrificial or a flaying knife." The relic is so described, although there is no indication of a blade.

By many, they have been called "idols," and strangely enough, have been seriously described and commented upon as "corn huskers," although their use as a husking peg would tend rather to retard than facilitate that work, as none have yet been found with a really sharp point, or one in any way available for piercing the husks, as the common hickory peg, now used, is expected to do.

Probably Messrs. Squier & Davis[140] correctly cover the whole ground concerning them, in stating that "it may reasonably be concluded from the uniform shape of these articles, and from their apparent unfitness as implements, as also from the wide range of their occurrence, that they were invested with a conventional significance as insignia, or badges of distinction, or as amulets. We know that the

[139] Hist. Condit., etc., N. A. I., vol. iv, p. 175, pl. xxiii, fig. 2.

[140] Squier & Davis, Anc. Mon. Miss. Valley, p. 239.

custom of wearing certain stones as preventives of disease, or as safeguards against accidents or the malice of evil spirits, has not been confined to one continent, or to a single age. It is not entirely obliterated among certain classes of our own people. Regal authority is still indicated by rich baubles of gold and gems. It matters little whether the index of royalty be a sceptre or a simple carved and polished stone, so that it is sanctioned with general recognition."

In a description of archæological specimens found in Michigan, Mr. Henry Gillman [141] has described one of these bird-shaped stones, "formed from a beautiful piece of variegated slate, of a grayish-green, interstratified with veins of a darker shade, and is neatly made and finely polished;" and further remarks, that "similar ornaments have been found throughout the United States; and, as there has been considerable discussion as to their use, I will here state that I have learned through an aged Indian, that in olden time these ornaments were worn on the heads of Indian women, but only after marriage. I have thought that these peculiar objects, which are always made of some choice material, resemble the figure of a brooding bird; a familiar sight to the 'children of the forest;' that thus they are emblematic of maternity, and as such were designed and worn."

This view of their significance has been met with considerable ridicule on the part of some, who, however, offered no better explanation of these objects, as a substitute. Their occurrence in graves, that were known to be those of females, by the fact that they were not associated with weapons of any kind, is certainly in favor of the view expressed by Mr. Gillman. In a local publication, the author [142] asked for information with reference to these bird-shaped stones; and, soon after, had the pleasure of receiving from Col. Chas. Whittlesey of Cleveland, Ohio, the following:

"Dr. E. Stirling, of this city, says, such bird effigies, made of wood, have been noticed among the Ottawas of Grand Traverse Bay, Michi-

---

[141] Gillman. Smithsonian Annual Report for 1873, p. 371. Washington, D. C.

[142] Abbott. Curiosity Hunter. January, 1878, Rockford, Illinois.

gan, fastened on the top of the heads of women, as an indication that they are pregnant.

"All of the stone bird effigies I have seen are perforated for attachment to some other object.

"No doubt all the ornamented stones of polished slate with holes for attachment had a meaning, and were significant of something personal to the wearer."

As further evidence that objects having this significance were not unknown to many of the Indian tribes, it may be mentioned that William Penn refers to a custom among the Shawnees and Delawares, with whom he formed his celebrated treaty, in 1682, that bears indirectly upon this subject. He says, "when the young women are fit for marriage, they wear something on their heads for an advertisement, but so as their faces are hardly to be seen, but when they please." (Harvey's History of the Shawnees, p. 14, Cincinnati, 1855.) While there is nothing to imply that the "something" that these Indians wore was bird-shaped, or was made of stone, wood or cloth; it does add to the probability, that the objects now under consideration, whether bird or mammal-shaped, or of so conventionalized a form that all trace of realism is lost, do have some such significance as mentioned by Mr. Gillman.

As an indication that these bird-shaped stones were not knife-handles, or corn-huskers, attention has been called[143] to the fact, that halves of these objects have been carefully ground smooth and polished on the fractured end, and a hole subsequently drilled for suspending them, which could be done more conveniently through the new hole, than through the two basal perforations common to all of these bird effigies. Easily pleased, as the Indian doubtlessly was, in the matter of decoration, it is hardly probable that a broken "husking peg" would ever have been used as a charm or pendant; but if the unbroken object had a significance, such as has been mentioned by Mr. Gillman and by Col. Whittlesey, then nothing is more natural than

[143] Abbott. Nature, vol. xii, p. 436, fig. 2. London, 1875.

that a piece of one should have been utilized in the manner described.

Fig. 352 represents the common form of these so-called bird-shaped stones, much reduced in size. It is a fraction over four and a half inches long. The body, or main portion, is very accurately sloped to the back, which is a narrow flat ridge, of a uniform width of one thirty-second of an inch. The "head" of the specimen is nearly square, and not unlike the head of a blunt muzzled mammal in shape. The knob-like protuberances stand out from the head one-third of an inch, and have a narrow neck, about one-half the width of the "knob" itself. The bottom of the implement, as the illustration shows, is flat. There is at each end of the specimen a small hole, drilled obliquely upward and outward from the flat base.

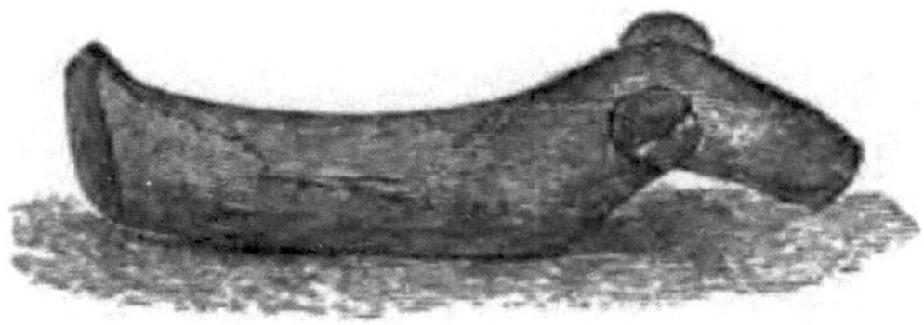

FIG. 352.—New Jersey. ½.

This specimen was found near Trenton, N. J. On the bluffs forming the eastern bank of the Delaware river, south of Trenton, N. J., on the site of one or more extensive Indian towns, fractions of these bird-shaped stones in great numbers have been found, and a few that are only "blocked-out." All are of brightly colored or handsomely marbled or striped stones, and none are without some degree of polish. The size varies exceedingly, the largest being about seven inches in length, the smallest scarcely three. In a series of eighty-four fragments, there were about equal numbers of each of the sizes mentioned, with perhaps a slight excess in the numbers of those of medium length, say of about five inches. None of those of the largest size were too heavy to have been worn upon the top of the head, an objection which has been urged as to their use in the manner suggested.

Fig. 353 represents an interesting specimen of one of these bird-effigies, made of striped Silurian slate. It was found in Cumberland Co., New Jersey, and is now in the archæological museum at Cambridge, Mass. Unlike most of the other examples of this class, this specimen has not the eye-like projections from the side of the head. Being found in the immediate vicinity of Delaware bay, and not far from the ocean, it has been suggested that it was intended to represent a "diver" or duck, and that the elongated "neck" was quite characteristic of these birds when rapidly swimming. On the other hand, it has been stoutly maintained that it was a "husking peg," and it must be admitted, that it is better adapted to this use, than either the preceding specimen, with the knobbed protuberances on its head, or the succeeding one with its broad circular base.

The fact that there is a marked individual difference in all these bird-

FIG. 353.—New Jersey. ½.

shaped stones is one of much interest, even if it has no bearing upon their significance. Of a very large number of specimens examined, no two can be considered as strictly alike, although most forms of stone implements can be very readily duplicated.

Fig. 354 represents one of these bird-shaped stones found in Vermont.[144] "It is made of a pretty breccia composed of light and dark material. It is finely wrought and very smooth, though not polished. The upper side is worked to a sharp edge, from which the sides round outwards towards the rectangular base, in which there is a hole at each end running obliquely through it. The length of the relic is four and one-half inches, and the height nearly two inches."

---

144 Perkins. Amer. Nat., vol. v, p. 16, fig. 6. Salem, Mass., 1871.

Fig. 355 represents another example of this class of objects, but is more like a mammal than any bird. Specimens with a broad circular base, like figure 355, are of more common occurrence in the west, than along the Atlantic seaboard. A perfect facsimile of this specimen,

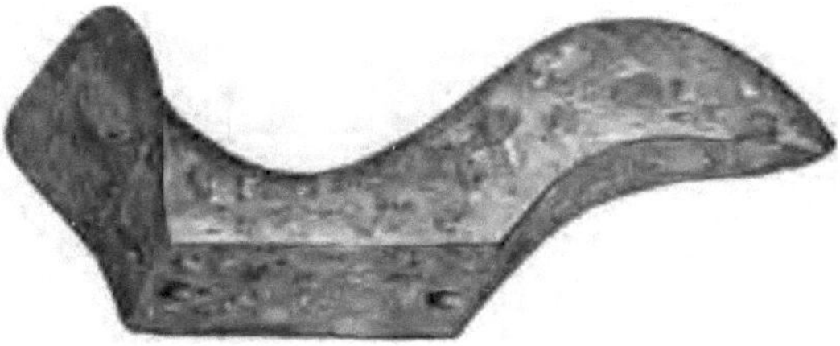

Fig. 354. — Vermont. ½.

except in having a pointed rather than blunt nose or beak (if it may be so considered), is among the interesting series of grave contents found near Swanton, Vermont. Another from Indiana differed only in being of a very dense granitic rock, of a dull gray color.

Fig. 355. — New Jersey. 1/1.

A variety of this form of ornamental stone, of which but few specimens appear to have been found east of the Ohio valley, consists in the supposed tail of the bird being repeated, as though the posterior halves of two of such specimens as fig. 352 had been placed end to end. This is certainly a highly conventional manner of representing a bird, but that such two-tailed, headless examples are the same object really, as the typical form, is, without doubt, a fair inference. (See Smithsonian Contributions to Knowledge, No. 287, fig. 211). To this latter type of these so-called bird-shaped objects must be referred the interesting specimen, figs. 356 and 357.

This ornamental stone bears but little resemblance even to those modified bird-shaped objects, which have only the "tails" of the supposed birds represented; but the characteristic oblique perforations at the ends, the flattened base and slightly curved upper surface, are well defined.

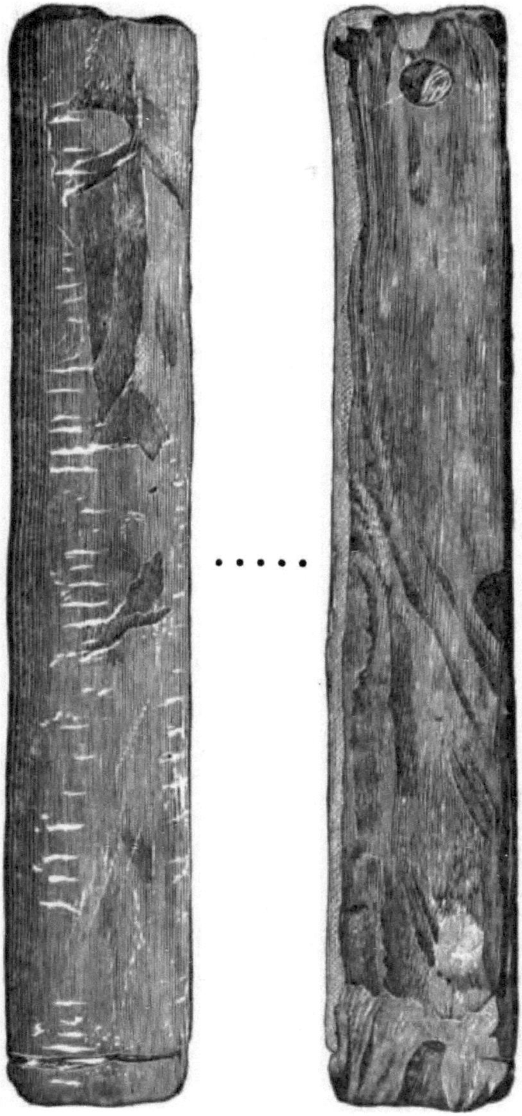

FIGS. 356 and 357.—New Jersey. ½.

This specimen is made of a fine grained sandstone of an olive-green color, and is carefully worked over its several surfaces. An interesting feature of this specimen consists in the series of narrow and now indistinct marginal notches extending along the edge of that side of the specimen shown in fig. 356. Other than these is a series of similar cuts or grooves of different lengths, extending down the middle of the back or upper surface of the specimen. That these are probably decorative only is questionable.

This rare form of ornament was found near Freehold, Monmouth Co., N. J., by Rev. S. Lockwood, to whom I am indebted for an opportunity to figure and describe it.

CHAPTER XXVII.

GORGETS, TOTEMS, PENDANTS AND TRINKETS.

INASMUCH as the name "gorget" given to a class of supposed decorative stones, or insignia of office or rank, has been so generally adopted by American archæologists, it is here retained as the specific designation of those large, thin slabs of stone, which, having two or more holes through them, are supposed to have been firmly attached to the clothing, and not, like the various trinkets, simply suspended.

Probably nothing in the whole range of the stone implements and ornaments of the Indian has been more discussed and written of than these simple slabs of slate and sandstone. That they were really insignia, as the designation "gorget" implies, is perhaps not known, but it is quite possible that such was the case. Certainly, if not badges, they were merely personal ornaments. Beyond this we are not warranted to theorize concerning them.

The following suggestion of Schoolcraft [145] is here offered as covering the whole ground; and, adopting it, they are classed as one form of several decorative objects described in this chapter. They are "one of the forms of those ancient badges of authority, to which, as a generic term, the modern Algonquins apply the name *Na-be-kow-á-gun*. The native tribes, from our first acquaintance with them, evinced their fondness for insignia of this kind. The modern medal is the result of a compliance on our part with this passion."

Fig. 358 represents a common form of gorget. This specimen is four and five-sixteenths inches long and one inch and five-eighths wide at the middle; it has been very rudely drilled in two places from each side until the depressions met, the distance between the holes

---

145 Trans. Amer. Ethnol. Society, vol. i., p. 401, pl. 1., fig. 2.

on one side being exactly four-fifths of an inch, a distance noticed particularly by Squier and Davis,[146] in several of the specimens they figured. They remark, "It is a singular fact that the holes in the three specimens first noticed, as also in some of those which follow, are placed exactly four-fifths of an inch apart. This could hardly have been the result of accident. These relics were found at different localities, several miles distant from each other." If this similarity of distance between the perforations was intentional, it would seem that the stone had some use other than that of a breast-ornament merely. Certainly, in such case, the mere distance separating the holes could have had no special use. In fig. 358 this distance is variable, inasmuch as one hole is obliquely drilled, and so produces a greater space between the two perforations on one side than on the other.

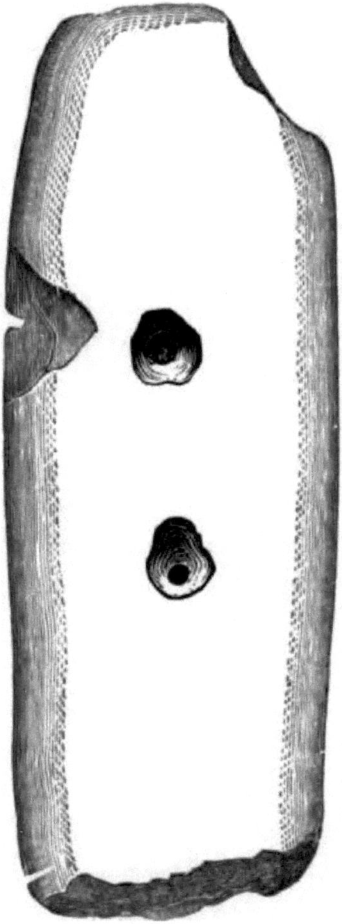

FIG. 358. — New Jersey. ½.

Fig. 358 is made of reddish sandstone of a fine grain, and is susceptible of high polish. Other specimens, the same in all respects, except material, from the same neighborhood, are made of variegated slate; and in one case the striped Silurian slate, of which so many of those found in Ohio are made, is used. The drilling in such of these gorgets, as are made of soft slate, is usually very irregular, when compared with that of specimens made of

146 Anc. Mon. Miss. Valley, p. 237.

a harder material. Why one series should be bored with great accuracy, and the other so indifferently, is indeed puzzling; all the more so, as the material that is easier to perforate is the more clumsily worked.

Figs. 359 and 360 represent the two sides of an exceedingly interesting specimen. As the illustrations so distinctly show, the entire surfaces are covered with incised lines, so closely arranged that their purpose is probably only decorative. As is common to a majority of gorgets and of allied trinkets, the margins of this specimen are cut into deep, closely set notches. The theory that specimens of this character were firmly attached to the dress of the wearer, and thus exposed to the gaze of others, only upon one side, is somewhat contradicted by the fact that an equal amount of decoration is found on the two sides. Were the combinations of straight, oblique and zigzag lines less closely placed, and fewer in number, portions, at least of them, might be considered as a record, rather than an ornament, especially as the lines are by no means as distinct as in the illustrations, which are reproduced from a photograph, after it had been carefully chalked, to bring out clearly the narrow, hairlike lines.

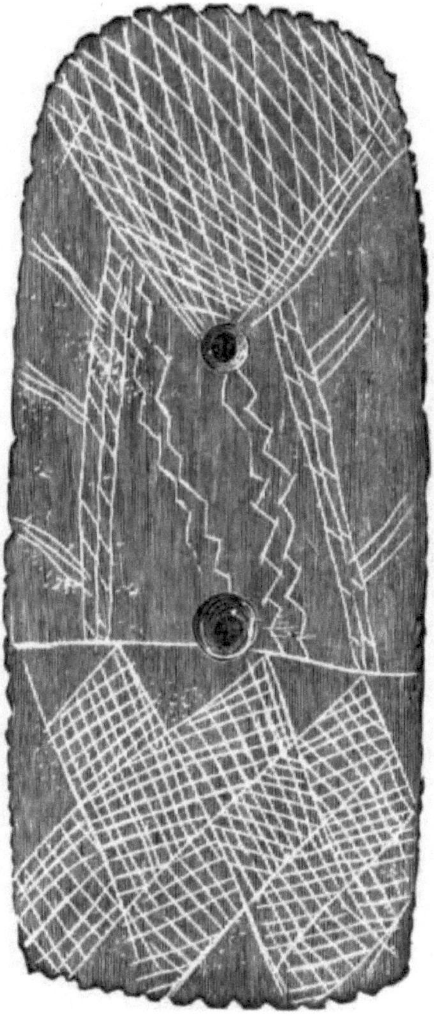

FIG. 359.—New Jersey. ½.

The surface of fig. 360 has a smaller amount of supposed ornamentation, is also a little smoother, and is nearly a perfect level, between the perforations. From these considerations, it would appear that fig. 359 represented the upper and fig. 360 the lower side.

FIG. 360.—New Jersey. ½.

Of a series of over one hundred of these gorgets from New Jersey, this alone exhibits any trace of incised lines, or other ornamentation, other than the little notches about the margins, which appear to be the rule, rather than the exception. This specimen was found near Freehold, Monmouth Co., N. J., and is now in the cabinet of Prof. Samuel Lockwood of that place.

Thin plates of native copper have been occasionally met with in New Jersey, which, although considerably smaller than the gorget here figured, were unquestionably ornaments, and used in essentially the same manner. They are mentioned here from verbal descriptions only, as no specimens have been preserved in our museums. In Ohio, copper gorgets of the usual size are frequent. A handsome specimen is figured among the illustrations of moundbuilder relics from Ohio, in MacLean's volume ;[147]

[147] MacLean. The Moundbuilders, p. 164, fig. 35. Cincinnati, 1879.

and another in the account of mound explorations in Ohio, by the late Prof. E. B. Andrews.[148]

Fig. 361 represents an interesting example of a gorget, which, although broken, was evidently not discarded. The lower, curved and unbroken end is two and three-fourths of an inch in width, and notched as usual. The narrower end, where the fracture occurred, has been carefully ground down, and now has as good a polish as the uninjured sides. The perforations, four in number, are very rudely executed. They have apparently been bored in pairs, those near the middle of the plate at one time, and the others at another. The latter are more evenly bored, and the holes are straight. In the central pair the perforations are slightly oblique. The marginal notches, in this specimen, are nine in number. The same number of notches are upon the under side, but they are not merely continuations of those seen in the illustration.

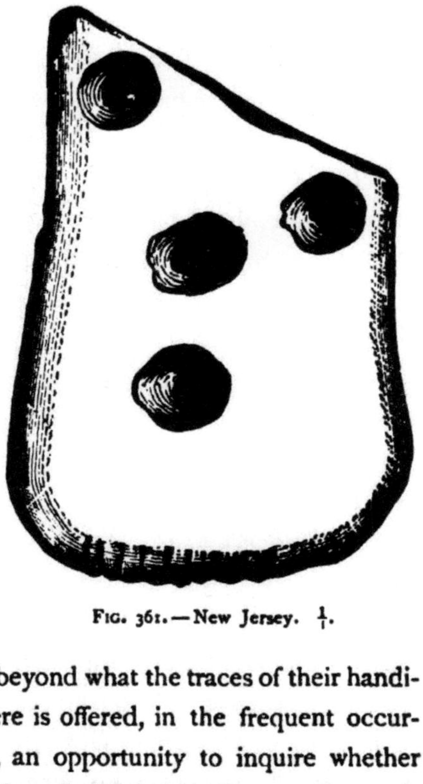

FIG. 361.—New Jersey. 1/1.

While far from being disposed to credit the native American tribes with any advance in culture beyond what the traces of their handiwork unquestionably warrant; there is offered, in the frequent occurrence of these marginal notches, an opportunity to inquire whether the early race of the Atlantic seaboard did habitually record prominent events, in the way indicated, by such carefully-cut notches as characterize the great majority of these gorgets. Have these notches a significance, or are they merely ornamental? We know, indeed,

[148] Andrews. Tenth Annual Report, Peabody Museum, Cambridge, Mass., 1877.

that the savage, like the child, is "pleased with a rattle, tickled with a straw;" but were ideas of ornamentation so primitive, that he could see any beauty in these marginal notches? Is the theory, that they were merely decorative in character, consistent with the fact that the same people, who cut these little notches, also shaped the truly beautiful ceremonial and bird-shaped stones, and carved lifelike portraits of both men and animals?

Gorgets are found in great abundance along the whole Atlantic coast. In New England they are as abundant as in the middle states, and perhaps the rich regions of the Ohio valley have not yielded a greater number. In the Champlain valley of Vermont "gorgets, with one hole or two, are found" everywhere.[149] "As is the case elsewhere, these are usually made of some compact, fine-grained stone that is capable of taking a smooth polish. Slate is the most common material in those that I have seen, sometimes a red roofing slate, often a dark-greenish talcose slate veined with black. The gorgets with one hole are less common and less elegantly made than those with two, and the material seems less carefully selected. Of the 'two-hole stones,' those of rectangular outline are most abundant; not that all these are rectangular, but with some modification of this form, as with corners cut off making an octagonal figure, or rounded more or less."

One variety of these gorgets, frequently met with in the west and south, is shaped like a boat, which, while very rarely found in New Jersey, is occasionally gathered with other relics from certain village sites and burial places of the Indians in New England. From ancient graves near Swanton, Vermont, a number of these boat-shaped gorgets have been obtained, associated, however, with others of plainer patterns, but of equally workmanlike finish. All are made of ornamental stone, and perforated.

Fig. 362 represents an example of this form of gorget found at Bradford, New Hampshire. This specimen, which is four inches in

---

149 Perkins, *l. c.*, p. 742.

length and but little more than one in width, is not so large, nor so deep as those found in Vermont, but may be taken as a connecting link between the flat or plain gorgets and those that are even more distinctly boat-shaped.

Some of the latter are not oval upon the convex side, but angular, giving them a triangular outline. An example of this kind, figured by Foster,[150] was found near Danville, Illinois; and others are in the Museum at Cambridge, from Tennessee. The perforations, in all of this triangular pattern, are near the ends, and are drilled in an oblique direction. In others, that are not so deeply excavated, the perforations are variously placed.

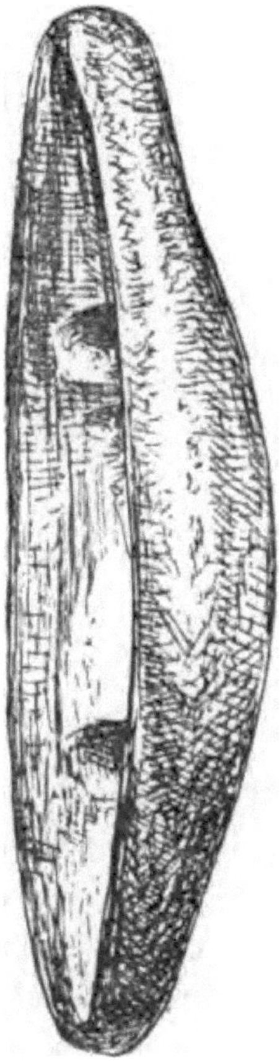

FIG. 362.—New Hampshire. ½.

A class of objects, closely allied to the preceding, but from their more elaborate character, supposed to have been invested with greater significance, are those perforated stones, which are either shaped to represent animals, or have representations of animals carved upon them. These carved and ornamented stones are here classed as "totems." Schoolcraft[151] has explained their origin and object, as being connected with "the system of names imposed on the men composing the Algonquin, Iroquois, Cherokee and other nations." With these, "a fox, a bear, a turtle, etc., is fixed upon as a badge or stem, from which the descendants may trace their parentage. To do this, the figure of an animal is employed as a heraldic sign or surname. This sign is called, in the Algonquin, town-mark or totem."

In this connection it may be well also to refer briefly to the tradi-

[150] Foster. Prehistoric Races of the U. S., p. 222, fig. 28. Chicago, 1873.

[151] History of Indian Tribes, vol. i, p. 52.

tional history of the Delaware Indians, as recorded by Heckewelder.[152] He mentions that, in accordance with their tradition, after they discovered the Delaware river, they explored "the Scheyichbi country, now named New Jersey." These migrating Indians finally settled on four rivers, "making the Delaware, to which they gave the name of Lenapè Wihittuck, the centre of their possessions." They here divided themselves into three tribes, two of them distinguished by the names of the turtle and the turkey; the latter settling nearest the sea.

FIG. 363.—New Jersey. ½.

The third tribe, the Minsi or wolves, settled in the mountainous region north and westward of Musconetcong creek.

Fig. 363 represents a highly polished, black hornstone pebble, oval in shape, flat upon the under side, and slightly convex upon the other, perforated at each end; and upon the upper side, as seen in the illustration, there is clearly and quite artistically engraved a turtle. This most interesting object was found near Titusville, Mercer Co., New Jersey, in the very neighborhood where the turtle tribe of the Delaware nation were settled. The inference is that this stone is not simply an ornament, as a bead, a bear's tooth or a perforated pebble,

---

[152] Heckewelder. Hist. Account of Indians: Vol. 1st, Transactions Amer. Phil. Soc., p. 33. Philadelphia.

but has that tribal or totemic significance which is mentioned by Schoolcraft as common to the Algonquin Indians.

This is not the only specimen of its kind that has been discovered in the valley of the Delaware. The turtle on the pipe (fig. 317), from Lewes, Delaware, may have had a totemic significance; and the heads of birds etched upon a stone-knife (fig. 43) and a bone implement (fig. 192), clearly show that animal representations were not uncommon, although probably none were of totemic significance, except fig. 363 and possibly the turtle pipe.

Objects made by carving the stone to represent an animal have frequently been met with in New England. Fig. 364 represents an interesting example found near Ipswich, Mass., and described by Mr. F. W. Putnam, as follows:—

"This stone was evidently carved with care for the purpose of being worn as an ornament, and was probably suspended from the neck. It is of a soft slate, easily cut with a sharp, hard stone. The markings left in various places by the carver, showing where his tool had slipped, indicate that no very delicate instrument had been used, while the several grooves, made to carry out the idea of the sculptor, indicate as plainly that the instrument by which they were made, had, what we should call, a rounded edge, like that of a dull hatchet, as the grooves were wider at the top than at the bottom, and the striæ show that they were made by a sort of sawing motion, or a rubbing of the instrument backwards and forwards. In fact, the carver's tool might have been almost any stone implement, from an arrowhead to a skin scraper, or any hard piece of roughly chipped stone.

Fig. 364 represents the stone of natural size, its total length being two and a half inches. It is of general uniform thickness, about one-fifth of an inch, except where the angles are slightly rounded off on the front of the head and on the abdominal outline, and the portion representing the forked tail, or caudal fin, which is rapidly and symmetrically thinned to its edges, as is the notched portion representing the dorsal fin.

The carving was evidently intended to represent a fish, with some peculiar ideas of the artist added and several important characters left out. The three longitudinal grooves in front represent the mouth and jaws, while the transverse groove at their termination gives a limit to the length of the jaw, and a very decided groove on the under side divides the under jaw into its right and left portions. The eyes are represented as slight depressions at the top of the head. The head is separated from the abdominal portion by a decided groove, and the caudal fin is well represented by the forked portion, from the centre of which the rounded termination of the whole projects. In this part there is an irregularly made hole of a size large enough to allow a strong cord to pass through for the purpose of suspension. The portion of the sculpture rising in the place of a dorsal fin is in several

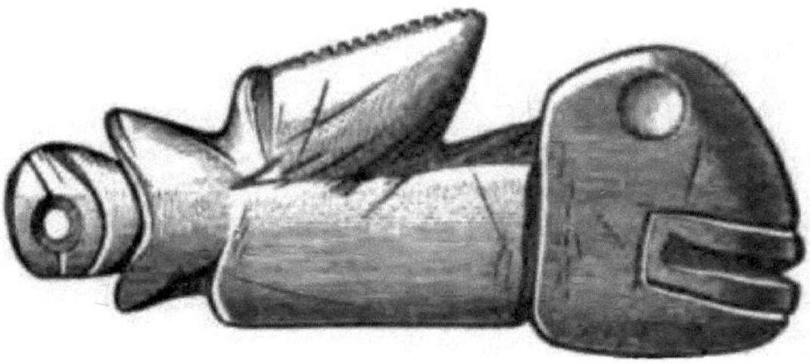

FIG. 364.—Massachusetts. ½.

ways a singular conception of the ancient carver. While holding the position of a dorsal fin, it points the wrong way, if we regard the portion, looking so much like a shark's tooth, as intended to represent the fin as a whole. It is very likely that the designer wished to show that the fin was not connected with the head and, as he was confined by the length of the piece of stone, after making the head so much out of proportion, he was forced to cut under the interior portion of the fin in order to express the fact. If we regard it in this light, the notches on the upper edge may be considered as indicating the fin rays; but the figure best shows the character of the sculpture, and persons interested can draw their own conclusions.

"The symmetry of the whole carving is well carried out, both sides being alike, with the exception that the raised portion at the posterior part of what I have called the dorsal fin is a little more marked on the left side than on the right, and the edge on the same side is surrounded by a faint, irregularly drawn line.

"The carving was, I think, unquestionably made by an Indian of the tribe once numerous in this vicinity; and, as it was almost beyond a doubt cut by a stone tool of some kind, it must be considered as quite an ancient work of art, probably worn as a 'medicine' and possibly indicated either the name of the wearer or that he was a noted fisherman."

Fig. 365 represents a carved stone, found at Seabrook, New Hampshire, which is supposed to "rudely represent a porpoise or, still better, a white whale or *Beluga*, as it had no protuberance representing the dorsal fin of the porpoise, and the Beluga is without the fin. The flippers or pectoral fins were represented by the protuberances on the sides, and the mouth was cut in and well indicated. The broad horizontal tail was decidedly cetacean in character, and the whole carving, though rudely done by picking the sienitic rock, from which it was made, with stone implements, was yet so characteristic as to indicate at once that a porpoise or Beluga was intended. A hole through the portion representing the tail shows that the object was suspended, but the stone is so large and heavy that it can hardly be classed as a personal ornament, though it is probably to be regarded as a totem. It measures ten inches in length by about two in depth at the pectoral fins, and is about two and a quarter inches wide across the pectorals as measured on the under side. This interesting specimen was found at Seabrook, N. H., and it is said that two other similarly worked stones have been found at the same place.

"The figures here given represent the 'object' in profile and from the under side."

Fig. 366 represents a typical specimen of a well-known class of perforated stones, generally called pendants, or gorgets. Whether any distinction really exists cannot now be determined; but the fact

---

[183] Putnam. Bulletin Essex Inst., vol. v. Salem, Mass.

that some of these stones have one perforation, whilst others have two or more, indicates a difference in the method of attaching

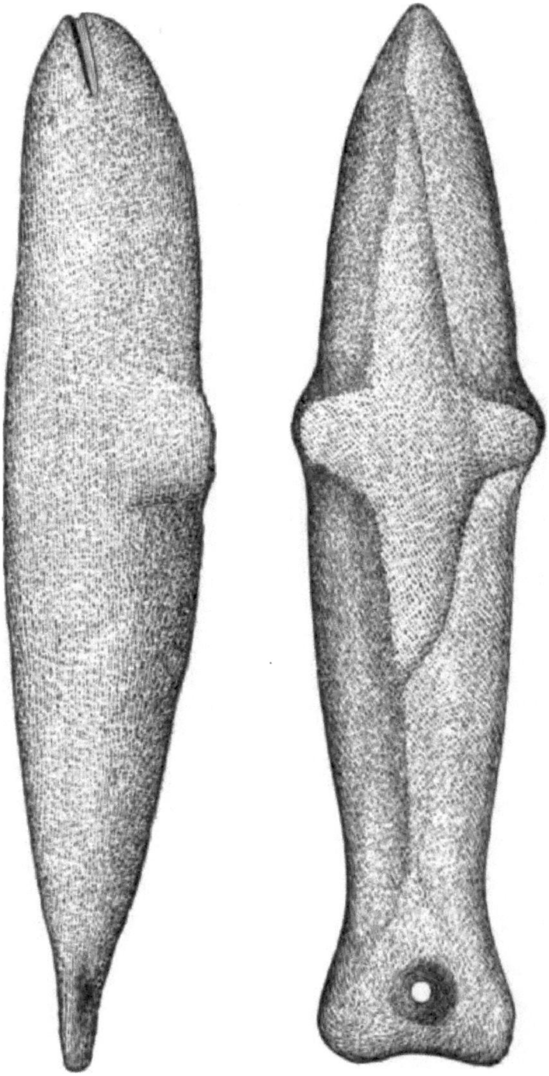

FIG. 365.—New Hampshire. ½.

them to the clothing. This difference may also be an indication that the two forms had different uses, the so-called gorget having been applied to some special purpose while the pendant proper was simply an ornament. Fig. 366 is made of serpentine, and is highly polished. In outline and finish, it is the counterpart of scores of similar articles gathered from every part of the Atlantic coast. Many, however, are much larger. In the Museum at Cambridge, Mass., are two examples of these large pendants, one of which (P. M. No. 602) measures nearly seven inches in length, by two and three-eighths inches in greatest width, with the perforation very near one end. The other is shorter and broader and has the perforation an inch from the end.

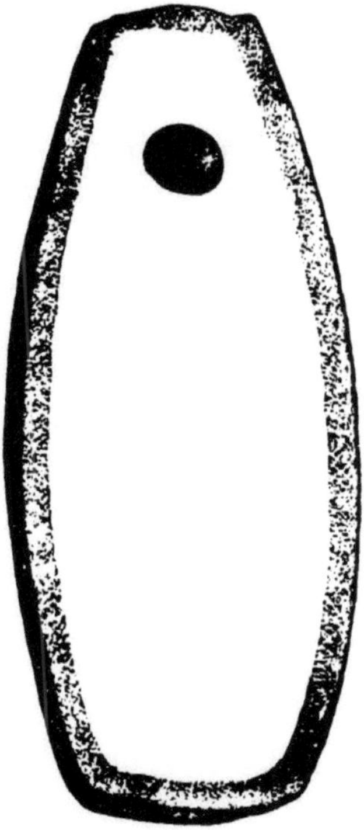

FIG. 366.—New Jersey. ½.

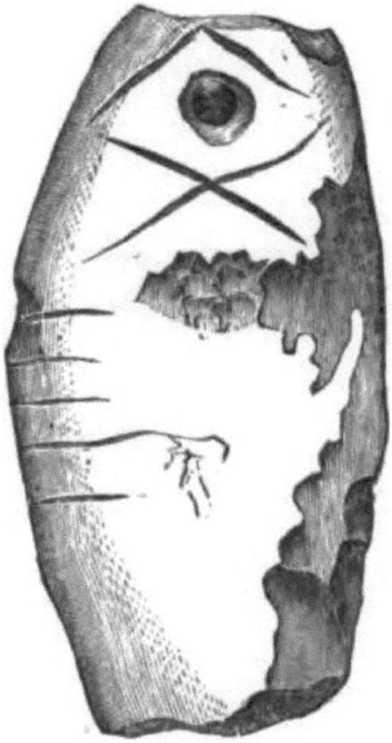

FIG. 367.—New Jersey. ½.

Among the contents of a series of ancient Indian graves, examined by Mr. Putnam,[154] were three of these pendants of large size. One is without a perforation, and another has a series of well defined notches on one end.

Fig. 367 represents a second example of a stone pendant, and

[154] Putnam. Bulletin Essex Inst., vol. iii, p. 123.

differs from the preceding in having a rude attempt at ornamentation. The pebble, from which this specimen is made, remains otherwise unaltered. The fracture on one side was probably caused by the plough. The lines are only upon one side, and are so far apart and so few in number, that they cannot *now* be considered as adding to its beauty. Indeed, so rude in many respects is this example, that it may have been intended for some other purpose,—possibly as a sinker for a fishing-line, to which use it had been applied before coming into the eager possession of a relic-hunter.

FIG. 368.—New Jersey. $\frac{1}{1}$.

Several pendants of the character and size of figs. 366 and 367, made of hematite and similarly perforated, have been found in Hunterdon Co., N. J. These presented no peculiarity whatever, other than in the character of the material, which was very seldom used by the New Jersey Indians, although it exists in the greatest abundance in the immediate vicinity of the place where these specimens were found. Ornaments and small celts of hematite seem to be principally found in Ohio, judging from the large proportion of objects from that state, made of this material, that are contained in our principal museums.

Fig. 368 represents an interesting specimen of the usual form, but made of a pebble and thicker than the usual pendants of slate. The remarkable feature of this example is the extremely rude representation of a human face cut upon one side. The stone is an accurate oval in outline, and near the smaller end a hole, for suspension, has been drilled. Unlike all other representations of the human face, this one is produced by a curious combination of straight lines. The eyes are shallow countersunk holes, enclosed in diamond-shaped figures, the nose is represented by straight lines, the surface of the stone on each side having been slightly ground down, so as to bring that feature more prominently into relief.

This specimen was taken from an Indian grave near Vincentown, Burlington Co., N. J., and, associated with it, were found two celts, a dozen arrowpoints, and a few glass beads. The presence of the last-mentioned articles shows that the burial was after the general introduction of articles of European manufacture among the native tribes; though it does not necessarily imply that the relic itself was of modern date. It is well known that many objects, the handiwork of their own ancestors, generations before, were held in veneration and preserved as relics by the recent Indians.

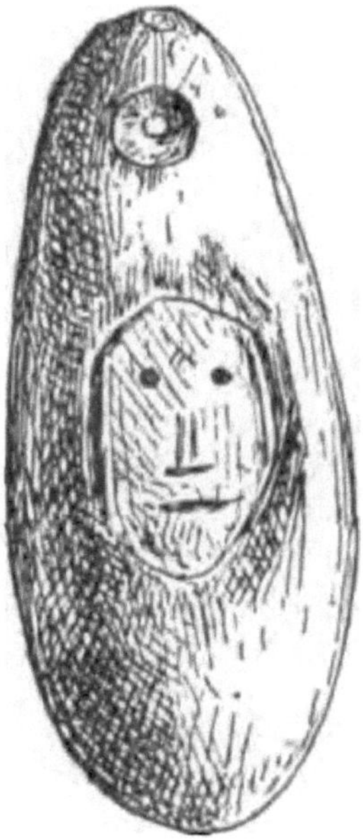

FIG. 369.—New Jersey. ½.

Fig. 369 represents a second example of a pendant, in this instance of much smaller size, with a still ruder representation of the human face. In this specimen, the features are delineated by a few lines, within an oval that marks the boundaries of the face. This face carving bears considerable resemblance to one found in Ohio, and figured in the Popular Science Monthly.[155] The workmanship in that case, however, is really artistic, and the stone on which it is carved is carefully and symmetrically shaped and polished. In fig. 369 we have simply a flat, elongated oval pebble, without any alteration of the surfaces other than the perforation and the rudely incised face.

In this, and the preceding specimens, we have examples of carving upon a very hard, unyielding hornstone, which may account for the

---

[155] Abbott. Popular Science Monthly, vol. vii, 1876. New York.

rude finish; as steatite objects, of a similar character, are much more artistically executed.

Pendants of this pattern are very rare, and none, as yet, have been found in New England. In the southern states their place seems to have been supplied in part by the shell-disks or totems, upon which are inscribed birds' heads, spiders, rattlesnakes, and in some instances, the human face. Examples of the latter, quite as rudely executed as the one figured above, have been taken from mounds in Virginia,[156] Tennessee and Arkansas.

Fig. 370 represents a small human face, carved in steatite, found many years ago, on the shores of Cayuga Lake, New York. This specimen is very interesting, as it is the only one of the series of illustrations given, wherein the Indian face is truthfully expressed or portrayed. While this specimen can scarcely, from its small size, be called a pendant, yet it is most appropriately considered here, in connection with the attempts at portraiture shown in the preceding examples. A specimen of a stone face, of somewhat smaller size, but of the same character of workmanship, and peculiar Indian cast of countenance, was lately found in a shell-heap, opposite Red Bank, Monmouth Co., New Jersey. The material is steatite of a dark greenish color. Associated with it, were fragments of pottery, and half of a "ceremonial object," which was profusely marked with the short, deep notches, so common to gorgets. This account is taken from a letter addressed to the Museum of Archæology, at Cambridge, Mass., by Chas. F. Woolley, Esq. (the gentleman who found the specimen), of Eatontown, Monmouth Co., New Jersey.

FIG. 370.—New York. $\frac{1}{1}$.

Representations of the human face upon stones are of common occurrence, not only among the relics of the modern Indians, but also of the western moundbuilders. In many cases, these carvings were

---

156 Carr. Tenth Annual Report of Peabody Museum, p. 87, fig. 3. Cambridge, 1877.

merely ornaments, though some of them appear to have been held in great reverence. In this connection, the following statement, transcribed from a Kansas paper, is of interest.

"About the 10th of July (1868), the Kiowas had a battle with the Utes, in which the chief, Heap-of-Bears, and seven other Kiowa braves were killed. Heap-of-Bears had on his person the medicine of the Kiowas, which was captured by the Utes, who still retain it. This medicine consists of an image about eighteen inches in length, carved to represent a human face and covered with the down and feathers of the eagle and other birds, and swathed in wrappers of different materials of value. Although I have been conversant with Indian habits and customs for a long time, I was surprised to find the value these people attach to this medicine. They begged and implored Col. Murphy to recover it for them, and promised to pay the Utes as many horses as they wanted, and also to make a permanent and lasting peace not only with the Utes, but also to refrain from further depredations on the Texas border, if this should be restored. Col. Murphy promised to endeavor to recover it, but I think his success in the matter will be doubtful, as the Utes also attach great importance to their capture, believing that while they retain it, the Kiowas will be powerless to do them harm."

The human face was not only represented in the several ways that have been described; but the Atlantic coast Indians appear to have occasionally attempted works of a more pretentious character, although their success in human portraiture cannot be commended.

Fig. 371 represents a stone mask found at Trenton, N. J.[157] Stone masks like this have been somewhat frequently found in the mound region of the Ohio and Mississippi valleys, but are not common eastward of these localities. The specimen here figured is probably the only one yet discovered in New Jersey, and thus far but few, if any specimens, have been found in New England. Fig. 371 is a hard sandstone pebble, and measures six inches in length by a fraction

---

157 Abbott. Nature, vol. xii, p. 49, figure not numbered. London, 1875.

over four inches in greatest breadth. It is concavo-convex, the concavity being shallow and artificial. The carving of the front or convex side is very rude, and certainly shows no evidence of the work of metallic tools. The eyes are simply conical, countersunk holes; just such depressions as the larger stone drills, so common among the surface relics of this neighborhood, would produce. The eyebrows or superciliary ridges are well defined, but are angular rather than curved, and in this respect the specimen resembles fig. 363. The nose is very flat and angular; the mouth merely a shallow groove. The ears are broken, but appear to have been formed with more care than any other of the features. The chin is slightly projecting. Until other specimens of like character shall have been found along the Atlantic seaboard, it is scarcely safe to conjecture even as to the significance of this specimen. Its resemblance to those found in the region occupied by the moundbuilders, and also to the Mexican masks, will at once be remarked; but, that it has any bearing on the relationship of the Red Indian to the moundbuilders, or the latter to ancient Mexicans, is very doubtful.

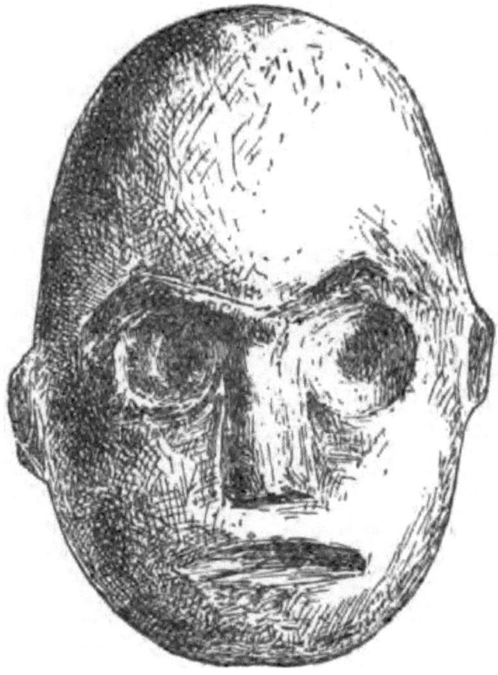

FIG. 371.—New Jersey. ½.

In the Fast collection of Alaskan antiquities, in the museum at Cambridge, Mass., there are two specimens of carved stone ornaments, one of which bears a marked resemblance to fig. 368, although finished much more artistically. The other, while of a different character, being a flat ring with eight human faces carved upon it, is also of interest, as the faces are quite similar to those upon the New Jersey specimens.

The Alaskan example, with a single face carved upon it, is a flat, oval pebble, two and one-half inches in width and three inches long. This ornament or charm has a hole through it, at the upper end, similar to the perforations in figs. 368 and 369. The surface upon which the face is cut is convex, but with a wide flattened margin ornamented with closely set parallel lines. The back of the stone is flat and without markings of any kind.

The similarity of these Alaskan specimens of stone carvings to those of the eastern coast Indians is not the only instance of this kind, of which there is evidence. Already, reference has been made to the identity of the semilunar slate knives of Alaska and those found so abundantly in the New England and middle states.

The sameness of the productions of distant and distinct people must be very cautiously taken as an indication of their former contact, or remote relationship, especially when there have always existed geographical barriers which were practically impassable. In the present instance, however, it is not strange that a marked similarity should be traced between the implements and ornaments of the Indians and the handiwork of the Arctic races. These two people have certainly been frequently in contact, and the belief that the Indian displaced the Eskimo over a considerable territory, far south of the present range of the latter, is founded on much strong evidence.

Fig. 372 represents a most interesting carving in stone that differs materially from all the others mentioned in this chapter. The others, with perhaps one exception, are representations of familiar objects, or are merely smoothed pebbles with decorative lines; but in this instance, by the use of effective lines, there is apparently a departure from the representation of natural forms towards conventionalism.

If this be the correct understanding of this carving, it is probable that it may be regarded as a conventionalized human face; and as such is of much interest from the rarity of such work in the relics o. the tribes of the eastern seaboard.

The material is steatite, now black from age and exposure. The upper surface and back of this object are perfectly plain and polished. The deep upper groove on the front surface is suggestive of a means whereby the stone could have been suspended or attached, but it is as likely that such objects were carried in pouches as attached to their persons. As will be seen by reference to the illustration, the upper portion is quite suggestive of the eyes and nose of a human face; but the lower portion can only be construed as representing cheeks, a mouth and chin, by closely regarding these lines in connection with those of the upper part of the carving.

FIG. 372.—New Jersey. ½.

For this interesting specimen, the author is indebted to the kindness of Master Herbert Coleman, of Recklesstown, Burlington Co., New Jersey, who found it on the bank of a small creek near that village.

Fig. 373 represents a very interesting example of a carving in stone, representing the head of a fox. In no one particular is the carving correct, yet the general resemblance is unmistakable, and no doubt can exist, but that the ancient sculptor intended to portray a fox's head. The small, knob-like protuberance at the neck, in connection with the deep groove that separates it from the head, shows that this little carving was suspended; and, either as a single ornament attached to the dress, or as the pendant and prominent feature of a string of beads, was used for personal adornment.

FIG. 373.—Penna. ½.

Animal carvings, like fig. 373, have been very rarely met with in the valley of the Delaware river, or anywhere in New England. In central New York, and along the valley of the Susquehanna, they are more frequently found.

This specimen of a carving in stone was found near Columbia, Lancaster Co., Pa., by G. W. Caley, Esq., and by him presented to the late Prof. S. S. Haldeman, from whom the author received it for purposes of description and figuring in this work.

A second example of these charms, pendants or trinkets, as the case may be, fig. 374, which was also received from my lamented friend, is a carved arrowpoint of steatite, which is quite elaborately ornamented. On one side of the specimen there is drawn, by incised lines, a spirited representation of our well-known "snapper" (*Chelydra serpentina*), and on the other side, there are closely arranged series of parallel, straight and oblique lines, which give a net-like appearance to this side. This is the only instance known to the author of well-known forms of weapons being reproduced, in miniature, as ornaments. The representation of the turtle, however, is of common occurrence.

FIG. 374.—Penna. ½.

In chapter V, attention has been called to the occurrence of birds' heads engraved upon a slate knife, and to the probability that they were intended to represent the heads of turkeys. In this instance we have the other "totemic" animal represented, of the three which were chosen by the Lenni Lenapè as the clan-marks of their nation: the wolf, the turkey and the turtle. Interesting as the specimen is, irrespective of the totemic significance of the animal engraved upon it, this historical knowledge of the political divisions of the Delaware Indians certainly adds to that interest very materially. Unlike many of the objects described in the present chapter, this specimen may not be simply an ornament; but what significance it had can now only be conjectured.

This unique specimen, fig. 374, was found by Mr. G. W. Caley, at Washington Boro', Lancaster Co., Pa., and presented to the late Prof. S. S. Haldeman.

Under the name "trinket" is included a miscellaneous series of small objects, which from their size, general appearance and material of which many are made, are believed to have been simply small ornaments worn either singly, or in numbers, as a string of beads; or carried about the person, as "charms." Almost all such objects are either perforated or grooved, and so were evidently attached to the dress or person by a cord. Many of these small ornaments were probably also attached to scalps; and, thus bedecked with small perforated stones, human finger bones and shells, these scalps were borne aloft in many of the well-known Indian dances. No object seems to have been too crude for the fancy of an Indian, and it is not strange that so many fragments of commonplace objects even are found, which show that they had been used in some such manner as described. These objects represent, in great part, the jewelry of the present day, and while not possessing the same intrinsic value, or any of the beauty of modern bijouterie, they certainly are no less meaningless.

Fig. 375.—New Jersey.

No classification of such objects can, of course, be made, and but little can be said with reference to the geographical distribution of the various forms. In some localities, perforated teeth and shell ornaments are more frequently found than in others, where stone trinkets are abundant; but in no locality does there appear to be any common form of trinket strictly peculiar to that neighborhood.

Fig. 375 represents a small, but symmetrical trinket, which has a marked peculiarity in the ornamentation upon one side. The specimen itself is short, being but one inch and three-quarters in length. The upper or perforated end is but three-eighths of an inch in width, and from this upper margin the specimen increases uniformly in width until near the bottom, when it rounds off in an almost regular curve. The hole appears to have been drilled wholly from the plain, or under side, being wider there at the surface than upon the opposite side, which has a slightly worn edge occasioned by the rubbing of the cord that suspended it.

The ornamentation of fig. 375 differs from any other specimen collected by the author. The surface of the stone has been smoothly worn off, leaving, a short distance below the perforation, a quadrangular figure that may be called a hollow square, there being a cleanly cut depression in the centre of the projecting "square," the width of which is just double the depth. Below this figure commences a second, which can be compared to an inverted pick-axe, with the iron arms straighter than usual. It is simply a "raised" ridge, the surrounding surface being cut away to leave it in bold relief. It is not exactly in the centre of the specimen, but near it, the upper ridge or handle of the pick being slightly inclined to one side. Below this, the specimen is smoothly polished and somewhat sloped to the end.

FIG. 376.—Massachusetts. ½.

Fig. 376 represents an ornament made of a metacarpal bone of a deer. One end is very much worked by being ground down and perforated, but the sides of the bone and the lower end (*i. e.*, lower end as seen in the illustration) are not altered in any way. Small bone trinkets or ornaments of this character are not uncommon, wherever the soil has been of such a character as to prevent the decomposition of the bone. The marked absence of bone implements generally, in some districts, is doubtless due to this cause. Fig. 376 was found in an Indian grave in Lagrange street, Salem, Mass. With it were several bone spoons, made from the jaw of a porpoise.

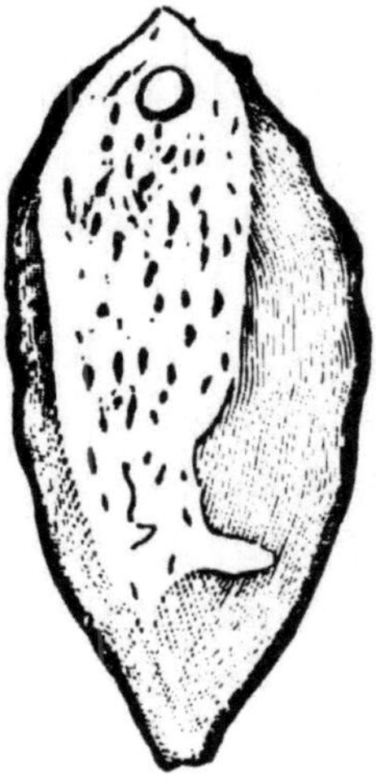

FIG. 377.—New Jersey. ½.

Fig. 377 represents an interesting specimen of this class of relics. It is a piece of black, well-worn stone, but with no polish; it is thin, but irregularly so, and has a greasy feeling which is most deceptive. One can almost smell the grease, now stale, with which the object seems to be saturated. The specimen is leaf-shaped, more pointed at one end than the other, and, when viewed horizontally, bears some resemblance to a fish.

So generally are these stone trinkets, such as fig. 377, found singly both upon the surface and in graves, that it seems probable that they were worn singly, to a great extent, as has been suggested, and not as one of many similar objects, as a string of beads, unless the beads have decomposed. Their occurrence in New Jersey, in ancient graves, is very unusual; and they are not mentioned as among the contents of ancient graves found in New England. These perforated stones certainly can be nothing else than ornaments, and as they are so abundant on many of our fields, it seems the more surprising that none should be found with skeletons, especially as glass beads and ornaments of European origin so generally occur in the more recent burials. It is possible, however, that all such ornaments and trinkets of every description are really from graves which have been obliterated by the destruction of the forest once covering these fields, and their long subsequent cultivation. If the Indians placed their dead in very shallow graves, it is possible that all the relics of this people, now found upon the surface in some localities, are really grave contents; but it is not certain that they practised this method of disposing of the bodies of their dead. They certainly had other mortuary customs also, as burying in tumuli, and in placing the body in a sitting posture in holes dug sufficiently deep to be beyond the reach of the ploughshare. The general levelling of our fields, however, through the action of winds and rain may have brought these, since the cultivation of the soil has been in progress, quite near the surface. It requires but a very short time, geologically considered, for a tract of country, which, while covered with trees, was quite rolling, to become comparatively level, when once denuded of its forest growth and put under cultivation.

Fig. 378 represents a more marked example of these thin, black hornstone pebbles, which has been utilized as an ornament. Whether the natural cordate outline of this pebble caused it to be chosen, or not, is doubtful. The fact that it is a smooth, jet-black, and thin pebble, was probably the sole cause of its being perforated, and made thereby an ornament. While this stone to us is a conventional heart in shape, it certainly had no meaning of this kind to the Indian who made it. "The two-lobed form is but a conventional device of civilized man to represent the human heart, and it is not at all probable that the North American Indian employed such a figure before he came into contact with the Europeans, especially as he does not use it in his paintings and etchings at the present time, but copies directly from nature."[158]

Occasionally this form has been found, of wholly artificial outline, which shows that it is a "type" of ornamental stones, the origin of which may have been the natural pebbles of this shape, which were chosen simply for their attractive appearance.

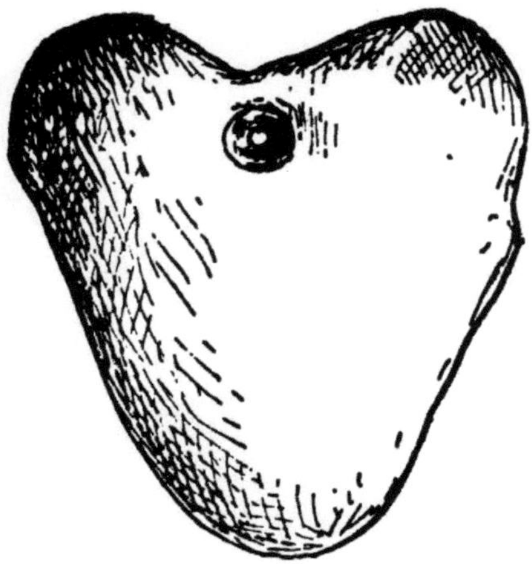

FIG. 378.—New Jersey. ½.

Of a later date are heart-shaped pieces of sheet-brass which have been found in considerable numbers in Indian graves, in the northern parts of New Jersey and in the New England states. As the material was derived from the Europeans, it is probable that these brass "hearts," like brass arrowheads, were made by the whites and sold to the Indians, and not generally designed by the Indians themselves.

Fig. 379 represents a split, water-worn jasper pebble, of somewhat irregular shape, with an extensive perforation through it. The hole upon the under or split side is about one-half the diameter that it is upon the upper. The under side, however, has an equally weather-

[158] Barber. American Naturalist, vol. xi, p. 45. Boston, Mass., 1877.

worn polish with the upper, indicating that the perforation was made subsequently to the splitting of the pebble, or that many years have elapsed since the "split" pebble was drilled, the peculiar gloss of the fractured surface indicating great age. The somewhat irregular outline of the perforation upon the "split" side of the pebble favors the belief that the fracture occurred *after* the drilling. This specimen is interesting from its resemblance to an African example figured by Sir John Lubbock.[159] This African drilled stone is square instead of pentagonal, and the drilling is of much less diameter at the junction of the two depressions which, together, make the perforation.

It may, perhaps, be questioned if perforated pebbles, as large as

FIG. 379.—New Jersey. ½.

fig. 379, were habitually used as ornaments or charms. May not such a stone as this have been used simply as a sinker for a fishing line? For this purpose it is certainly well adapted; and, on the other hand, possesses no one attractive feature to suggest its use as an ornament. This is judging such a perforated pebble, however, from our own standpoint; and every one will surely be misled in such matters, if he base his opinion of the use of any object, or its value in the mind of a savage, upon his own experience. The similar specimen from Africa, to which reference has been made, was used as an ear-ornament, and the weight and size of fig. 379 are not ob-

---

159 Jour. Anthrop. Inst. of Great Britain, vol. i, pl. 1.

jections to the supposition that this specimen may have been worn in like manner. Earrings, so called, of greater weight and bulk, are still worn by many of the western tribes.

Numerous specimens, of larger, naturally perforated pebbles, have been found in New Jersey, especially on village sites, and it is highly probable that all these had been carefully gathered and mostly worn as ornaments. When much larger than fig. 379, their use as weights for nets and lines is probable; as large, artificially perforated pebbles were so used, as described in chapter XVIII.

Fig. 380 represents a small perforated disk of steatite, which varies much from the preceding, and notably in being wholly of an artificial shape. Steatite rings, or disks of this size and smaller, are quite abundantly met with in many parts of Chester Co., Pa., from which locality this specimen was derived. That such specimens were probably used only as ornaments is indicated by the size and the lightness of the material of which it is made. It is possible, also, that these small, perforated disks were used as gaming-stones. Whether the larger specimens were used as spindle-whorls, mace-heads or weights for digging-sticks, uses ascribed to them as found in other countries, cannot be determined; but the relative scarcity of these objects is, of itself, an indication that their use, except as ornaments, was with the Atlantic coast tribes wholly exceptional. Steatite rings, of a quite different character, being more like modern napkin rings, are not uncommon in the region of the mounds, not only in Ohio, but southward and westward of that state. A single fragment of such a ring has been found in New Jersey; a second whole specimen was found in a grave in Lancaster Co., Pa. From the position in which it was lying when found, it appeared to have been attached to the hair

FIG. 380.—Pennsylvania. $\frac{1}{1}$.

of the person buried; a third specimen, nearly entire, was ploughed up in a field near Bushkill, Pike Co., Pa., in the spring of 1879.

Fig. 381 represents a very handsomely designed steatite bead, of which very many have been found in New Jersey and Pennsylvania, and in fewer numbers in New England. The illustration so clearly conveys the character of this pattern of bead, that a detailed description is not required. This specimen was found with a small string of shell beads in an Indian grave in Mercer Co., New Jersey. From this it might be inferred that objects of this pattern were only used singly in association with small beads of other kinds; but such was not always the case; as several strings of beads, all of the size, pattern and material of fig. 381, are reported to have been taken from ancient Indian graves in New Jersey.

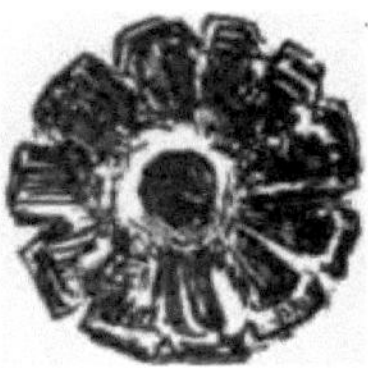

FIG. 381.—New Jersey.

With the native tribes of the Atlantic seaboard, as with the Indian everywhere, beads were the commonest form of all personal ornaments. They were made of stone, bone, shell and baked clay, and present a greater range of patterns than do even the arrowheads. Those of stone and baked clay were probably never as abundant as the shell and bone beads, and as the date of earliest European contact is that of the abandonment of their manufacture, they are not now very frequently met with.

A simple form of a bead is a small pebble that has a natural perforation. Many such occur in our tertiary gravels. Occasionally, a series of these have been found in a grave. May not such pebbles have been the starting-point, from which were developed the manufacture and general use of elaborate beads of all patterns?

Figs. 382 to 386, inclusive, represent a pretty series of small disks of sandstone and other more compact minerals, which, by a single perforation and, in some, a notching of the edges, have been converted into veritable trinkets. These, it can readily be seen, are by fact of being thin disks not available as beads; but as additions to a string of beads they might be used, and also as small pendants with which their pipe-stems were decorated. Indeed, there is scarcely a limit to the methods of utilizing these objects for decorative purposes.

In many localities these small perforated stone disks are very abundant, and they have been found by the score where beads were rarely, if ever, met with. That they really took the place of beads is, however, exceedingly improbable. As a series these small perforated stones, both with worked and unworked edges, bear a strong resemblance to the beautiful shell ornaments from southern California.[160]

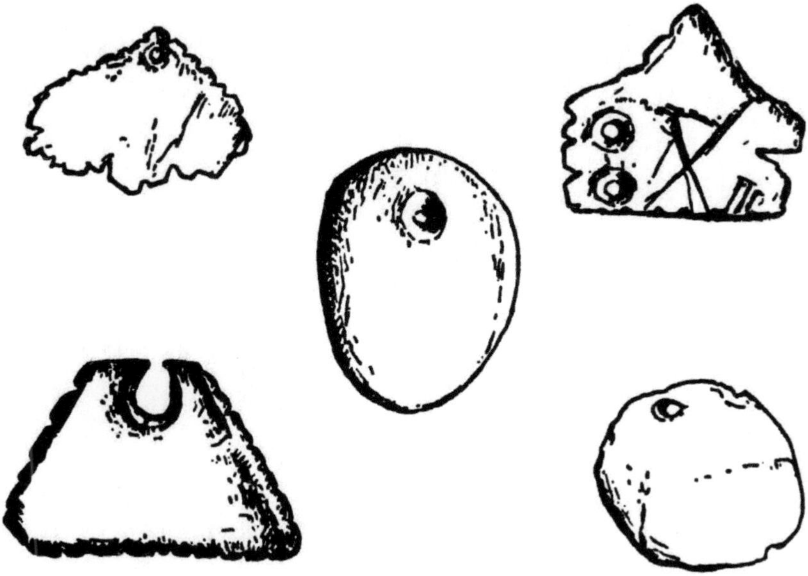

Figs. 382 to 386.—New Jersey. ½.

Fig. 387 represents a common object in all series of Indian ornaments found in New Jersey. The fossil sharks' teeth, that occur so abundantly in the cretaceous formations of New Jersey, did not escape the quick eye of the Indian, and they were used, when perforated, as in fig. 387, as ornaments, and very probably to some extent as arrowheads. These are probably the "fishes' teeth, fastened in with

[160] U. S. Geog. Survey West of 100th Meridian, vol. vii, Archæology, pl. xii.

glue," to which Holm refers.[161] They do not appear to have been in common use, in any manner, much beyond the valley of the Delaware river.

Fig. 388 represents a canine tooth of a bear, with a perforation near the base. This is one of the most common objects of all the Indian trinkets, and is to-day as much in use as even in ancient times. Frequently a dozen have been found in a single grave.

FIG. 387.—New Jersey. ½.

While canine teeth of other animals were also used as ornaments, as of the wolf and wild-cat, those of the bear were generally chosen, or, at least, largely outnumber the teeth of other animals that have been thus utilized.

Fig. 389 represents a curiously-shaped clay ornament, such as are of common occurrence in some localities, but are rarely met with in others. This specimen is made of nearly pure clay, and has been burnt to nearly a black color. There are two holes or perforations in each end of the globular portion of the object, but these do not extend through. A cord therefore was not passed through it.

FIG. 388.—New Jersey. ½.

Objects of this character, and others that are rude representations of animals' and birds' heads, are quite rarely met with in southern New Jersey, but are common in the neighborhood of the Delaware Water Gap and throughout the eastern portions of New York state. In New England they are less frequently found. Mr. Frey[162] figures two examples of these clay ornaments,

---

[161] Holm, *l. c.*, p. 129.

[162] Frey. Amer. Nat., vol. 12, p. 782, figs. 10 and 11.

one representing an owl's head, the other that of a fox. These, as he remarks, "are rude in style * * * but by no means inferior to similar terra-cottas from Mycenæ and Cyprus."

Some of the Atlantic coast pottery had the margins of the vessels ornamented with objects, either meaningless like fig. 389, or representing animals' heads, as is so commonly the case with the mound-builders' pottery of the southwestern states. Objects of this kind, therefore, when found in a fragmentary condition, and showing by the fracture that they had been detached from a vessel, can be readily distinguished from those that were made as separate ornaments.

FIG. 389.—New Jersey. ½.

FIG. 390.—New Jersey. ½.

Figs. 390 and 391, which conclude the series of objects described in this chapter, may be thought to be misplaced, when considered as ornaments. That these small, cylindrical pebbles, with the groove near one end, are possible sinkers for fishing-lines, is a very natural view to take of them probably, but it is one that cannot be shown conclusively to have been the case. While they are in shape, the minimum size of the well-known "plummet" of New England, they are so small and made of so light a stone, that they would be of little use as sinkers; and the fact that one, identical with the larger of these two, was found in a grave with a series of shell beads, and *three brass buttons*, made it evident that *it* had *finally* been used as an ornament, if it had had other uses in the lifetime of him in whose grave it was found.

FIG. 391. New Jersey. ½.

The custom of wearing and carrying about the person small trinkets, such as have been described in this chapter, is by no means confined to the Indians of this continent. Mr. Ludwig Kumlein, in his valuable notes on the Eskimo of Cumberland Sound (Bulletin of the National Museum, No. 15, p. 45, Washington, D. C., 1879), remarks of this people: "Among their many superstitious notions, the wearing of charms about the person is one of the most curious. These are called *amgoouk*, or *amusit*, and may be nothing but pieces of bone or wood, birds' bills or claws, or an animal's teeth or skin. To these charms they attribute supernatural powers, and believe them to be able to keep the wearer from sickness or misfortune. It is a common custom for the wife to throw a piece of seals' blubber on her husband's kyack when he is about to go hunting; this will give him success. Little strips of deer-skin are hung about the person in different places to insure success in some undertaking or to ward off some misfortune, real or imaginary. We discovered one of these charms, which seemed to possess unusual interest. It was worn by a little girl about eight years old. She had a small envelope of seal-skin that was worn on the back of her inside jacket. We succeeded in bribing her grandmother to show us the contents of the envelope, which proved to be two small stones, the one a bluish flint, the other *apparently* meteoric iron. The tradition connected with these stones, the grandmother said, is that a very long time ago an Eskimo, from whom she was a lineal descendant, had discovered the iron, and had picked up a stone to break a piece off and take home with him; but when he struck the iron fire flew from it, and he soon learned how to make use of this accidental discovery, and became a great man among the people. At this point we lost the thread of the old woman's narrative, and all we could further learn was that these two small pieces had been preserved in the family for successive generations, and were inherited by her from her mother, and that she had now given them to her grandchild, the child's mother being dead. The child will in turn give it to her children. She thought this charm of inestimable value, and could not be induced to part with it, for, she said, 'no one has yet died while wearing this charm.'

"Another charm of great value to the mother who has a young babe is the canine tooth of the polar bear. This is used as a kind of clasp to a seal-skin string, which passes around the body and keeps the breasts up. Her milk-supply cannot fail while she wears this."

CHAPTER XXVIII.

COPPER IMPLEMENTS.

If we are to judge of the extent to which copper was used by the native populations of the northern Atlantic seaboard of this country, by the number of objects made of it, which have been discovered, it is evident that we must look upon the use of this mineral as an exceptional occurrence, interesting in itself but of no ethnological significance. Indeed, the character of the few copper implements found, judging from their size and shape, although apparently indicating that the value of this material for certain useful purposes had been recognized, does not really establish the fact, inasmuch as they certainly are not as serviceable as their counterparts of polished stone. Native copper, worked merely by hammering, as was done by these people, is not sufficiently hard to retain a cutting edge. To this end, it must be converted into bronze. As pure copper, it is not so valuable for cutting purposes, as newly chipped or even polished stone.

It is quite probable that the copper "celts," made by the Indians of the Atlantic coast, were never designed for use as weapons or implements, but were intended for display upon special occasions; as for instance in their various dances, when much ceremony was observed, and various objects were displayed, that at other times remained hidden in the custody of their fortunate owners, or of the appointed keepers, if tribal property.

In the description of the white-deer dance of the Hupâ Indians of California, to which reference has already been made on page 307, Mr. Powers remarks that there are many articles paraded and worn in their various ceremonial dances, that are held in great esteem, as "rendering their possessors illustrious in the eyes of all men." Among them "is the flake or knife of obsidian or jasper." In this manner the copper celts, found along our Atlantic coast, which when new were bright, lustrous and attractive looking, were possibly used; and hence they might be classed with the ceremonial objects described in a preceding chapter.

The several forms of small copper implements, such as the awls, needles and spoon-shaped objects, met with in some western localities, and particularly in Wisconsin, have not been found on the Atlantic coast; and, indeed, unless the implements of this material have been so far destroyed by chemical action (through exposure to the atmosphere and soils unfitted for their preservation), it is strange that there should be so very few examples within the limits of New Jersey, considering the amount of material accessible to the native tribes.

The late Prof. Lewis C. Beck,[163] in an able account of the occurrence of copper in New Jersey, remarks that "small pieces of this metal have been found on the surface of the ground in various parts of New Jersey. In the vicinity of Somerville, specimens weighing from five to ten pounds, have been obtained. The largest mass which has, to my knowledge, been found in New Jersey, is now in possession of James C. Vandyke, Esq., of New Brunswick. Its weight is seventy-eight pounds; but a large piece has been detached, and it is said to have weighed when first obtained, one hundred and twenty-eight pounds. It was ploughed up by a farmer near Somerville. On examining this specimen, pure metallic copper is visible in various parts; but with it is mixed the lead-gray oxide, and it is generally incrusted with the green carbonate of copper." Prof. G. H. Cook[164] also reports copper as occurring in the city of New Brunswick. He remarks that "flakes of metallic copper, from one-sixteenth to one-eighth of an inch in thickness, and one or two feet across, have been found in cutting the street east of the college, and also in digging a cellar in Somerset street."

---

[163] Beck. Amer. Jour. of Science; vol. xxxvi, p. 107. New Haven, Conn.

[164] Cook. Geology of New Jersey, p. 678. Newark, 1868.

Kalm mentions (Travels in N. A., vol. i, p. 300, 2nd ed., 1772) that "the *Indians*, before the arrival of the Europeans, had no notion of the use of iron, though that metal was abundant in their country. However, they knew in some measure how to make use of copper. Some Dutchmen who lived here (Philadelphia) still preserved the old account among them, that their ancestors at their first settling in *New York* had met with many *Indians*, who had tobacco pipes of copper, and who made them understand by signs, that they got them in the neighborhood; afterwards the fine copper mine was discovered upon the second river between Elizabethtown and New York (*i. e.*, in New Jersey). On digging in this mine, the people met with holes worked in the mountain, out of which some copper had been taken, and they found even some tools which the *Indians* probably made use of when they endeavored to get the metal for their pipes. Such holes in the mountains have likewise been found in some parts of *Pennsylvania*, viz.: below *New Castle* towards the seaside, and always some marks of copper are along with them. Some people have conjectured that the Spaniards * * * * made these holes in the mountains: but * * * * it is * * * * almost undoubted that the *Indians* dug these holes."

It would seem from the above that the Indians had access to a great deal more copper than they appear ever to have made use of, limited, as it was, in comparison to the supply obtainable in the Lake Superior region. Indeed, it is not improbable that all the copper articles, found along the Atlantic coast, were brought from western localities; and that the metal that was at hand in New Jersey was not recognized, or, at least, not utilized. Covered in part by earth, discolored by oxidation, and lying among rock of many descriptions in a densely wooded country, it might well have escaped even the notice of the keen-eyed redskin.

Whether the copper of New Jersey, or that from other localities, was utilized, it is certain that there was enough in use, when the Europeans first visited these shores, to attract their attention. Robert

Juet,[165] who served under Hendrick Hudson as mate in the Half-Moon, relates in his journal that the Indians "had red copper tobacco-pipes, and other things of copper they did wear about their necks."

The use of this mineral, it would appear from the account given by Brereton, was much more common among the Indians of Massachusetts, than among these tribes occupying the territory between the Hudson and the Susquehanna rivers. Brereton's statement is, as given in his Brief and True Relation of the Discovery of the Northern Part of Virginia; London, 1602, that "they have great store of copper, some very red and some of a pale color: none of them but have chains, ear-rings or collars of this metal: they head some of their arrows here with * * * * * * broad arrowheads, very workmanly made. Their chains are many hollow pieces cemented together, each piece of the bigness of one of our reeds, a finger in length, ten or twelve of them together on a string, which they wear about their necks: their collars they wear about their bodies like bandeliers a handful broad, all hollow pieces, like the others very fine and evenly set together. Besides these, they have large drinking cups made like skulls, and over them thin plates of copper, made like our boar spear blades, all which they so little esteem, as they offered their fairest collars or chains for a knife or such like trifle; but we seemed little to regard it, yet I was desirous to understand where they had such store of this metal, and made signs to one of them (with whom I was very familiar) who, taking a piece of copper in his hand, made a hole with his finger in the ground, and withal pointed to the main from whence they came."

Celts made of copper have been occasionally found in Maine, Massachusetts, New York, Pennsylvania and New Jersey. Many unrecorded specimens, of course, have been gathered, and are lost to science, but their number is probably not sufficient to affect the statement, as to their comparative rarity in the territory mentioned.

---

[165] See Rau in Smithson. Annual Report for 1872 for an able essay on Copper, from which the above reference is derived.

Mr. A. F. Berlin[166] has described a specimen found near Millbach, Lebanon Co., Pa., which is much smaller than usual, measuring but two and a quarter inches in length.

Mr. L. W. Brodhead[167] has recorded the finding of "a copper axe, made however from the raw material (*i. e.*, hammered into shape) and ground down to the required size and form."

Mr. Elias Lewis, jr.,[168] of the Long Island Historical Society, has kindly sent me the following information respecting copper implements, in the cabinets of that institution. He writes that among various other implements "are two copper axes, one very rude, the other well formed; both obtained from one spot, with a polished axe of hard jasper; surrounded by twenty large flint arrowheads setting upright. They were two feet below the surface."

Mr. F. W. Putnam[169] has recorded copper celts as found at Wakefield, Mass.

Fig. 392 represents the more common form of the copper celts found along the northern Atlantic seaboard. This specimen measures three and one-fourth inches in length, and has a cutting edge of two inches in extent. The upper and lower margins are almost twice as thick as the blade. The marks of the hammer, by which this celt has been shaped, are plainly seen over most of the surface except at the edge and on the adjoining portion of the blade from which all hammer-marks have been removed by subsequent grinding. It was found near Damariscotta, Maine, and is preserved in the museum of Archæology, at Cambridge, Mass.

Mr. Squier mentions, in his Aboriginal Monuments of New York (Smithsonian Contributions to Knowledge, vol. ii, p. 78), the discovery of a copper celt, similar to figure 392, "ploughed up somewhere in the vicinity of Auburn, Cayuga county." He further says,

---

166 Berlin. American Antiquarian, vol. ii, p. 154. Chicago, 1879.

167 Brodhead. Delaware Water Gap, p. 89. Printed for author. Philadelphia, 1870.

168 Elias Lewis, jr. Official Correspondence of Long Island Historical Society. Brooklyn, 1877.

169 Putnam. Bulletin of Essex Institute, vol. i, p. 90. Salem, Mass., 1869.

"it appears to be pure copper," and does not express the belief that it was "cast" but that it simply had that appearance, due to "the granulations of the surface;" unless, indeed, it was made from molten copper, either by the Europeans, or by the Indians after they had acquired the art of moulding copper, from the white settlers. It is now well ascertained, however, that the granular surfaces of many of the copper objects found are due to the oxidation of these surfaces. Still, it should be here mentioned that Roger Williams (Key to the Indian Language, p. 55) says, of the Indians of Rhode Island, that

FIG. 392.—Maine. 1/1.

"they have an Excellent Art to cast our Pewter and Brasse into very neate and artificiall Pipes."

Through the kindness of Rev. W. M. Beauchamp of Baldwinsville, New York, I have been able to gather some important facts with reference to the occurrence of native copper implements, to all appearance the handiwork of the Indians, and am indebted to him for the following illustrations of characteristic forms.

Fig. 393 represents a "copper gouge belonging to Albert Hopkins of the town of Phœnix, on the Oswego river, New York. It was found in Oswego Co. of that state." This specimen is "convex on the lower side, nearly flat upon the upper, with long ridges. The cutting edge is hammered into a hollow on the upper side, and is round (convex) on the lower." This specimen has recently been somewhat mutilated.

Mr. Beauchamp has kindly informed me of other specimens of celts and gouges found in the same neighborhood, one of which is remarkable for its size. This specimen, of which I have a beautifully executed drawing, made by Mr. Beauchamp, is described by him as a "large copper implement found by Mr. J. Schultz, in May, 1880. One side is nearly flat, very slightly hollowing; the other ridged, as in like implements, and a little hollowing on each side of the ridge. There are the usual flattened rough lines, which I now see are probably part of the original surface, the rest being corroded and granulated. It has a dull, chisel edge. Its weight is two pounds, fourteen ounces." This specimen measures one and three-eighths inches in width, at the upper end; one and three-fourths inches wide at the cutting edge, and eleven and one-half inches in length.

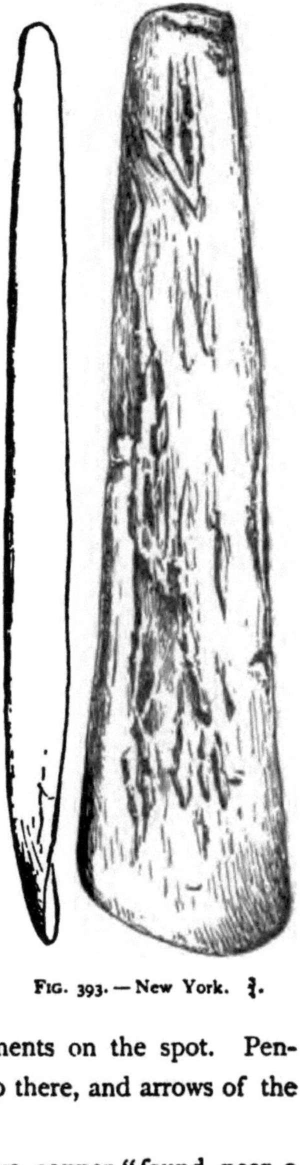

FIG. 393. — New York. ⅓.

In reply to enquiries concerning the evidence still existing of the use of copper by the Indians formerly occupying western New York, Mr. Beauchamp further says "shreds of sheet copper are common at Indian Hill, in Pompey, having apparently been used to make ornaments on the spot. Pendants of thin copper may yet be picked up there, and arrows of the same are found."

Fig. 394 represents a spearpoint of native copper "found near a stockade site on Seneca river. The site is prehistoric, and I found a rolled copper bead there. This spear is in the collection of Mr. Otis M. Bigelow, of Baldwinsville, N. Y."

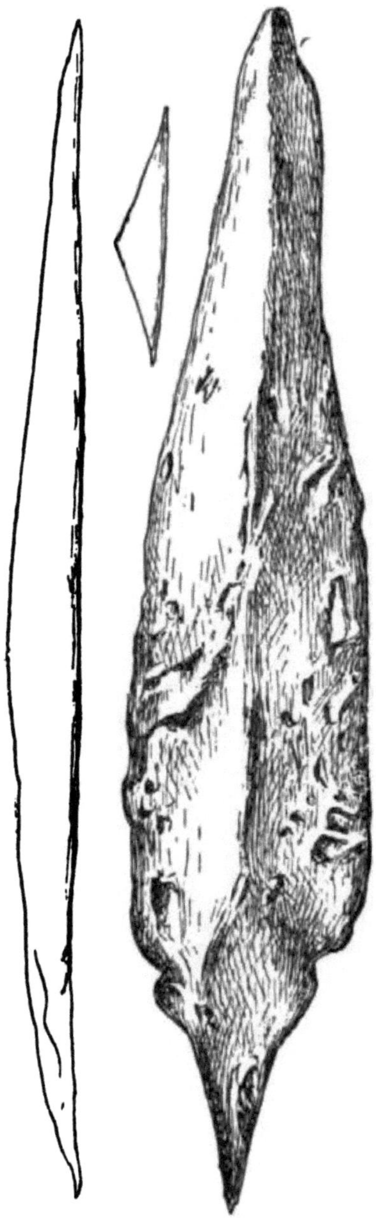

FIG. 394.—New York. ⅔.

Fig. 395 represents a "copper arrow or spear found near the river, six miles west of here (Baldwinsville) but not on a village site. Another of the same pattern, certainly pure copper, hammered, was discovered under the roots of a large tree which had been felled. This specimen, also, is in the possession of Mr Bigelow."

Mr. Beauchamp further informs me that "the other copper finds in that neighborhood, so far as he can learn, are "one large arrow, two gouges, two large and unique celts, and a copper bead or sinker."

Arrowheads and other objects, both of copper and brass, are also found in this same locality, which are of European manufacture, but Mr. Beauchamp considers them readily distinguishable from similar articles, made of hammered, native copper. It is not at all improbable that copper weapons were in quite general use, at the time of European contact, and the early voyagers seeing these objects of copper, simply introduced a better finished article, just as they did glass beads, and so led to a discontinuance of the manufacture of native copper articles, by the Indians. That the two patterns (as well as a third form, which is that of Indian-made objects of European sheet copper) should now be found associated is quite natural, for the introduction of European copper would not have led to the discarding of home-made objects, of the same material. Judging from the number of objects of brass found in many localities near the seacoast, it is probble that this material, rather than pure copper, was mostly used in manufacturing such objects as were used in barter with the coast tribes. On the other hand, objects like figs. 393 and 394 may have been made in Wisconsin, where so many similar specimens have been found, and brought as far east as western New York, by that system of trade, which, it is well known, existed between the many tribes or nations, occupying the whole North American continent.

FIG. 395.—New York. ½

The ornaments of copper, as yet discovered along the Atlantic coast, do not differ in pattern from those of stone. So far as ascertained, there have been a few examples of plain gorgets, disks with one or more perforations, and a few globular and long cylindrical beads, found in graves; and less frequently single specimens have been ploughed up; but all the material now preserved in our museums would not suggest to any one, that any of the New England tribes or those occupying the coast so far south as New Jersey, ever had "great store of copper."

The tobacco pipes made of copper, to which Juet refers, may possibly not have been copper, but merely wrapped with a thin sheet of that metal. This suggestion is based upon the fact that clay pipes covered in this manner have been taken from Indian graves in Massachusetts.

In this connection, I desire to call attention to an unusually long, cylindrical bead or tube made of sheet brass, recently found by Mr. Wm. Wallace Tooker of Sag Harbor, New York, at "Sabonic Neck, Shinnecock Hills, Long Island;" and which has been kindly loaned for illustration, fig. 396. It measures four and one-quarter inches in length and one-fourth of an inch in diameter. It is neatly rolled and the diameter is nearly uniform throughout its entire length. Judging from the smallness of the perforation, and the character of the material of which this specimen is made, it may be safely said that it was used as a bead or pendant, and not as a pipe.

FIG. 396.—New York. ½.

In the collection of the same gentleman, there is also an interesting specimen of a brass arrowhead, fig. 397, from Long Island. It is of the ordinary triangular pattern, with the sides and base slightly convex in outline. At a distance from the base, of about one-third the length, there is a small, circular perforation. The surfaces have apparently been hammered, and not rolled, although they are quite smooth, and the specimen has a uniform thickness of one-fortieth of an inch. It is identical in size and pattern with a series of brass specimens, now in the archæological museum at Cambridge, Mass. It was found in an ancient grave, at Revere, Mass., and does not differ materially from metal arrowheads found by the late Prof. Haldeman, in Pennsylvania. In his posthumous publication, "On the Contents of a Rock Retreat in southeastern Pennsylvania" (Transactions of the American Philosophical Society, vol. xv, p. 351), that author figures a small brass arrowhead, of which he remarks, "fig. 35 represents a thin regular metallic arrowhead of a coppery appearance, but yellow on a new surface, and presumed to be European brass, therefore within the historic period, with brass dishes occurring in graves. It was found outside of the retreat." The specimen described in the above quotation is about one-third smaller than fig. 397 and is without the perforation.

Irregularly shaped fragments of sheet brass, also, and in some cases of silver mixed with copper, have been found in New Jersey, associated with the ordinary surface-found stone implements. In the course of a few years' collecting, the late Prof. Haldeman found similar fragments quite frequently. In one of the small islands in the Susquehanna river, near Columbia, Pennsylvania, he found many "torn bits of sheet brass," associated with stone implements.

These specimens, as well as all others of the same material, are of comparatively recent origin, as all the evidence goes to show that the Indians of the Atlantic coast at the time of the arrival of the whites, were ignorant of the manufacture of brass, and only knew copper as a mineral that could be rolled and hammered, but not smelted.

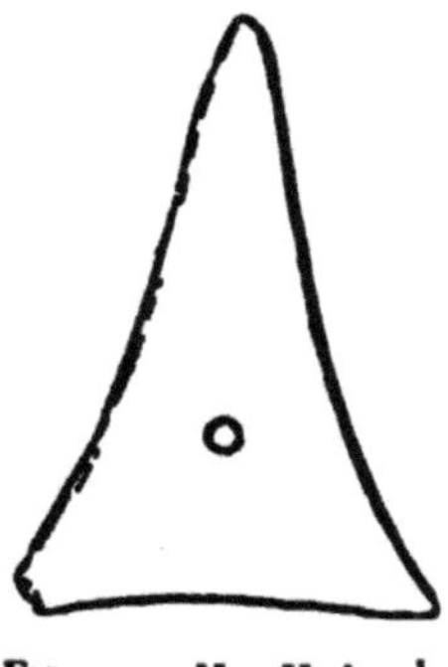

FIG. 397.—New York. ½.

The Jesuit missionary, Claude Alloüez, says of the Indians at Lake Superior, that they "respect this lake as a divinity, and offer sacrifices to it because of its size, for it is two hundred leagues long and eighty broad. * * * * It happens frequently that pieces of copper are found weighing from ten to twenty pounds. I have seen several such pieces in the hands of savages; and since they are very superstitious, they esteem them as divinities, or as presents given to them to promote their happiness, by the gods who dwell beneath the water. For this reason they preserve these pieces of copper wrapped up with their most precious articles. In some families they have been kept for more than fifty years; in others, they have descended from time out of mind—being cherished as domestic gods." (Quoted in "Report on the Copper-lands of Lake Superior, by Foster and Whitney, Washington, D. C., 1850, p. 7.")

The preceding quotation renders it the more probable, I think, that the plain, so-called copper celts found on the Atlantic coast were more likely to have been regarded in this light, than used as cutting instruments; but the variety of forms of metallic implements and weapons found in western New York, on the other hand, are to be looked upon more as objects in daily use. To a certain extent also, they may be of comparatively recent origin, as there is sufficient evidence to show that in the historic period they made their own arrowheads from metal derived from the European traders. Thus, we find in Underhill's history of the Pequot War, that a Dutch trader was prevented bartering with the Pequots, on the ground that they were to be supplied in part with "kettles or the like, which make them arrowheads." (Mass. Hist. Collections, 3rd series, vol. vi, p. 17, Boston, 1837.)

CHAPTER XXIX.

HAND-HAMMERS AND RUBBING STONES.

In picking up a pebble and striking a blow with it, we do but repeat an act of primitive man when he chose a rounded pebble, and used it as a hammer. Had the pebbles, that required no preparatory alterations in order to be made available, been used only to strike upon substances more yielding than themselves, they could not now be recognized as relics of a vanished race. Indeed, those that were used only for cracking nuts, or as weapons, are now nothing but pebbles to us, although many of them have had a history, as a weapon or hammer, which, if known, would dispel every doubt that envelops the dawn of civilization.

In the present chapter, these small pebble-hammers, here called hand-hammers to distinguish them from the hafted hammers or mauls already described, are associated with a very dissimilar class of stones, which were used largely in connection with them. It is supposed that the "pecking," or process by which stone implements, such as grooved axes, were brought to the desired shape, was done with them. This pecking process would appear, from experiment, to be as destructive to the hammer, as it was effective in removing the surface of the stone operated upon, and hence it is probable that more than one such hand-hammer was required to remove the inequalities, and reduce the stone chosen for an axe, pestle or other implement, to the required form. By reference to fig. 6, p. 18, it will be seen how much of the original stone it was necessary to remove in order to produce the slender, conical head of the specimen represented. To accomplish it by pecking away the surface, several of these small hammers were probably required, as their value for such purpose apparently decreases in proportion as their surfaces become battered.

The associated objects referred to, called "rubbing-stones," are supposed to have been used in rubbing or smoothing the slightly rough surfaces which remained after pecking, and also in grinding and polishing the edge and adjacent sides of the blade of the implement.

Fig. 398 represents a pretty little hand-hammer, made from a small cylindrical quartzite pebble. The sides retain their natural surface, but the ends are much battered.

Specimens of this simple pattern are not as widely distributed as might be supposed. In many localities, where other implements are in great abundance, these simple objects are often entirely absent; while, on the other hand, the writer has frequently gathered a hundred or more specimens on a single village site.

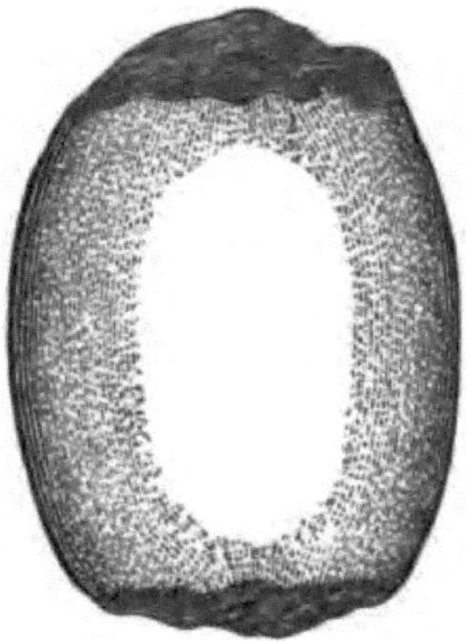

FIG. 398.—New Jersey. ½.

These implements vary much in size, many being fully five and six times as large as the one here figured. The weight also varies, but not always in proportion to the size; as many of the smaller ones are of such compact mineral, that they equal in weight others of more than double the size. None are found, however, which cannot readily be used with one hand. The battered condition of the entire surface of some of them cannot readily be explained, as it is difficult to see how the sides of small cylindrical pebbles could have been used for pecking the surfaces of other stones; but our knowledge of the processes by which the Indian fashioned his stone implements does not always justify us in forming or rejecting conclusions, on the subject. Hand-hammers with the entire surface battered from usage are, however, but seldom met with, in comparison with those in which only the ends show the effect of use.

Fig. 399 represents another form of hammer-stone which is of common occurrence not only along the Atlantic seaboard, but in Europe. These hand-hammers, as they are called, are usually flat pebbles, circular or oval in shape, with a well-marked depression in the middle of each side, generally known as "thumb and finger pits." Many of these have no trace of battering about the edges, nor other marks of hard usage, while others distinctly show traces of use as hammers. Fig. 399 represents a specimen of these finger-pitted hand-hammers frcm the valley of the Susquehanna. The material is "a tolerably hard stone,

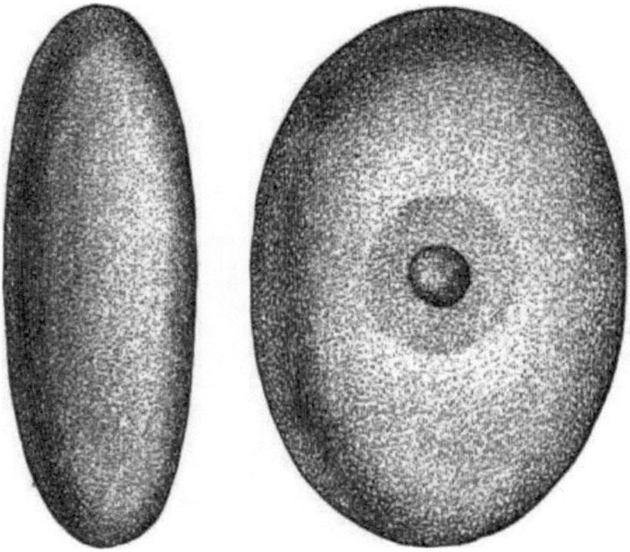

FIG. 399. — Pennsylvania. [illegible].

consisting of rounded quartz grains, apparently a metamorphic quartz or quartzite." In a large series of these implements, the size varied from five to less than three inches in diameter; and the weight, from one pound and ten ounces to half a pound. It is worthy of note that of the hundreds of these objects collected in various localities, particularly along the Delaware river, but very few reach the maximum weight of those found in the valley of the Susquehanna. In New Jersey, the heaviest specimens are all of irregular shape, and are but seldom even comparatively flat and thin. It would not, however, be safe to infer from this fact alone, that the larger examples were used for some other purpose than the one mentioned, as the battered condition of the edges shows that they had been subjected to violence that could scarcely have proceeded from any other cause, than violent contact with some material equally hard or harder.

Fig. 400 represents a second specimen differing only from the preceding in being of much smaller size. This specimen was found on the banks of the Delaware river, and, like that from the Susquehanna

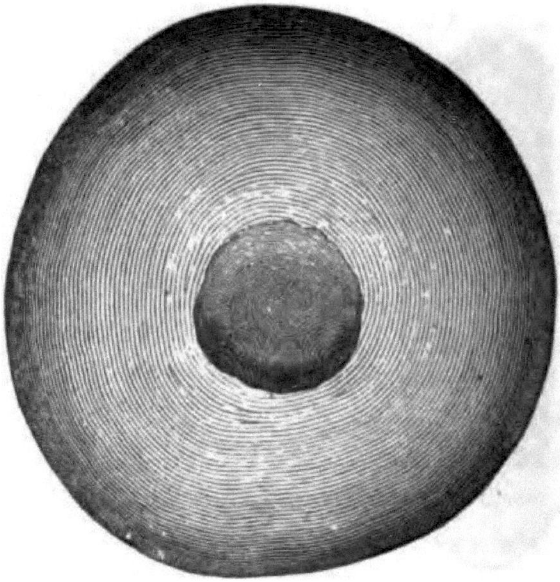

FIG. 400.—New Jersey. 1/1.

valley, may safely be considered as the handiwork of the Lenni Lenapè or Delaware Indians.

For whatever purpose this specimen may have been intended, it is evident that it has been but little used, and was never subjected to any such violence as is implied in the word "hammer." As will be seen by reference to the illustration, this hammer is nearly circular in outline, and though not polished, its entire surface is very smooth. The lateral depressions or finger pits have been drilled instead of pecked, and are now very smooth. As this specimen is so nearly accurate in outline, and free from blemish of every kind, it led the writer[170] to believe that it had been probably used as a "chungké stone," as they are usually called. The absence of a flattened margin, however, and the fact that specimens of typical discoidal stones have been found in New Jersey and Pennsylvania, render it quite doubtful, whether it should be classed with this characteristic implement of the southern and western Indians.

It appears from the verbal accounts of several collectors, that occasionally stone hammers of this pattern have been found in Indian graves. If it could be ascertained that in all that are so found there are no traces of use as hammers, it would indicate that they were intended for some other purpose. This is not improbable.

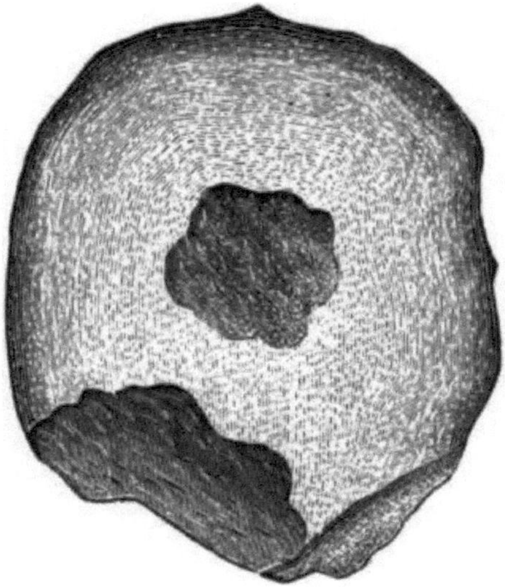

FIG. 401.—New Jersey. ½.

Fig. 401 represents a pitted hand-hammer which is so battered along one portion of the margin, that it is quite evident that it has been used for hammering stone or some other equally hard material. The battered and badly broken condition of so many of these objects is generally considered as an indication that one of their chief uses was to break off the flakes of jasper, that were subsequently worked into spears and arrowpoints. Not only is this apparent from their battered surfaces, but hammers of this pattern are found in numbers mingled with the chips, cores and discarded specimens that are found in such profusion in many localities, and mark the sites of "open-air workshops." In a subsequent chapter, refer-

[170]Abbott. Smithsonian Annual Rep. for 1875, p. 366, fig. 209.

ence will be made to a series of these hammers found associated with the refuse of a long occupied workshop. The late Prof. Haldeman[171] refers to them in his description of the contents of a rock-shelter discovered by himself, at Chickies Rock, Lancaster Co., Pennsylvania. He says, "y compris les pierres rondes provenant du gravier de la rivière, et non seulement les marteaux ou percuteurs portant d'un côté ou de deux une dépression artificielle destinée à loger le pouce et les doights, marteaux dont les bords ont été brisés par l'usage, mais encore les spécimens exempts de toute marque artificielle et paraissant avoir été colligés en vue d'un emploi ulterieur." The other contents of this shelter were of such a character, that it fully justifies the conclusion of the author quoted, that "l'abri semble avoir été occupé par une succession de fabricants de flèches."

The jasper cores, to which reference has been made, are bowlders of this mineral, from which have been detached pieces suitable for making arrowpoints. These detached pieces or flakes are sometimes six or eight inches in length; and a large number of these were frequently obtained from one bowlder. A hammer of some kind must necessarily have been used for detaching these flakes, and the fact of the cores and these finger-pitted hammers being found together, as described by the late Prof. Haldeman, renders it quite evident that the small hammers, such as fig. 401, were largely used for this purpose.

Occasionally there are found specimens of these hammers made of a soft sandstone, which would of course be valueless for flaking jasper. The object of these is difficult to determine, as they are too small and friable for use, even as nut-crackers; and indeed, their shape is not fitted for this or any ordinary purpose. It is possible that they may have undergone material decomposition since used as hammers; but if so, then these soft specimens must be of very ancient date, and thus furnish evidence of the antiquity of the Indian on the Atlantic coast.

It is of course probable that, to a considerable extent, all these

---

[171] Congres des Américanistes: Luxembourg Session, 1877. Tome Second. Un Abri en Pensylvanie: S. S. Haldeman, p. 324.

hammers with pitted sides were used for pecking stone implements, as was suggested of fig. 398. But to determine the particular purpose of any or all hammer-stones is a task as vain as it is unnecessary.

Fig. 402 represents an implement in which the edges, or a portion of them, have been worn away by rubbing, instead of having been chipped or splintered by hammering. Implements like this are as evidently rubbing or grinding stones, as those that are battered are hammers, and

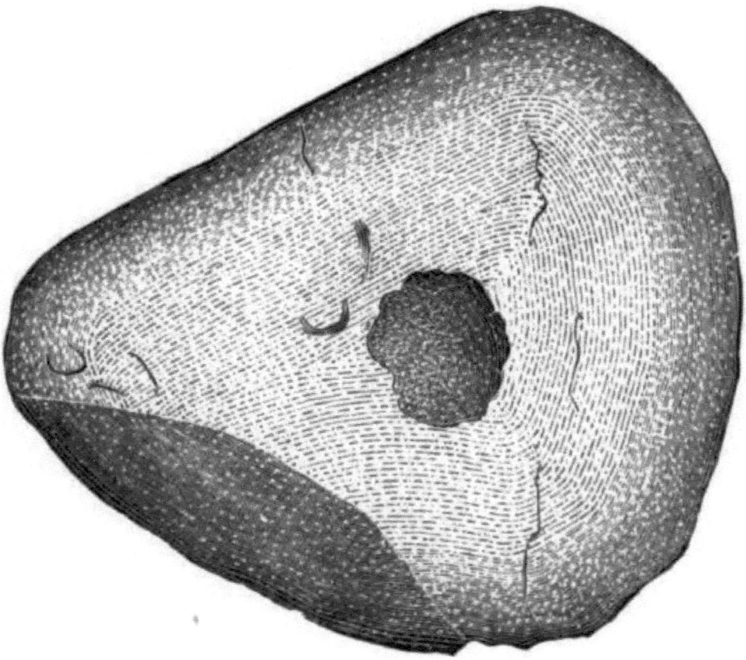

FIG. 402.—New Jersey. ½.

they have been used in polishing the beautifully finished celts, which are almost as abundant as these supposed tools, wherewith in part they were made. These rubbing stones are of all shapes, and are about equally divided between those that have the finger-pit depressions, and those in which the natural surface of the pebble, on the sides, is still retained. They are usually of a finely grained sandstone, but not always. A few specimens of jasper and other equally hard stone are also found. In New Jersey, these rubbing stones are found in abundance on the former sites of Indian villages, but they are not so frequently met with singly.

An exhaustive examination of a limited locality near Trenton, N. J., known to have been formerly a village site of the Delaware Indians, resulted in the discovery of a number of these implements, having both a worn and a battered edge, as though used for hammering as well as rubbing. If these specimens were really used in both ways, the hammer portion was doubtlessly used in pecking implements like celts, axes, pestles and many of the carved ceremonial and ornamental stones, which were shaped by this process, before being polished. It is not improbable, however, that a hammer, like fig. 401, may have been subsequently used as a polishing or rubbing stone, and *vice versa*.

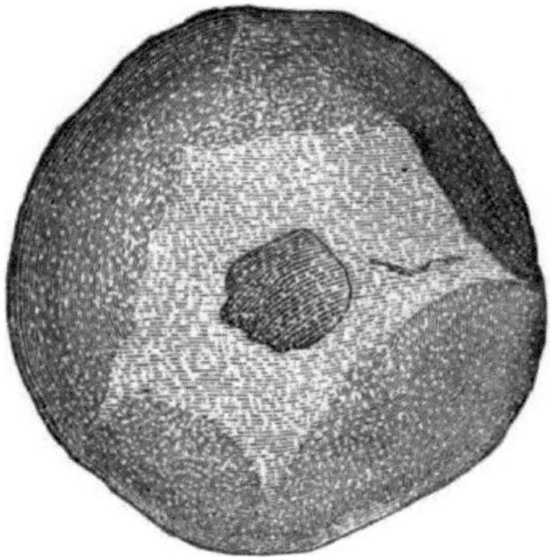

FIG. 403.—New Jersey. ½.

Fig. 403 differs from the preceding in that its entire margin has been worn away by rubbing against other stones. The finger pits in it are deep and smooth. Specimens with so little of the natural surface of the pebble remaining are not as abundant as those irregularly shaped pebbles in which limited portions of the periphery only are worn. It would appear from this, that in most cases, these rubbing stones were used for a short time and then discarded, as very many are found with but a single worn surface, and that of very limited extent. Those that have been least worn by use are, as a rule, of quartz, jasper or chert, materials that are not so well adapted for polishing purposes, as is a sandstone of fine grain.

Fig. 404 represents a second specimen of these rubbing stones, which is even more worn than the preceding. It is a quartzite pebble that has been altered in shape, by rubbing, until every vestige of the natural surface has been removed. In it the finger pits have been pecked out, and subsequently worn smooth. They are of unusual size and depth. In outline, this rubbing stone is more irregular than fig. 403, and has one obtuse point on the margin. This blunt point is quite a common feature, not only in those specimens which are worn entirely around the edge, but in those which are large, angular, and worn only over a limited extent. The point is simply the result of the continual wearing off of adjoining portions of the margin of the stone, and has not itself been used as a rubbing surface.

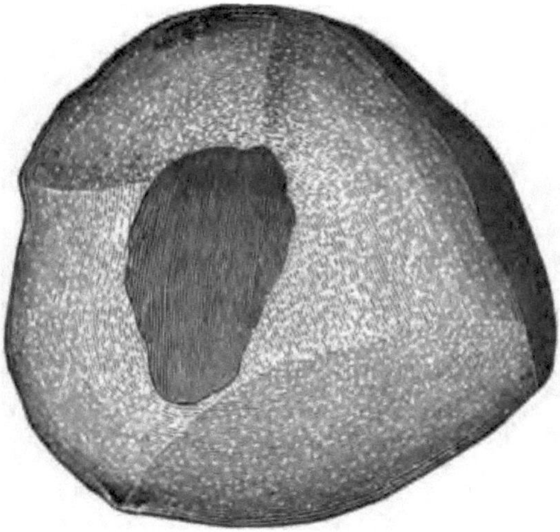

FIG. 404.—New Jersey. ½.

These rubbing stones vary indefinitely in thickness, and many are almost globular in shape. Indeed, it is difficult to draw a line of demarcation between them and those small artificial globes of jasper and quartz which are occasionally met with. These latter are seldom so smooth as to warrant the belief that they were rubbing stones; they rather appear to have been pecked to a truly globular form and subsequently worn smooth by constant handling.

Fig. 405 represents a pattern of rubbing stone that differs in one important particular from those already described. Instead of being worn along the edge, this portion is only smoothed so far as this may be effected by constant handling, while the sides are worn down until they are perfectly level, and highly polished. In this respect, this specimen closely resembles the upper millstones already described.

Rubbing stones of this pattern are far less common than the preceding, and were probably used for some one purpose not now determinable. For polishing the curved surfaces of celts and cylindrical implements generally, they are not as well adapted as are those like figs. 402 and 403; nor is it probable that so large a pebble was used as a pottery polisher.

Fig. 406 represents a second example of this form of rubbing stone, which, like the preceding, has had the sides worn down until they are very smooth and even polished. Unlike fig. 405, however, this speci-

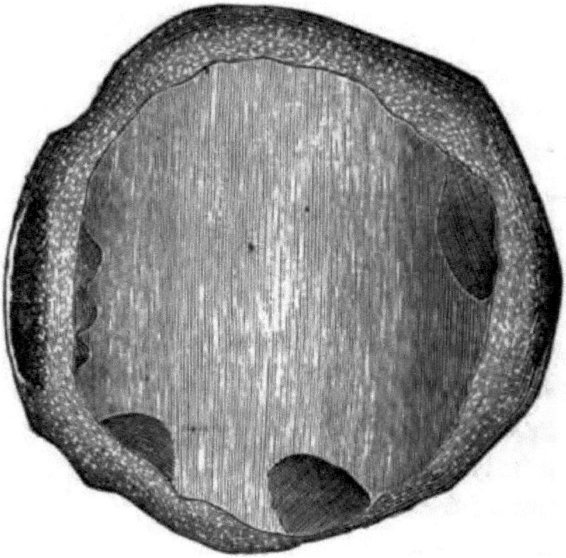

FIG. 405.—New Jersey. ½.

men has the margin badly battered, and thus gives evidence of having been used as a hand-hammer. A small portion of the margin is somewhat smooth and shows that it has been pecked. It is therefore probable that, when first in use, this rubbing stone had a pecked margin, as in fig. 405, which was afterwards destroyed by the use of the implement as a hammer.

It is perhaps questionable whether these implements were the only tools in use among the Indians, for polishing stone, from the fact that in many localities, where objects of stone of various patterns are abundant, there is often a total want of these rubbing stones; and this is true not only of New Jersey, but of the other Atlantic coast states. To a certain extent they may, of course, have been overlooked, but this will scarcely explain their absence, as proved by careful search instituted particularly for them. On the other hand, it is difficult to explain the remarkable abundance of these objects in such a limited locality as a single township in Mercer Co., New Jersey.

The foregoing examples show how indefinitely these small hand-

FIG. 406.—New Jersey. ½.

hammers and rubbing stones vary. In fig. 407, we have represented another form of rubbing stone, which shows many traces of having been long in use, but precisely in what manner, or for what purpose, is not altogether clear. Judging from the more marked features of the specimens gathered, of which fig. 407 is an average example, this pattern of these implements should be called a "sharpening" rather than a "rubbing" stone, as the character of the worn surfaces appears to be such as would be produced in the process of making and subsequently sharpening the cutting edges of celts and axes. This would not explain some of the deep and wide grooves, worn obliquely across the margins of many of these specimens, nor would it account for certain other features found in many examples, otherwise identical with fig. 407; but the examination of a very large series leads me to infer that the principal use of these irregularly worn rubbing stones was to give a cutting edge to the celts and axes—a sort of finishing tool, used only after the body of the implement had been shaped and smoothed. This, however, it must be borne in mind, is wholly conjectural.

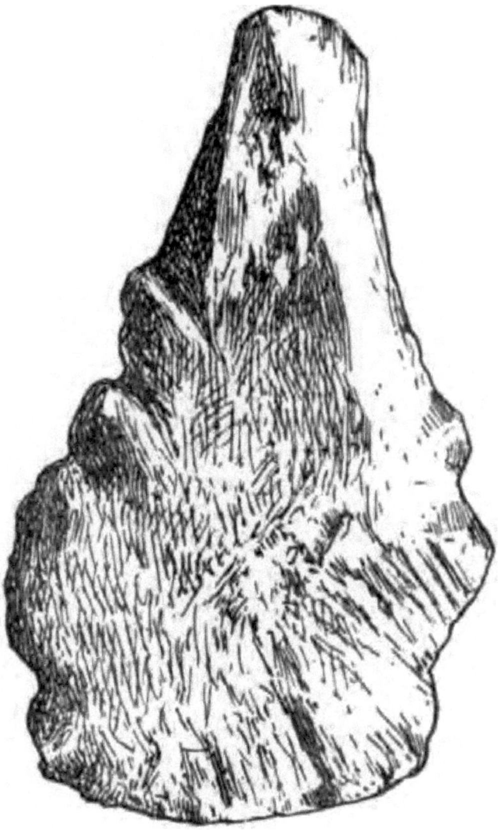
FIG. 407.—New Jersey. ½.

Originally an oval, flat pebble, this specimen, fig. 407, has been used as a rubbing or sharpening stone, until the outlines are wholly altered. The straighter portions of the two sides have been ground off until they meet in a point. The slope of these worn edges shows that the motion of rubbing was always in one and the same direction,—in this case, from left to right. In this, as in nearly all the specimens collected, the base is rounded in outline, and has a number of small, worn surfaces of different shapes and characters. Some are flat and circular, while others run into broad and shallow notches. There are also a few narrow and deep notches of the same character as those on the so-called sinew-dressers, described in a preceding chapter. These may indicate that specimens like fig. 407 were used for several purposes; as it is very doubtful whether the edges of polished celts could be smoothed or sharpened by being drawn through deep grooves, such as are seen on this specimen. This, however, is the prevalent impression, though it is probably erroneous.

This form of rubbing or sharpening stone, like the preceding, is common in the neighborhood of Trenton, N. J., but it does not seem to be very well known elsewhere. In the collections of New England forms of stone implements, in the Museum of Archæology at Cambridge, Mass., there is not a specimen that can be classed with them, nor are they mentioned by collectors of stone implements in the western and southern states. To a certain extent they may have been overlooked, but it is improbable that so marked an object as fig. 407 should not have been noticed and preserved. That they are probably comparatively scarce in most localities is inferred from the fact that in several large collections of Pennsylvanian and northern New Jersey stone implements, not a single specimen of these greatly worn sharpening stones is to be found. Possibly of this same character, are the "small pieces of grit-stone, almost the size of a hen's egg, with deep grooves on all sides," which Dr. Potter has mentioned as among the relics of the moundbuilders of Missouri. That author believes these objects to have been used for sharpening knives of bone and deer's horn. (Contributions to the Archæology of Missouri: pt. 1, Pottery, p. 19, Salem, Mass., 1880).

Besides the several forms of hammers for flaking stone, for "pecking" it into shape, for rubbing down implements until their surfaces were polished, and for grinding to a sharp cutting edge, the celts, chisels and axes, there are often found long, cylindrical pebbles, which have been used as whetstones, for sharpening the cutting tools of the Indians.

Stones of this character appear to be abundant wherever polished stone implements are found. In many, there are several worn surfaces, as in fig. 408; while in others, there is but one. When but a single worn surface is upon one of these whetstones, it is usually of such large area, that it may have been used for other purposes than for sharpening stone cutting-tools. Fig. 409 represents an example of this form. This specimen is of convenient form for grinding bone im-

FIG. 408.—New Jersey. ½.

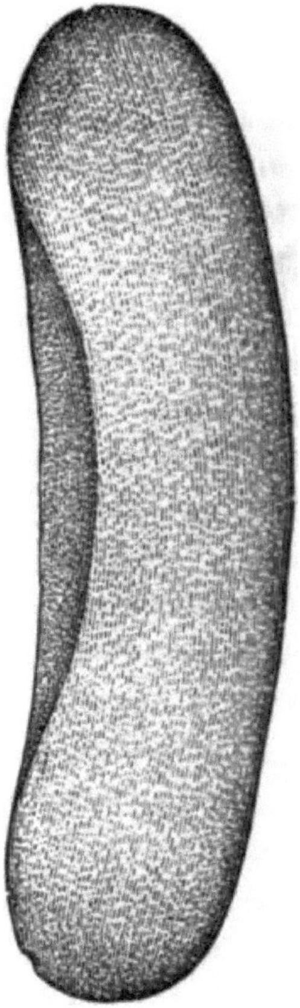

FIG. 409.—New Jersey. ½.

plements, for rounding fragments of steatite for beads, and for many similar purposes. A very fine grained slate was generally chosen for this class of implements, and usually they were so well selected, that when they are now found, they are carefully preserved for sharpening metallic tools.

CHAPTER XXX.

SHELL-HEAPS.

SACO BAY.

So extensive has become the literature treating of those artificial accumulations of shells, and bones of animals used for food, that but little can be said that will not appear to be a repetition of the published accounts of various archæologists.

The term "shell-heap" has generally been applied to these traces of the earliest occupants of our coast, not because they are exclusively what this name implies, but from the fact that the shells of different edible mollusks are the principal or characteristic feature of the accumulations. The term "kitchen-middens, or "kjœkkenmœddings," has also been used, especially in Europe, and is admirably descriptive of those that occur in northern Europe, and in the New England states. This term implies the accumulations not only of shells, but also of the bones of fishes, birds, and mammals used as food. Throughout this mass are scattered the implements of stone, bone and clay, which were made and used by the people who dwelt upon this ever-increasing pile of refuse from their daily repasts. In New Jersey, however, there are to be found not only such kitchen-middens, which are in fact, village sites, but there are also shell-heaps proper, or the accumulations of shells made at various points where, periodically, during centuries, different mollusks were gathered in vast quantities, and preserved by drying over a fire.

Kalm (Travels in North America, vol. i, p. 240, 1771) refers to heaps of shells near New York, made by resident Indians, who fished for oysters and other mollusks, not only as a food supply, but for sale. He says, "The Indians, who inhabited the coast before the arrival of the Europeans, have made oysters and other shell-fish their chief food; and at present, whenever they come to a salt water where oysters are to be got, they are very active in catching them, *and sell them in great quantities to other Indians who live higher up the country:* for this reason you see immense numbers of oyster and muscle shells piled up near such places, where you are certain that the Indians formerly built their huts." Such heaps contain neither implements nor pottery, nor bones of any mammals, fishes or birds; and, except for their size and present position with reference to the sea level, offer but little attraction to the archæologist.

In none of the shell-heaps of the northern Atlantic coast have we any evidence of that succession of deposits of different remains, such as has been so graphically described by Major W. H. Dall,[172] as occurring on the Aleutian Islands. This archæologist has determined that in these islands, "the stratification of the shell-heaps shows a tolerably uniform division into three stages, characterized by the food which formed their staple of subsistence and by the weapons for obtaining, and utensils for preparing this food, as found in the separate strata; these stages being—

I. The Littoral Period, represented by the Echinus Layer.

II. The Fishing Period, represented by the Fish-bone Layer.

---

[172]Dall. Tribes of extreme Northwest, p. 49. Contributions to N. A. Ethnology, vol. i, Washington, 1879.

III. The Hunting Period, represented by the Mammalian Layer." Somewhat analogous to this is the evidence of the succession of the shell-heaps in Florida, as determined by the late Prof. Wyman,[173] who mentions as one of the ten "conclusions" that close his invaluable monograph, that "fragments of pottery exist in the later *but not in the oldest mounds*. The pottery was in all cases of a rude kind."

Marine shell-heaps are still abundant along the entire Atlantic coast from the Bay of Fundy to the Gulf of Mexico. Some of the largest are at Cape Henlopen, Delaware;[174] along the greater part of the New Jersey coast, from Cape May[175] to Keyport;[176] on Long Island,[177] and along the coasts of Maine and Massachusetts.[178]

The shell-heaps on Long Island are probably more extensive and more numerous than elsewhere along the coast, southward of Massachusetts. These heaps are all true kitchen-middens, and contain bones of birds and mammals, implements and pottery mixed with the shells. Mr. W. W. Tooker of Sag Harbor, Long Island, has kindly furnished me with the following descriptive note of one of these heaps, which may be considered as fairly representative of the whole series.

"West of the Otter Pond is a shell-heap of considerable extent, that covers nearly three acres. On its surface have been found hundreds of arrowpoints. Part of this deposit is still hidden under the leaves and soil of the woods and has never been disturbed. Along the cove beyond, for a distance of about one mile and a half, is one almost continuous shell-heap. It is thicker at some places than at others. Back on the southern slopes of the hills, near swamps and springs, are others, some being an acre in area. At Payne's Creek, there is found one of the largest and most compact shell-mounds on this part of Long Island. At the time the shells were deposited, the creek evi-

---

[173]Wyman. Fresh-water shell-mounds of Florida, p. 86. Salem, Mass., 1875.

[174]Leidy. Proceedings of the Academy of Nat. Sciences for 1866. Philadelphia, Pa.

[175]Cook. Geology of New Jersey, p. 362. Newark, New Jersey, 1868.

[176]Rau. Smithsonian Annual Report for 1864, p. 370. Washington, 1865.

[177]Elias Lewis, jr. Popular Science Monthly, vol. x, p. 436. New York, 1877.

[178]Wyman. American Naturalist, vol. i, p. 561. Salem, Mass., 1868.

dently flowed in front of the deposit, but now it is filled up and a sandy country road extends along its front. This deposit covers about three acres, and is fully four feet in depth. In some spots on this shell-heap, are remains only of the oyster; in another, of the clam; and a third of the scallops, and then the various shells will be found to be about equally abundant, and mingled together. In this deposit have been found bones of the raccoon, bear, otter, fox, deer, and rabbit. Almost all the stone implements used by the natives have been found in this shell-heap. Also awls or pointed instruments of bone, and one large bone fish-hook (chap. XV, fig. 193). Fragments of their pottery, made of pounded shells, clay and sand, are seen strewn in every direction."

Fresh-water shell-heaps are also of common occurrence in the valleys of those rivers in which the larger Unios or mussels live. These inland kitchen-middens do not differ materially from those found upon the seacoast, except that, generally, they are much more limited in extent, especially in the New England and Middle States. In the latter locality, indeed, they are quite insignificant in comparison with the deposits of oyster and quahaug shells at Beesley's Point, Cape May Co., and at Tuckerton, Burlington Co., New Jersey. Prof. Wyman[179] has described an extensive shell-mound on the shore of the Concord river, in Massachusetts, made up of valves of the *Unio complanatus*, a species still existing in that river. Charcoal and stone and bone implements were found in this deposit. A small Unio shell-heap formerly existed on the shore of the Delaware river, at the mouth of Crosswick's Creek, in Burlington Co., New Jersey, which was composed of valves of several species of Unionidæ; *Anodonta purpurea*, and *Unio viridis* being most numerously represented. A few chipped stone implements were found in it. The building of the Delaware and Raritan Canal obliterated this deposit.

A second and much smaller mussel shell-heap, on the banks of Crosswick's Creek, four miles from its mouth, approached most nearly

[179] Wyman. Proceedings, Boston Society of Nat. History, vol. ix, p. 243. Boston, Mass.

the Danish kjœkkenmœddings. Within a few square yards, there was a layer of Unio shells and charcoal, with a few fish and bird bones, nearly two feet in thickness. In this mass were several strata of fine sand and mud, which suggested that it had been several times abandoned and subsequently re-occupied. In it were found a small portable mortar and an oval "crusher" or upper millstone, a score of leaf-shaped arrowpoints of large size—possibly knives—two rude axes, and two remarkable "flame-shaped" or zigzag chipped implements, which were supposed to be spearpoints. These specimens, unfortunately, were lost or stolen, but drawings were published in Nature, London, vol. xi, p. 190. The larger one, measuring five inches in length, was too delicate to have been used as a weapon, and doubtless was used only upon ceremonial occasions. Whatever its purpose, its pressure in a shell-heap was remarkable. No duplicates, nor indeed any crooked arrowheads, approaching these unique forms, have since been found.

Peter Kalm, the Swede, to whom I have so frequently referred, mentions that in southern New Jersey "*Mytilus anatinus* (*Unio*), a kind of mussel-shell, was found abundantly in little furrows, which crossed the meadows. The shells were frequently covered on the outside with a thin crust of particles of iron, when the water in the furrows came from an iron mine. The Englishmen and Swedes settled here seldom made any use of these shells; but the *Indians who formerly lived here broiled them and ate the flesh.*"

In a recent letter from Mr. Ernest Ingersoll, who is known to archæologists in connection with the early discoveries of the now famous cliff-ruins of the far southwest, I am furnished with an account of some inland shell-heaps in central New York. Mr. Ingersoll writes:

"I heard of several shell-heaps along this part of the Susquehanna, but during my brief stay had only opportunity to examine one personally. This was on the high northern bank of the river, just in the outskirts of the village of Owego, Tioga county, N. Y., and had previously been much disturbed. Tradition says it was once a hundred yards or more in length, along the edge of the bluff. These dimensions had been vastly reduced, but enough remained to give me a day's labor. The shells were from a few inches to about two feet in depth and without intermixture of dirt, which only slightly covered the whole. They were all of the two or three species of *Unio* and *Anodon* which inhabit the river there, and the exterior coat of every shell had disappeared. Many of the shells could be taken out entire, but required handling with extreme care to be preserved. A few land shells (*Helix*) were also seen, but they may have crawled there and died; that is, I would not care to assume they were eaten by the Indians. Associated with the shells was very little of value. Evidences of fire occurred in the shape of blackened strata, cinders and calcined shells. Pebbles occasionally were found; but the only stone implements disclosed was one hammer-stone, three netsinkers and a possible awl or drill of argillite. Fragments of bones of fishes, birds and chiefly deer, were abundant, however, scattered through the mass. All of these fragments, except the joint-ends (which were broken short off), were split,—I suppose for the marrow,—and were of small size. Upon one of these pieces of bone, which was about two inches long and half an inch wide, with smooth edges, a series of transverse notches had been cut, as though to tally some score or series of events.

"This locality, where the Owego—or *Ah-wa-ga*—creek empties into the Susquehanna, was a warm and fertile valley much in favor with the Indians who had a permanent village here. Many local traditions and memories remain of their occupancy. They are asserted to have been a band of Massachusetts Indians who retreated to this region and procured, from the Iroquois tribes who held it, permission to settle. I have not investigated the truth of this statement. In the centre of the present town was the Indian graveyard, tradition asserts, and the whole region abounds in stone relics, which in large number and variety have been saved by half a dozen persons interested in the matter, throughout the county. During a visit of some weeks, I collected as many stone implements as possible.

"Flint and slate 'chips' are extremely abundant all along the river-bank and throughout the valley; a good deal of jasper also occurs, and I found one large 'core' of this substance, which is all imported. The arrowheads found are of various shapes; rude, unsymmetrical forms predominating. I found no large spearheads, but heard of a cache of thirty or more planted close together and upright, which were ploughed up some years ago; what became of them, I failed to learn. Hammer-stones, some of them elaborate, I found very common, and as for netsinkers, of all sizes, shapes and degrees of value, they were almost innumerable. Evidently these Indians (who no doubt antedate the occupancy of the Massachusetts strangers) were great fish eaters.

"Some skin-scrapers, long pestles, celts, grooved axes of syenite (?), corn-mortars (one biconcave) and flints of various forms, either found by me or by others in this locality, are not worth special remark."

Dr. Daniel G. Brinton,[180] in a brief notice of various inland shell-heaps, more especially those of Florida, and along the Tennessee river and its tributaries, has called attention to the fact, that "the exclusively artificial character of many of these deposits, even of very considerable size, was first prominently brought before the scientific public by Mr. Lardner Vanuxem, in the proceedings of the American Association of Geologists and Naturalists for 1840–42 (pp. 21–23). The existence of enormous accumulations of the shells of the *Ostrea virginica*, and *Venus mercenaria*, on the shores of the Chesapeake and its affluent streams, on the Jersey shore and Long Island, was discussed, and various proofs of their formation by the aboriginal tribes pointed out;" and he further states, that "these proofs may be briefly summed up as follows: First. Valves of the same animal are rarely found together. Second. Arrowheads, fragments of pottery and charcoal are mixed with the shells *in situ naturali*. Third. The shells are broken and frequently charred. Fourth. The substratum of the deposit is the same as the surrounding soil. Fifth. The deposits are at the mouth and shores of water-courses, where the shell-fish abound. Sixth. There is absence of stratification and older fossils."

---

[180] Brinton. Smithsonian Annual Report for 1866, p. 356. Washington, D. C.

Dr. Brinton subsequently remarks: "It seems hardly necessary to adduce evidence from the old voyagers to show that in the commissariat of the native coast tribes, ésculent shell-fish constituted an important item. Cabeza de Vaca describes the accolents of the Gulf of Mexico as dwelling in houses of mats, 'built on heaps of oyster shells' (Ramusio Viaggi, tom. iii, fol. 317), and the first settlers of Maryland record with pleasurable recollections, 'oisters, broil'd and stewed,' that the savages offered them in profusion.—(Relation of Maryland, 1634, p. 18, in Shea's Southern Tracts).

The late Prof. Wyman,[181] in the American Naturalist, has given somewhat detailed descriptions of various marine shell-heaps in Maine and Massachusetts, and these may be taken as typical of such deposits wherever found upon the New England coast. These deposits are in Frenchman's Bay, and at Crouch's Cove, on an island in Casco Bay, Maine; and at Eagle Hill, in Ipswich; in the town of Salisbury; and at Cotuit Port, in the town of Barnstable, Massachusetts. The shell-heap at Frenchman's Bay was "examined only near its border, where a pit was sunk, showing a deposit of clam shells about two feet in thickness. Among these were found the bones of several animals, including those of the deer, elk and beaver, but no implements of any kind." Another deposit on an unnamed island was more carefully examined. "A section through the heap at its thickest part showed that it belonged to two different periods, indicated by two distinct layers of shells. The lowest, a foot in thickness, consisted of the shells of the clam, whelk and mussel, all much decomposed, and mixed with earth. Above this was a layer of dark vegetable mould, mixed with earth and gravel, and from six to eight inches in thickness. Above this was a second layer of shells, of the same species as those just mentioned, but in a much better state of preservation, and with less intermixture of earth; this deposit was in turn covered by another layer of earth and mould, and these now sustain a growth of forest trees, but none of them of large size."

---

[181] Wyman. American Naturalist, vol. i, p. 560, *et seq.* Salem, Mass., 1868.

The shell-heaps at Crouch's Cove also proved, on examination, to have been "deposited in two different layers, very much as on the island in Frenchman's Bay."

In all these heaps, the increased size and solidity of many of the shells, as compared with the same species now living in the adjacent waters, were noticed; and to some extent, certain species, notably the quahaug (*Venus mercenaria*) which is now very scarce and local, north of Cape Cod, were abundant.

Dr. Chas. Rau[182] has given a detailed description of shell-heaps near Keyport, Monmouth Co., New Jersey, which may be considered as typical of the larger deposits found along the New Jersey coast. These heaps of refuse were made up of the common oyster (*Ostrea borealis*), and the hard-shell clam (*Venus mercenaria*), with a small percentage of periwinkles (*Pyrula canaliculata* and *P. carica*). Evidence of the occupancy of these places by the Indians is to be seen in the presence "of numerous fragments of pottery and stone implements of the usual kind, otherwise very scarce in this part of New Jersey."

On the extensive meadows immediately back or west of Atlantic City, Atlantic Co., New Jersey, are many small shell-heaps, some only a few yards square. These heaps are made up of shells of the oyster, hard-shell clam or quahaug, periwinkles and soft-shell clams, all of which show the effects of exposure to fire. Careful examination of many failed to yield a single stone or bone implement. These heaps, which are "in the marsh, and extend down to the hard ground," were made when gathering the shell-fish for winter use, and were probably occupied only temporarily. Dr. G. H. Cook,[183] mentions the existence of these heaps in many localities along the New Jersey coast, and states that "there is every indication that the marsh has grown several feet about them since they were deposited."

During the summer of 1879, in company with Mr. F. W. Putnam,

---

[182] Rau. Smithsonian Annual Report for 1864, p. 370. Washington, D. C.

[183] Cook. Geology of New Jersey, p. 362. Newark, New Jersey, 1868.

the author visited and carefully examined an extensive shell-heap on the west bank of Dowdy's Creek, near Absecom Inlet, Atlantic Co., New Jersey. This deposit measured one hundred yards in length, and varied from three to ten yards in width. At present it is about four feet above high water, and nearly as many below low water.[184] These measurements, with that of the rise and fall of the tide make the deposit, at present, fully twelve feet in thickness. There have, however, been many shiploads taken from the surface, and when finally abandoned by the Indians, it was doubtlessly half as deep again as at present. It is composed in great part of hard-shell clams, with about ten per cent. of oysters and periwinkles. A few fragments of charcoal were found, and all the shells and fragments showed traces of fire. No stone or bone implements were found.

This extensive shell-heap, like the many smaller ones in the immediate vicinity, was evidently simply a favorite spot for gathering the shell-fish, and in no respect a village site or even a camping ground. Situated in a marsh meadow on a navigable creek, and but a short sail from high ground, the Indians appear to have passed to and fro, from the heap to the high land, where there are abundant traces of their village sites. Failing to find any implements or handiwork of any kind in the shell-heap, we hunted over various ploughed fields on the higher ground, and gathered numerous fragments of pottery, and a few arrowheads.

Why the shell deposits and village sites, in this locality, should be thus separated, is not readily understood, especially when it is seen that these deposits, now surrounded by marsh, extend to the hard ground. Then, as now, there was a deep, navigable stream between them and the present main shore, but this of itself would seem to

---

[184]This measurement was made from the present bed of the creek, which has cut a channel upon one edge of this deposit, and shows a fine section of the heap. It is a homogeneous mass: the lowermost portions being more decomposed than those nearer the surface. Whether the "fast ground," upon which the shell-heap is built up, is much lower than the bed of the creek was not determined positively, but it probably is two or three feet below it, and thus gives the deposit that additional thickness.

offer no objection to the locality as a village site. We can only account for it by the probability that, when the deposit was commenced, this hard ground was then so near the level of the ocean, that occasional tides and the sea, during storms, swept over it. If this were an exceptional occurrence, it would offer no obstacle to the temporary occupation of the place as a station for collecting clams and oysters, but it would be an insuperable objection to the locality as a permanent abode. This, too, would explain the absence of the bones of edible mammals, birds and fishes, and also of implements, all of which are found in the New England shell-heaps, and in those of New Jersey, which are built up upon the main shore, or upon ground that could be permanently occupied.

Mr. Chas. F. Woolley, in a communication to the American Antiquarian (vol. i, p. 225), refers to the existence of "many shell-heaps of aboriginal origin along" the New Jersey coast, and adds: "but few of these reward the seeker with anything except a few chippings, and in many, even these are wanting, * * * * the majority having been made by the Indians in casting away the valves of the shell-fish, after stringing or otherwise preparing them, to carry back to their more permanent habitations in the interior; for our state was traversed by well-defined trails leading from the Delaware to oyster-producing inlets of the Shrewsbury, Squan and other streams. One of these heaps near Tuckerton, known as the "Hummock," has its base upon the Salt Meadows, a half-mile from the firm land, and is very conspicuous as it can be seen from out at sea; it is a solid mass of clam shells (*Venus mercenaria*), eleven feet high, twenty-five feet long, and an average width of six feet: upon the top are growing several red cedars, six inches in diameter; no whole shells could be found and but few entire valves, they all showing the marks of having been opened with a rude instrument. This has been opened several times, and tons of shells carted away, but not a chipping or fragment of pottery has been seen, though it is of undoubted aboriginal origin, and upon the mainland opposite and a half-mile away, ground axes, celts and other implements have been found."

The New England shell-heaps, after most critical examination by Prof. Wyman,[185] exhibit no indications of an antiquity as high as those of the old world. He remarks: "The materials of them present some variety in the degree of decomposition, which has resulted from time and exposure, the lower layers being much more disintegrated and friable, the shells in fact, falling to pieces, while those of the upper ones generally preserve their original firmness." Dr. Rau[186] observed this same peculiarity in the heaps he examined at Keyport, Monmouth Co., New Jersey. He says "that considerable time was required to heap up these shells is evident, and, moreover, indicated by the chalky, porous appearance and fragility of many of the valves, while those that were cast away at later periods exhibit these signs of decay in a far less degree." Prof. Wyman also notes the fact of "a disintegration of the shores, the seas undermining and destroying the deposits," and adds, "there can be no doubt that these (deposits) were once much more extensive than now, and that the water has worked its way into their places. Lastly, these deposits contain the remains of animals, as of the elk, not known at present to exist, to the eastward of the Alleghany mountains; of the wild turkey, now virtually extinct in New England; and of the great auk, which * * * * has receded almost, if not quite, to the Arctic regions,"[187] and concludes that there is as yet "no proof of great age, or high antiquity."

The same is applicable to the New Jersey shell-heaps, the age of which can only be estimated by the indications of the subsidence of the entire coastline, now in progress.

Prof. G. H. Cook[188] states that "there is an abundance of evidence to show that a slow subsidence of all the land along the tide-waters not only of New Jersey, but of the whole eastern coast of the United States, has been going on for several hundred years past, and there is evidence that it is still in full progress. This movement is one of a

[185] Wyman, *l. c.*, p. 571.

[186] Rau, *l. c.*, p. 372.

[187] The auk is now believed to be extinct.

[188] Cook, *l. c.*, p. 362.

series which has occurred on our coast, by which the line of water-level has been alternately elevated and depressed. The extent of the movements is quite limited, the whole range being comprised within twenty feet." Prof. Cook was able to make some measurements, from which he concluded the rate of subsidence to be "about two feet in a century, or one quarter of an inch in a year. The whole amount of this subsidence is not known; it must, at least, equal the whole depth from high water-mark to the lowest points at which stumps and roots of trees have been found in their places of growth. This from the evidence, * * * * is seventeen feet, and it may be more."

This, in brief, would indicate that if the shell-heaps now resting, as they all do, upon "fast ground," but surrounded by marsh-meadow, were begun at the commencement of the subsidence, then at the rate of two feet per century, the depression being about twenty feet, they would date back one thousand years. If, however, we consider that there have been a series of these movements, and that the elevation may not be perfectly regular in its movement, there is no reason why they should not be much older. Possibly, as the movements themselves have been oscillatory there may have been earlier shell-heaps that were destroyed by a preceding period of submergence; as many of the present heaps are now being washed away by the encroachments of the sea.

A thousand years is not time enough to explain many facts connected with the archæology of the inland districts. If it could be shown that there were no shell-heaps of greater age, then it would be necessary to class them among the later traces of the former occupants of this continent, which no one would willingly admit. Hence it is evident that a considerable antiquity must be assigned to them — an age greater by far than ten centuries.

Dr. Cook[189] mentions that "the enormous piles of clam and oyster shells, which were accumulated by the Indians, are all in the marsh,

[189] Cook, *l. c.*, p. 362.

and extend down to the hard ground." It is obvious from this, that this "hard ground" was in the immediate vicinity of the oyster and clam beds, when the Indians first tarried here. There is absolutely no reason for supposing that it was after the subsidence had commenced, that the shells were first thrown down in a heap. It is even possible that at that time, the land might have been slowly rising to a higher level, and it is clearly as probable that the Indian was wandering along the coast at the very earliest practicable date, as that he first trod these sandy beaches at some very recent period.

Furthermore, some of the fresh-water shell-heaps have now an accumulation of peat and alluvial deposit over them, that is itself indicative of a considerable antiquity. One of these has been found at a depth of six feet below the present surface, resting upon the tertiary gravels exposed by the shrinking of the old glacial river to its present narrow channel. This refuse heap is of an identical age with the argillite fish-spears already described.

Shell-heaps built upon "fast ground," and now oceanward of our coastline, have long since been destroyed by the steady encroachment of the sea, and so our study of the antiquity of those that remain intact, can give us no clue to the earliest of these accumulations. Man therefore is older than the oldest of the shell-heaps that remain.

Furthermore, in basing any estimate of antiquity upon shell-heaps, it must be borne in mind, that the savage race or races, when they first appeared on this coast were certainly few in numbers and widely scattered; and it must necessarily have been long subsequent to the time of their arrival, that their numbers so increased and their habitations were so far permanent as to result in the formation of these enormous heaps of shells.

When we consider the millions upon millions of shell-fish that were gathered, and from them pass to our river valleys, and see there too, thousands of the relics of the Indians dotting every field, and further, trace the growth of the flint-chipping art from the relics of the deeper to those of the superficial soils, is it unreasonable to ascribe a "high antiquity" to the dusky savage whom the pale-faced intruder found in possession of these lands?

CHAPTER XXXI.

FLINT-CHIPS.

UNDER this general term may be considered the various flakes, splinters, chips, cores, and the refuse of those minerals that have been used in the manufacture of such chipped implements, as knives, scrapers, spearheads, arrowpoints, and drills or perforators. Wherever any of these or allied forms of finished implements occur, it is usual to find chance specimens of this refuse material, and, when thus met with, they have much more archæological significance than is generally supposed. When, however, vast quantities of such chippings occur, in a very limited area, they indicate, without doubt, that the various forms of weapons and implements were there made in quantities; and, doubtlessly by a system of trade[190] among adjoining tribes, they were in time dispersed over a large extent of country.

Such accumulations of this refuse material are met with under two quite different circumstances: as where a suitable exposure of living rock occurs that is adapted to the manufacture of the various forms of implements, which are, in such cases, necessarily of the same mineral, and present, curiously enough, a very uniform appearance in the pattern adopted, although the mineral is readily chipped into other and more delicate forms, as shown by broken specimens in the refuse heaps where a bowlder of this same rock has been utilized; and again, we have such accumulations, where selected minerals, in small masses, have been brought together, and from this little storehouse of crude materials the various forms of implements and weapons have been formed.

---

[190] Rau. Smith. Ann. Reps., for 1872, p. 348, and 1877, p. 291.

In the upper valley of the Delaware river, where the rocks frequently afford such shelter of which the Indian was glad to avail himself and which are, at the same time, suitable for making many of the forms of implements which his ingenuity had at that time devised, these traces of an early occupancy of the country by savage man, shown by the accumulations of flakes and splinters,—these "open-air workshops," as they have been called,—are frequently met with. The abundance of small flakes or chips, and of broken and unfinished specimens, at once indicates the character of the locality. In such as have been carefully examined, there has been a marked absence of the small pebbles and indented oval stones that are supposed to have been used in flaking the chert, jasper, and quartz, which were so largely selected for the manufacture of stone implements. Whether they had been removed and the places abandoned while the Indians were still in possession of the country or not, cannot now be determined; but in all the workshop sites that were visited where the living rock was utilized, there was in every instance an absence of several features that characterize the same sites when found in more southern localities, where there is no living rock, and all the material and the tools themselves, when of stone, were transported from more or less distant points. The former have always seemed to imply that the locality, being accessible to all, the Indians came and went as their needs suggested, and fashioned for themselves what implements they desired, and in this case they naturally took away with them the tools they used. This would account for a marked absence of even the simple hammer-stones. On the other hand, where all the material is foreign to the spot, it was probably the property of one or a few individuals; and if left at any time, all connected with it would be left behind, unless it were the finished implements.

At first sight it might be thought that where a rock *in situ* could be utilized, portions of it could be selected, such as would be available for the desired forms of implements, and hence that comparatively few failures would occur. In other words, that the detached mass would be in such shape and so far free from weathered surfaces, that any defect in its constitution, as the presence of a foreign mineral, could be detected, and no trial chipping need be made, to determine its availability. No such discrimination, however, was exercised by the Indians; and in a rock-shelter, near Belvidere, New Jersey, where arrowpoints were once made in vast numbers, there was an immense accumulation of chips that had every appearance of being simply failures, being, for the most part, blocked-out spearheads, which had been broken into halves, or otherwise so fractured, as to render them useless. In all such cases there were to be traced such variations from the characteristic constitution of the mineral as explained the cause of the failure to produce a finished implement on the part of the ancient worker.

While, therefore, one cannot but admire the beauty of workmanship, and marked display of taste, as shown in the finish and design of the thousands of arrowpoints that have been gathered from our fields, it must be admitted that, notwithstanding all their skill, the makers experienced much unnecessary trouble from their inability to judge of the qualities of a given mass of mineral, prior to expending considerable labor upon it. It was to this, rathei than to want of skill, when good material was used, that we must attribute the quantities of "failures," as they have been well called, which characterize many a former site of an arrowmaker's labors.

It is desirable now, to refer in detail to an accumulation of chips, cores, hammers, and other material, found in a limited space, and which indicated very clearly that at that spot some one or more redmen had, for a long time, been accustomed to manufacture all the forms of weapons and domestic implements collectively known as *chipped* implements; for it is a curious fact, that, so far as it was practicable to determine, not a trace of a polished implement or polishing tool has been met with in these open-air workshops.

In Hamilton township, Mercer Co., New Jersey, the author discovered, during the summer of 1878, a large series of splinters and irregular chips of jasper, mingled with the soil of a newly drained swamp. Immediately over these fragments large trees had grown, flourished to maturity, and died of old age. Judging from the depth at which some of the jasper chips were buried, it was evident that these trees had either been mere saplings, or had not yet sprouted, when the arrow-maker here pursued his calling. Attention having been called to the spot by the surface indications, the ground over an area of several square yards was examined and a large quantity of interesting material gathered which has since been placed in the Cambridge Museum. It consists, first, of masses of jasper and allied minerals, gathered from the gravel beds that form the eastern bank of the Delaware river, at a point some two miles distant, as the crow flies. Secondly, of cores, or the remnants of the selected masses above mentioned, which were too small or too irregular in shape to be further available. Thirdly, of large flakes, which being similar to those that often occur associated with relics in our fields, may or may not have been considered and used as finished implements. These flakes, however, show little secondary chipping, and were intermingled with chips, splinters, and other refuse material. A noticeable feature of these flint-like masses and chips is the wonderful range of color they exhibit; not only are there different shades of red, purple, blue, green, brown, and yellow, but many are most beautifully variegated. While there seems much evidence to show that attractive coloration was prized by the aborigines, it is found that the implements made from such material were either retained by them on relinquishing this territory to the white settlers, or the mineral thus brightly colored is more easily fractured, when used as weapons. Of the thousands of arrowpoints already gathered, those still perfect, of pale green, bright yellow, blue, and the lighter shades of red, are quite rare; yet flakes and broken implements of these colors are represented, not only in the fields, but in the workshop accumulations, as already mentioned. Fourthly, of blocked-out and subsequently discarded specimens,—the failure to finish them being evidently caused by a flaw in the mineral not detected at the outset. Fifthly, of specimens that were nearly finished but irreparably injured by some unlucky finishing touch. These are often arrowheads, with the points broken off, or such as have a barb or a basal corner detached. A very considerable number of these are certainly quite as useful as weapons, as perfect specimens, and why discarded, as they were, for some very trifling defect, is difficult to understand. Are we to conclude, that the readiness with which adepts manufactured these chipped implements was far greater than has been supposed, and, therefore, that these objects were really sold at a very cheap rate, and that the Indian purchaser could afford to be very particular? Sixthly, of chips and splinters of every size and shape, being the ordinary refuse that would necessarily accumulate in the course of chipping jasper into arrow and spearpoints. Seventhly, of a series of oval, of nearly square, and of some irregularly shaped pebbles of sandstone, jasper, hornstone, and porphyry, mostly with shallow depressions, one on each side, and with the ends, if oval, and the angles, if square or of irregular outline, so battered as to show that they had been used in striking mineral as dense or more resisting than that of which they consist. Considering the circumstances under which they were found, their shape, and the evidence of hard usage which they exhibit, it is very evident that they are rude implements used, at least in part, in the manufacture of arrowpoints and other forms of chipped implements. The use of such hammerstones was doubtlessly quite limited, and other less uncertain means were taken to produce the delicate finish of the smaller implements and weapons. Although the size of these hammerstones varies greatly, it is difficult to see how minute flakes could be detached by their aid. Probably bone implements, that have since decayed, were used as finishing tools.

This simple form of hammer, it is well here to observe, is very common wherever the ordinary patterns of relics are met with, and in every series gathered by collectors in various parts of the state (New Jersey) there are several examples, varying considerably in size and shape. It can scarcely be held that they had any use as a weapon, while as hammers they would be useful in many ways.

Eighthly, of a few flat slabs of stone of small size, with an occasional trace of hammering on either side. These were possibly used as lap-stones in part, and may have been used, also, in connection with bone implements, or flakers, as a breastplate, whereon rested the base of the bone flaking-tool, when, by pressure, series of small flakes were detached. This, however, is altogether conjectural.

In all this large mass of material consisting of about one thousand pieces, there were no traces of charcoal, nor any indication whatever of fire; no pottery, nor any whole or fragmentary weapon, nor domestic implement of any description other than those mentioned.

The apparent absence of fire from this unquestionable workshop site deserves a few words of comment, as many suggestions naturally arise. It may, indeed, be thought that the absence of charcoal is mere negative evidence, and does not show that fire was not used at this spot; but when we consider that fire here would always be at or very near the same place, whenever kindled, and would be kept continually burning during winter, it is evident that some trace of it would remain in or on soil not subsequently disturbed, as was the case in this instance. Very frequently in examinations of different known localities where stone implements abound, there have been found traces of fire that were clearly evidences of the occupation of the locality before the first visit of the white settlers. Assuming, then, that the absence of all indication of fire is evidence that one had never been kindled there by the Indians, the fact seems to imply that the spot was not occupied in winter, a season when we should naturally suppose the Indians would be far busier in this industry than at any other time. The supposed absence of fire also indicates that the flint, prior to use, was not heated, as has been stated was often done. What the effects of moderate exposure to heat are, so far as facilitating the fracture of jasper and allied mineral, is questionable; but the author's experience in arrow-making leads him to conclude that the varieties of jasper and quartz represented in these several series are quite readily fractured, by either percussion, or pressure; and the art of arrowpoint making consists wholly in the acquired skill in governing the size and direction of the flakes detached, after the implement has acquired, in general outline, the desired shape and size.

On the other hand, while a site, such as this, was apparently only occupied in pleasant weather, it is noticeable that all such sites, so far as discovered, lie in close proximity to a spring or rivulet of good water.

The entire amount of refuse material, and number of tools found on this workshop site, do not throw any light upon the length of time this spot was occupied; but, as it is not demonstrable how many chips were necessarily stricken off in making an arrowhead, and as the earth was here thickly strewn with thousands of very small flakes, looking very much like coarsely crumbled shells and as much of this refuse, as well as some of the larger pieces, was quite deep in the soil, it seems evident that the spot was occupied for a long time. The coarse refuse may indeed at times have been gathered up and removed, but even supposing that this was never done, the thickly and deeply bestrewn condition of the soil with very minute chips indicates a prolonged occupation of this particular site. We find just such fragments about the known sites of Indian settlements, and flint chips are recorded among the contents of graves.[191]

Considering all the circumstances, the story that this accumulation tells is this: here, shaded by dense woods, on a slightly elevated knoll, in the midst of a meadow-like expanse of low-lying ground, through which trickled a sparkling spring brook, had tarried, at times, for years, an arrowmaker, shaping with marvellous skill those varied patterns of spearpoints and delicate drills, which are still gathered from the adjoining fields. Unlike localities of many acres in extent, where the traces of former occupation are scattered throughout the whole area, and indicate that manufacturing had once been in progress simply by the abundance of chips, we have in this workshop site the evidences of the toil of, probably, a single skilled workman, who, in the quiet ot his forest retreat, spent the greater portion of a long and useful life.

There is one feature of this interesting find to which it is desirable to call particular attention, inasmuch as it probably has an

---

191 Eleventh Annual Report of Peabody Museum of American Arch. and Eth., p. 313.

important bearing on the age of the supposed *pre-Indian* arrow and spearpoints, to which allusion has been made elsewhere.[192] This is the entire absence of argillite, or that material of which nearly the entire series of palæolithic implements from the gravel-beds are made. There have been several descriptions already published of the character of workmanship and position of a determining majority of the rude argillite arrowheads met with near the surface, and if they are, as a class, to be considered as having the same origin as that of the more elaborately formed specimens of jasper, chalcedony and quartz, then it would be natural to expect to find in the refuse of an open-air workshop, such as this, an abundance of flakes, splinters and cores of that mineral. Such traces, however, do not occur; nor, as yet, have otherwise similar sites been discovered where that mineral only was used. It can scarcely be held that, as argillite occurs in the valley of the Delaware as a living rock, there only should we expect to find traces of the localities where the mineral was worked up into arrowheads. It occurs also in the drift in the lower portions of the same valley, and is as readily obtained as the pebbles of quartz and jasper, with which it is associated. Flakes of argillite do occur quite frequently in the fields, just where we find the finished relics of the same material, and also some rude examples of what may be blocked-out or unfinished implements; but why may not these have been lying on the surface and in the soil before the advent of the Indians?

While the two classes, or those stone implements made of argillite and those of quartz, occur on the surface intimately associated, and it is obviously impracticable to dissociate them with anything like scientific accuracy, yet there is evidence that these argillite implements were frequently worked over by the Indians. However this may be, there is no reason why the rudely fashioned and weathered argillite implements may not be far older than those made of other material;

---

[192] Eleventh Annual Report of Peabody Museum, p. 254, 1878. See also Chapter XIX, p. 277, of this work.

and, when a considerable extent of the territory is carefully examined, it becomes evident, as elsewhere shown,[193] that such implements, whether of pre-Indian or Indian origin, antedate the jasper and quartz specimens with which they are now associated. Unquestionably, the occasional occurrence of argillite implements of exceeding delicacy of form, accuracy of outline, and comparative freshness of surface, has an important bearing upon the question of the date of the *general use* of this mineral,—for such specimens of savage handiwork are probably the productions of the Indians; but when we remember that there have been no arrowpoints made in New Jersey for at least two centuries, and that argillite chipped some time prior to that has undergone no appreciable alteration, we surely have something of a guide as to the rate of weathering of those other forms that are so deeply altered over their entire surfaces.[194] No jasper specimens have been found showing great alteration of the surface. They are as fresh as though chipped but yesterday, and is it not probable, that, although argillite will be affected by exposure sooner than jasper, if the two minerals were in common use from the date of the arrival of the Indians, there would be some difference detected in the surfaces of jasper arrowheads when thousands are examined and compared?

Until we meet with the chips and other refuse of argillite associated with those of jasper and quartz, or discover a workshop site where it was solely used, is it not safe to conclude that, from the great degree of weathering which the vast majority of the implements made from it have undergone, such specimens are of an earlier origin than those of jasper and quartz; the exceptions being referable, first, to the fact that outcroppings of this rock, where found in place, were sometimes

[193] Nature, vol. xi, p. 215, Jan. 14, 1875, London; and American Naturalist, vol. x, June, 1876, p. 329.

[194] Arrowheads of slate and shale are quite commonly met with, throughout the entire area of the state. These, very generally, have "weathered" to a greater extent than those of argillite, and having much the same appearance of the surface, are readily mistaken, for such as are made of the latter material. Arrowheads of argillite are not abundant—in many localities are very seldom found.

utilized, though not to a great extent, and also to re-chipping by the Indians, of weapons, that to them, when they were gathered, were veritable relics of a by-gone time. Finally, as the palæolithic implements proper, as found deeply embedded in the drift deposits are, as a class, argillite, it is rational to ascribe the origin of these rude weapons of the same material to the descendants of the same people who fashioned them; and their absence from the open-air workshops seems confirmatory of this supposed earlier origin of these ruder, and much weathered implements, in which we have a trace of an industry, once world-wide, on the part of a people ruder even than the Indians, whose productions as a class are as primitive and uniform as those of the later race are, in comparison, elaborate and varied.

The relationship which the argillite implements bear to those made of jasper and quartz is probably still further demonstrated by their abundance in those places where the two are found commingled; for it is in the places which are believed to have been occupied, first by the users of argillite, and subsequently by the Indians, that we can gather those facts which give us some insight into the relative antiquity of these different peoples. There are no known localities where argillite has been made use of exclusively, and, if there were, it would be difficult to determine whether the implements occurring under such circumstances were of Indian origin or earlier. On the other hand, as these implements are scattered promiscuously over the country, never occurring in such workshop sites as have been described, and are found in virgin soil at an average depth, greater than that at which jasper and quartz generally occur, there is in these facts alone, a strong indication of their greater antiquity. Where thoroughly commingled, as has been the case with the great majority of surface "finds" of relics of pre-European races in North America, the question of the abundance of the implements of the one mineral, as compared with those of the other kinds, becomes all-important.

As an indication of what is supposed to be true of the Atlantic coast of North America, it may be mentioned that of a series of twenty thousand objects gathered by the author in Mercer Co., New

Jersey, forty-four hundred were of argillite, and of such rude forms and in such limited varieties, as would be expected of the productions of a less cultured people than the Indian of the Stone Age. Of this series of forty-four hundred, two hundred and thirty-three are well designed drills or perforators and scrapers; the others being spear-points, fishing-spears, arrowheads and knife-like implements—not so great a variety of patterns as are now made by the Eskimo, nor of such excellent workmanship. Although it is true of these implements that they are of more primitive forms and therefore probably older than the objects made of quartz and jasper, the argument does not rest so much upon this greater simplicity, as upon their decomposed condition, their occurrence at greater depths in the undisturbed soil, the greater adaptability of the spears for fishing purposes, and the absence of all indications in the deeper soils, of the utilization of the minerals *habitually* used by the later Indians.

For these reasons, it is claimed that we find sufficient evidence, in them, of a pre-Indian people—believed to be the Eskimo—who, it is further claimed, are the direct descendants of that still older race, the fabricators of the palæolithic implements of the river drift.

A second find of this character also deserves a detailed notice, although it has already been referred to by Professor Wyman. At the time of its discovery, the importance of carefully noting everything connected with it was not recognized. Professor Wyman [195] was furnished with a few brief notes with reference to the specimens gathered and forwarded to him. This "open-air workshop" was discovered near Trenton, N. J., in 1872, and is less than a mile distant from the one previously described. It is situated on the brow of a hill, or rather of a plateau margin, where it is broken by a ravine through which flows a considerable brook. Originally surrounded on two sides by a dense forest, here always has been an open spot, with an extensive southern outlook over a broad expanse of meadow extending from the foot of the uplands to the river. When discovered, there

[195] Fifth Annual Report Peabody Museum, p. 27.

was simply a shallow depression to be seen, nearly circular in outline and about ten feet in diameter. On removing a thin layer of vegetable mould, through which projected a few irregular masses of yellow jasper, there were found a large quantity of thin flakes, chips, and a number of broken arrowpoints, especially of the triangular pattern. Of the latter the proportion was much larger than in the preceding instance, and this fact may indicate that the workman who had operated here was either less skilful, or that this pattern is more liable to breakage, which seems improbable. The accumulation of refuse, in this instance, was on a level floor of compact clay-earth, about which I could not discover a trace of fire. Separated from this refuse, by a layer of earth nearly a foot in depth, there occurred a somewhat similar deposit, except that cores and large chips only were found, with no trace of either hammer-stones, or broken or unfinished implements. There was also but little variety of mineral, the deposit consisting exclusively of dark, yellow-brown jasper. My impression is that there was no connection between the two finds, but that the deeper one was just so much older as it requires years for some eleven inches of soil to accumulate in a forest, where the growth of vegetable mould, from the decay of the annual fall of foliage, is steadily in progress.

A third deposit of flint chips consists of a series of some fifty specimens of brown jasper of quite a uniform size (P. M. 14,706), but which in the character of the fracture differs materially from both cores and flakes. They are too irregular and small for the former, and exhibit no regularity in the flaking. Still, when we consider that they were found closely packed together, and buried nearly a foot deep in a meadow, which was originally swampy ground, it is evident that they were designedly buried, but for what purpose it is difficult even to conjecture.

It may be well here to state that the three deposits just described are all in the immediate vicinity of the extensive deposit of finished implements to which reference has been made in other publications.[196]

---

[196] Annual Report of Smithsonian Institution for 1875, p. 272.

These were all of the same material, and identical with the jasper fragments here mentioned. If the former were made at the spot or near where they were found buried, this later find of fragments may indeed be a portion of the refuse accumulated in their manufacture; having been selected for converting into small arrowpoints, and afterwards forgotten.

An interesting feature connected with these accumulations of refuse is, that while the minerals there found are the same as that of which the great bulk of arrowheads and other implements are made, there will occasionally be met with a specimen of an arrowpoint, or fragment of a spear, of different pattern and material, such, for instance, as the extreme point or barb of an arrowpoint of obsidian; or, it may be, a fragment of jasper, of a color not occurring here, yet common in distant southern or western localities. These instances are numerous and show clearly the contact of different and distant tribes; for what better evidence can be produced to show that an implement has been brought, either through the vicissitudes of warfare, or through barter, from some far-off point, than to find that all trace of the material, of which it is made, is wanting in the heaps of refuse? The range of patterns of all our chipped implements is infinitely varied, and it needs but a glance at the material I have gathered from this one spot to show how little can be inferred from the shape of any specimen. Scarcely a European pattern except, perhaps, the most delicate arrowpoints from Denmark, that does not find its counterpart here, at least in a fragmentary state. In some instances the shapes seem to have been determined by the particular use for which the implement was intended; but in vast numbers the outline was determined by the shape of the chips, and thus originated the non-symmetrical specimens that we frequently find. Some of these are so crooked that their availability as arrowpoints is questionable, and they may have been used as knives. In the territory of every petty tribe (and every creek appears to have had one such tribal community dwelling in its valley), there is seen a family likeness, so to speak, extending through the whole range of chipped implements, and especially among the arrow and spearheads so that it is not difficult to pick out an occasional specimen, and set it aside as something foreign.

Having already incidentally referred to the chance occurrence of flakes and chips upon the surface of the ground, it is well, in conclusion, to refer to them more particularly, as it is possible that, collectively, they may have more significance than might at first be supposed. Jasper flakes, such as fig. 411, occur not only in "open-air workshop" sites, but on the surface of our fields; while flakes of other minerals, more especially of argillite, also occur whenever we find arrowpoints and spearheads of the same mineral. On the other hand, no deposits

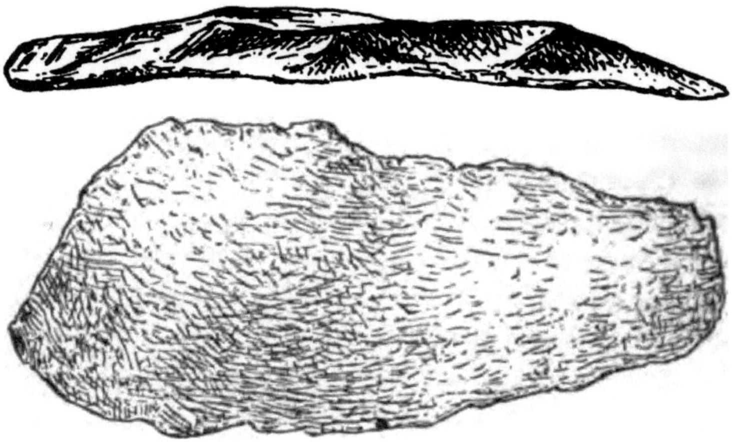

FIG. 411.—New Jersey. 1/1.

of argillite chips and cores have been discovered. From these facts the inference may be drawn, I think, that as argillite flakes, mostly exhibiting a great degree of weathering of their surfaces, are associated with finished implements of the same material, but never in such numbers as to indicate the spots whereon the former were fabricated, it is to be supposed that as flakes they were put to some use, such as knives, or if very small and not too irregular in outline, as arrowpoints. Figs. 412 and 413 are examples of flakes of argillite (P. M. 9,008, 16,315) such as are found singly on our fields. In general outline, fig. 412 does not materially differ from flint flakes, as found in Europe; and with the sharp edge that this mineral presents, when freshly fractured, it certainly is well adapted to cutting such yielding substances as the skin and flesh of small mammals, and to scaling and cleaning fish. Whether fig. 413 could be used as an arrowpoint, or is to be looked upon as a small knife, is altogether conjectural; and yet there are many ways in which just such a fragment of stone, provided the edges be sharp, might be utilized. The same character of flakes (fig. 411 P. M. 14,629) with small chips of jasper, and finished implements of the same mineral being likewise scattered singly over the entire area of the state, must therefore, if the argillite examples are not misinterpreted, have a like insignificance. Considering the absence, as yet, of

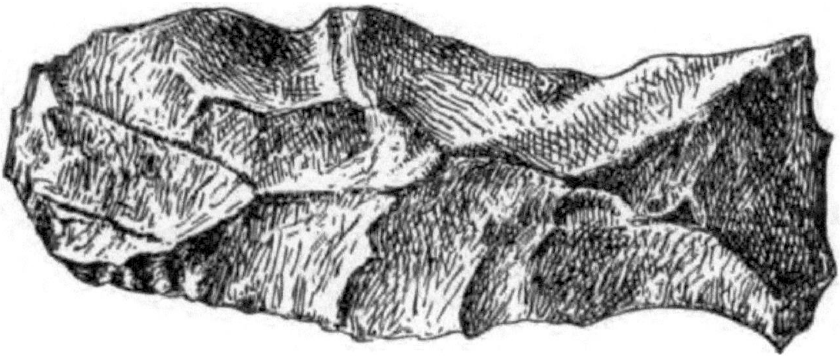

FIG. 412.—New Jersey. ½.

sites of arrowmakers' workshops, where argillite was exclusively used, and that chipped implements of this mineral are, when compared with jasper, *characteristic of the deeper soils* wherever the virgin earth has been examined; and bearing in mind also that argillite is a living rock in the vicinity, while jasper and the allied minerals, of which the bulk of the chipped implements are fashioned, occur only as bowlders in the drift, and require more labor to gather than it would take to visit a ledge of living rock,—it is safe to conclude that the argillite spears, arrowpoints and flakes, as a class, are of an earlier time than the same implements of the other minerals, and doubtlessly bear a closer relationship to the still ruder—the *primitive* implements found in the river drift gravels of the valley—than they do to those made of jasper. Certainly, if fabricated by a former people, or even by their own ancestors, these discarded or lost implements would not escape the keen vision of the Indians, few of whom lacked sufficient skill to repoint, and render available these ruder specimens of the same weapons to which they were accustomed.

The conclusion might readily be drawn from the foregoing that weapons and implements of all kinds, chipped from stone, were made by comparatively few persons, who supplied the people of their respective tribes with such implements as they required. While the manufacture of the finest specimens was very likely confined to adepts, who

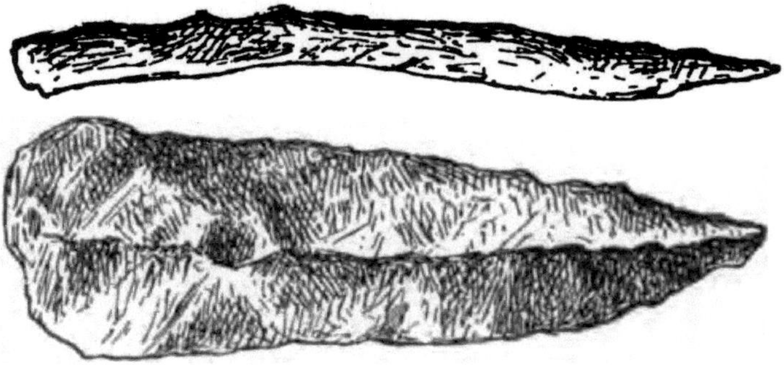

FIG. 413.—New Jersey. ½.

made it the business of their lives, it is probable that those of inferior finish, which are found scattered over the state and mingled with others of artistic workmanship were made by hunters or warriors, as the case might be, who subsequently lost them. However occupied, whether on the war-path or in the chase, it would scarcely be possible for a warrior or a hunter to supply himself with as great a number of arrowheads as he would need, even for a few weeks. Vast numbers unquestionably were lost or broken when first discharged from the bow; and when we consider the various conditions under which these same arrowpoints now occur, it is evident, that to a certain extent, every Indian was his own armorer. Ornaments and stone implements, whether weapons or for domestic purposes, were of careful or careless finish—and we find both patterns—as their owner happened to be proud or indifferent, or possibly neat or slovenly. A ground edge to an oval pebble being required, the subsequent shaping of the other portion of the implement might be gradually accomplished, if the implement itself were used without a handle or could be readily separated from it. Such shaping, by grinding away all irregularities, was at best a slow process, and one very unlikely to be followed as a means of livelihood. When therefore, we find a beautifully polished and symmetrically fashioned celt we probably have a proof of the patience and skill of its original owner; and any one, with the exercise of the same patience and skill, could soon learn to chip from flint his own arrowpoints, knives and spears.

CHAPTER XXXII.

PALÆOLITHIC IMPLEMENTS.

We have now to consider a class of objects, which, though bearing marked resemblance to the ruder forms of Indian or neolithic implements, from their uniformity of manufacture, their identity of material, their slight variation in design, and their occurrence in an older geological formation than the present surface soils, are believed to be veritable traces of a people, who inhabited the northern Atlantic seaboard of America, prior to the advent of the Indian; if we accept the current opinion that the latter was a comparatively recent comer to our shores.

Before passing to a detailed description of a series of these earliest traces of man, it will be necessary to refer briefly to the physical character of the river valley in which these implements have been found. This is the more necessary because having been seriously misled[197] by the various geological reports that purport to give, in proper sequence, the respective ages of the several strata of clay, gravel, bowlders and sand, through which the river has finally worn its channel to the ocean level, I have probably, in previous publications, ascribed too great an antiquity to these implements, although what is now known to be a substantially correct history of the various deposits in the river valley does not dissociate these traces of man from a time when essentially glacial conditions existed in the upper valley of the Delaware

---

[197] American Naturalist, vol. vii, p. 204, figs. 37 and 38, Salem, Mass. 1873: and vol. x, p. 329 fig. 21, Boston, Mass., 1876.

Also: Tenth Annual Report of Peabody Museum, Cambridge, Mass., p. 30, figs. 1, 2 and 3; and Eleventh Annual Report, same institution, p. 223, figs. 1 to 4, inclusive. Cambridge, Mass., 1878.

river, though they occurred subsequently to the existence of the great continental glacier, when at its greatest magnitude.

It was not until the surface geology of the Delaware river valley was carefully studied by Mr. Henry Carvill Lewis, of the 2nd Geological Survey of Pennsylvania, that we were in possession of all the facts necessary to enable us to recognize the full significance of those early traces of man, discovered in one of the latest geological formations of this valley.

The results of Mr. Lewis' studies are embodied in two communications[198] to the Academy of Natural Sciences of Philadelphia; but to these, but brief reference will be made, as their author has kindly prepared for this volume, an able abstract of his earlier papers, and added thereto, the results of his later investigations.

It is sufficient here to state that, according to Mr. Lewis, "the last and newest of all the gravels"—in the valley of the Delaware river—"is one which, at Philadelphia, seemed to be of little importance. It lies close along the river; and, rising a few feet above it, extends but a short distance back from the river bank. It covers the flat ground of Camden and the lower part of Philadelphia, and forms islands in the river. It was called *the River gravel and sand.* It is this alluvial gravel, the latest, except the recent mud-flats, of all the surface formations, * * * * which, from its great development farther up the river, is now named *the Trenton Gravel.* It is in this * * * * gravel only, that traces of man have been found."

This deposit of gravel has also been described in full, by Prof. Geo. H. Cook,[199] who says: "The beds of stratified drift, at various places in the valley of the Delaware, south of the line of glacial drift, bear marks of having originated from the action of water. The bowlders and cobble stones are all water-worn, and round, and are not scratched

---

[198] The Surface Geology of Philadelphia and vicinity: Proceedings of the Mineralogical and Geological Section, Academy of Natural Sciences, Philadelphia, Pa., Nov., 1878.

Also the Trenton Gravel and its Relations to the Antiquity of Man. Philadelphia, Pa., Nov., 1879.

[199] Annual Report of the State Geologist of New Jersey for 1877, p. 21; Trenton, N. J., 1877.

or streaked. They have all come from places farther north in the valley and have been moved and deposited by powerful currents. There are to be seen in the railroad cuts near Trenton, where the exposure of this kind of drift is very fine, bowlders of gneiss, from the rock near; of red sandstone from the country just north; of trap from Lambertville; of altered shales from the near trap; of conglomerate from New Milford; of magnesian limestone from the valleys of Warren county; of conglomerates from the Blue Mountain, and of cherty and fossiliferous limestones from the Delaware valley north of the Water Gap. The gravel consists largely of quartz, but it contains numerous fragments of red shale, and black slate."

This is a correct description of this drift material, except in one most important particular. It is an error to state that all the bowlders and smaller pebbles are water-worn, and that none are angular. There is, it is true, but a small proportion of scratched pebbles or bowlders, but there are a large number of sharp, angular stones. No such uniformity of size, shape and smoothness of surface as characterizes the pebbles of our ocean beaches, is found to exist.

Prof. Geikie has described, as characteristic of all the river valleys in England, that the upper levels consist of coarser material than the lower, "and frequently contain large blocks of stone which could only have been transported by river-ice;" while the lower level gravels are usually of finer grained character, and that these "seemed to point to a milder condition of things—to a time when the rivers were less liable to flood, and the ice-rafts were uncommon" ("Great Ice Age," American ed., p. 435). An examination of the bed, as now exposed on the bank of the Delaware river, shows that no such distinction can here be drawn. By actual count, in a section measured off, there did not occur more large bowlders above the line equally dividing the bluff, than below it; but, on the contrary, there were three more beneath, not counting those that had rolled from the bed down to the water's edge. This of itself would be by no means conclusive; but upon frequent inquiry of men who had sunk wells and excavated cellars, it became evident that these bowlders were met with in greater abundance, at considerable depth, rather than near the surface. This irregularity of their occurrence and position is of itself indicative of the agency of ice in the origin of the deposit. Mr. Lewis, however, has informed me, that he thinks, if the entire area of the Trenton gravels be considered, that this gravel "is certainly characterized by bowlders *on top*, resting upon finer material, as an examination of the railroad cut at Trenton will show." My own impressions were that this was not the case, and I therefore give the reader the benefit of the conclusions, on this point, of a competent geologist, allowing my own statements, as originally published (Peabody Museum Reports, vol. ii, p. 226), to remain unaltered.

These large bowlders are also met with upon the surface. Many of them have been very carefully examined *in situ*, and it was evident that they were not deposited with the gravel beneath them, and the latter subsequently removed by rapidly running water, for in nearly every instance, there was a foot or more of sand between the lower surface of the stone and the gravel, and this layer of sand extended so far beyond the limits of the bowlder, that unquestionably it must have been slowly accumulated prior to the deposition of the bowlder; as evidence of this, a well chipped, spear-shaped implement was found in such a stratum, immediately beneath a stone that would weigh at least half a ton. These surface bowlders undoubtedly have been dropped from ice-rafts, together with sand and gravel, the ice then floating over a broad expanse of country, in comparatively quiet waters.

In the Annual Report for 1877, of Professor Cook, State Geologist of New Jersey, we find an excellent map, and a detailed account of the glacial drift that covers the northern portion of the state and consists of unstratified bowlder clay and ice-scratched, angular pebbles. Where the débris of the ancient glacier ceases to be of this character, Professor Cook considers the glacier terminated, and all the material lying to the southward as a modified deposit due exclusively to water action.

Of the great glacier itself, Dr. Cook remarks, in the report alluded to, "even in New Jersey, it covered the tops of the highest mountains.

"This immense mass of ice had a slow movement from the north towards the south, in which it scraped or tore off the earth and rocks from the rocky mass under it, grinding, grooving and smoothing down the rocky surface, and pushing forward, tumbling and rounding the fragments of stone and rock, and finally leaving them at the southern edge of the glacier, or wherever breaks in it may have allowed the loose materials to rest.

"The terminal or southern edge of the drift is well and very plainly marked by a line of hillocks of mixed clay, sand, gravel, rounded stones and bowlders of large size."

Of its extent, geographically considered, he further remarks of it, as "beginning on the eastern side of the State on the north side of the Raritan, at Perth Amboy, the line of Short Hills extending from that place to the First Mountain, and passing just north of Metuchen, Plainfield and Scotch Plains, marks the southern edge of the drift.

"From there, it extends to the Delaware below Belvidere. The portion near the Delaware shows the gravel and bowlders very plainly, but it appears to have been washed and otherwise modified by floods or great bodies of water descending in that valley. The whole line of this moraine is remarkably plain and well defined.

"Across New Jersey the line is not exactly east and west, but appears to deviate towards the north, the deviation being greater somewhat in proportion as the ground is more elevated.

'The hillocks of stones, gravel and earth, which together made this long chain, have every appearance of piles of débris which have been thrown down without order, and without the presence of water to sort or arrange the various materials."

Nowhere, as here described, does the terminal moraine of the great glacier approach the bluff at Trenton nearer than sixty miles, or following the valley of the river, fully seventy miles; but these distances are really of little moment, in connection with the subject of man's presence here during the maximum severity of glacial conditions in North America. With the existence of a glacier filling the entire valley of the Delaware, sixty miles northward, and extending across the state to the Atlantic coast, there must necessarily have been a widely different physical condition of the entire territory extending southward. A large part of this area, now constituting the southern, low-lying portion of the state, was submerged; and Mr. Belt[200] has pointed out, that over such low-lying and submerged areas, there would be spread out a vast amount of material, by the agency of sub-glacial torrents, consisting of the true glacial débris, borne still farther southward by the currents caused by the melting of the glacier at and near its base. Such swift-flowing currents might readily, through long periods of time, being charged with sand and small pebbles, wear away much of the ice-scratching that is so characteristic of the pebbles in the more northern drift; but to such sub-glacial rivers we cannot well refer the enormous bowlders scattered promiscuously through the gravel deposits, as seen at Trenton, that must be ascribed to the more powerful agency of floating masses of ice detached from the glacier existing farther to the north.

What has been here ascribed, by the late Mr. Belt, to the action of sub-glacial streams, doubtless took place at a much earlier date than the deposition of the gravel, through which flows the shrunken river of to-day. These accumulations of gravel are, it is much more probable, as Mr. Lewis has shown, of a character that requires the aid of floating ice to transport the larger bowlders.

These masses of floating ice, as Mr. Lewis believes, having displaced the older clay and bowlder deposit, in great part, from the valley at Trenton, and southward, and worn a deep basin through it, the present river, then a glacial stream, brought by the aid of floating ice enormous quantities of material from the terminal moraine of the continental ice-sheet and re-filled the valley created by the removal of the older clay and gravel. During the accumulation of this later gravel, man must have occupied the adjacent land, and there existed no physical reasons why he should not have done so. However recent, the careful studies of the geologist may ultimately determine these

---

[200] Quarterly Jour. of Sci., Jan., 1878: London.

gravels to be, there need be no relinquishment of the belief in the archæologically great antiquity of the traces of man found in it.

Admitting, without doubt, that the sequence of conditions resulting in the various superficial deposits from the later tertiary to past-glacial times, as determined by Mr. Lewis, there is to be considered the important fact that the minimum time required for the deposition of the Trenton gravels was not necessarily the actual length of time during which these accumulations were completed. In all such cases, it must be borne in mind, that rapidity and regularity of action are not constant factors of the conditions that result in geological formations. Further, the supposed second glaciation of the Delaware valley, limited as it may have been, and comparatively of short duration, was yet an event of remote antiquity, as man measures the events of time. Not simply a few centuries ago, was the river the mighty stream that Mr. Lewis describes, when the vast beds of this "recent" gravel were deposited. Long vanished, indeed, may have been every vestige of the earlier, continental glacier—but what of the interim, between the disappearance of the former and the appearance of the supposed latter? Can we assume that the events of the accumulation of gravel on gravel, clay on clay, removal of earlier and deposition of later strata, occurred without a break? We read of them as following each other in a given order, but seldom pause to think how long a time might have elapsed between these several occurrences.

Finally, as bearing on the one important question of man's antiquity, it is here strenuously maintained that the forces that caught up these later gravels also gathered, in part, the rude implements that now give such interest to the deposit. It is evident from the condition of some, and the depth at which many are found, that they were made prior to the formation of the containing bed, and were lost or discarded when the floods swept down the valley. Surely, we are without warrant in assuming that only after the last pebble was in place, that man appeared, and dropping these implements in the water, they sank into the gravelly bed of the river, even to a depth of forty feet.

Admitting that man was not interglacial, and is more recent than the epoch of the brick clays described by Mr. Lewis, he surely may have been an occupant of the Atlantic seaboard before the total disappearance of the glacier farther north. Whether this glacier is a first or second occurrence of this phenomenon cannot yet be positively determined.

Whatever the changes that may have taken place during the glacial epoch, there is no evidence that the land now constituting the southern portion of New Jersey was uninhabitable, long prior to the deposition of the implement-bearing gravels. There is abundant evidence, on the contrary, to show that it was inhabited in preglacial times by a number of the larger mammals, including the mastodon, which, it should be remarked, lived long after the disappearance of every vestige of glacial conditions. There seems no reason, therefore, for excluding man.

No floods arising from the melting of the great glacier or continental ice-sheet, submerged the portion of the state, which extended seaward miles beyond the present boundary,[201] as Mr. Lewis has shown beyond question, and the absence of traces of man in the older glacial

[201] It is well known that there is now in progress a gradual subsidence of all the lands along the tide-waters of New Jersey. How long ago this commenced is not readily determined. This depression and elevation is believed to be within a range of twenty feet; but even this amount of elevation would materially increase the area of the state. Not unfrequently fragments of cedar trees which are derived from forests now covered by the ocean, are thrown upon the beach, during violent storms, and remains of the mastodon have likewise been found, that are, with reason, believed to have been washed from these same sunken forests. In an interesting lecture by my friend, Dr. Maurice Beesley of Cape May Co., I find the following statements, which have a direct bearing upon this subject, and clearly show how vast an area has been lost in the last few centuries by the encroachments of the sea, all of which, in palæolithic times, was habitable land.

"We find, likewise, that Egg Island, twelve miles from our shores, was surveyed by Budd and Worlidge in 1691, and taken up, as per their draft for Thomas Budd, a Quaker gentleman of Burlington, N. J., for 300 acres full measure. My father, Thomas Beesley, has related to me that when he followed the water from 1790 to 1800 it contained about 60 acres. Here we have a loss of 240 acres in a century. In 1830 gunners went from our place to Egg Island to kill geese, which frequent those meadows in great abundance, and were very successful. This Island made a final disappearance about twenty years ago, and vessels now sail over the entire area it formerly occupied. To throw its former area of 300 acres into a circular form we find the wash has been a rod a year since 1691 up to the time of its final disappearance. Therefore, the loss on the east side of Maurice river cove having been one and a half rods a year, and on the west side one rod, we

drift indicates, it may be, that he never dwelt in preglacial times upon that portion of the country over which this drift was carried, and northward of the terminal moraines, but, of itself, offers no reason why he should not have lived, so long ago, in the territory beyond the reach of glacial conditions.

Whether the view taken by Professor Cook, that the Trenton gravels are of glacial origin and derived from floods caused by the melting of this great continental ice-sheet, or are an indication of a post-glacial condition, subsequent to the great glacier, as considered probable by Mr. Lewis, the fact remains, that these implements are indicative of man's presence, and have been placed in their present posi-

---

have a loss per annum of two and a half rods between those shores, and as the distance between Egg Island light and the mouth of Dennis creek is now estimated at twelve miles, we find it must have been solid land or meadow that whole distance, with the exception of creeks and rivers, 1536 years ago, or A. D. 343. That the whole extent of Maurice river cove was originally meadow seems plausible, from the fact that a muddy bottom exists throughout its widespread limits.

" From the present mouth of East creek, when the tides fall out low, a range of oyster beds can be traced for more than a mile into the bay, being the original bed of the creek. In fact Christopher Ludlam, father of Charles Ludlam, Esq., located 40 acres of meadows in 1801 on the western side of said creek, which from the minutes of the deed ran up it a straight course 100 perches to a ditch passing westward from the creek at right angles. In 1869 I visited the spot and found the whole tract of 40 acres had been swallowed up by the capacious maw of the bay, and exists only in the shape of atoms. Therefore, if 100 rods has been washed away in sixty-eight years, we find the average loss one and a half rods a year, as before stated.

"Dennis creek, according to a survey made in 1767 by Aaron Leaming, the 2d, and re-surveyed by Leaming M. Rice and the writer in 1867, has lost 150 rods during the century. As the " reach " that disappeared in that time ran a southwest cousre, running diagonally with the direct course of the 150 rods loss, the mouth of Dennis creek must have been about three-fourths of a mile nearer to the mouth of Goshen creek than at present. As this loss of soil still continues, where no gravel exists, it is ėvident the farmers of East creek neighborhood will have the waters of Delaware bay dashing against their homesteads in the course of three or four centuries more.

" This wash is not confined exclusively to the bay shore. At Cape May City by ancient deeds made in 1689, the distance across the Island, from the beach to the creek, was 265 rods. Is it half that distance at the present time? For many years past, barriers have been erected to prevent further encroachments. Those of you who are familiar with the seaboard, will have seen along our beaches after heavy storms and tides the meadow-land jutting out oceanward with the salt grass still adhering to it, as it was two or three centuries before, when first enveloped by the westward movement of the beach. I well remember, likewise, large creeks upon our beaches that have been submerged and lost to view, probably forever, through this gradual change of soil, and through the intervention of the winds, and the waters of old ocean constantly encroaching upon and frittering away our territorial bounds."

tions, varying from three to forty feet in depth, by the same agency that laid down the gravels; and in either case we are carried back to a time when a bed of bowlders, sand and gravel, of many feet in thickness, was laid down, and subsequently a channel worn through it, until the waters of the river reached the level of the sea, and all accumulation of other material than sand and silt practically ceased. From that time until the present, how many centuries may have elapsed! Although brought to the near present—to but yesterday, geologically speaking, we have yet to consider, that since the last bowlder found its present resting place, there have accumulated in the present flood plain of the river, vast beds of sand and mud, charged and recharged with vegetable growths, that overlie much of this later gravel, and this also contains more highly wrought evidences of man's presence, themselves suggestive of a considerable antiquity.

Professor Cook [202] very accurately describes the appearance of the river drift. In a late annual report he remarks, "there are a number of terraces, on ground open towards the sea, which have their upper surfaces almost at equal heights above tide level. The terrace on both sides of the Pennsylvania railroad, near Trenton depot, is one of them. It is between fifty and sixty feet high. The gravel bed at the railway depot at New Brunswick and also that on Bayard street are nearly the same level. The terraces at the Narrows in New York harbor are also on the same level, and others along the west bank of the Passaic, in Newark, are of nearly the same height. It can also be recognized along Bergen Hill, both on the North river and the Newark bay sides. Up the North river, near Peekskill, similar terraces are very conspicuous, but their height above tide is about ninety feet. In New Jersey, in the more southern part, terraces are seen, but they are at a somewhat lower level. The whole of them taken together indicate that at some former time, perhaps at the close of the glacial period, the ocean level was somewhat higher than it is now, enough higher to bring it just over the tops of these terraces, and water

---

[202] Cook. Annual Report of State Geologist for 1878, p. 22. Trenton, N. J., 1878.

bringing stones and earth from the higher ground would deposit them, when it reached sea level, in these steep flat-topped terraces. The terrace at Trenton is just where the valley of the Delaware comes down to this level, and where it widens out to allow room for such a deposit to accumulate. These are good examples of terraces of the Champlain Period."

In this brief extract from the report on the surface geology of New Jersey, its author, it will be seen, refers all the evidences of combined ice and water action to the glacial epoch, or not later than a period constituting its close; and further shows that his view is that the Trenton gravel is an *ocean terrace*, made by water which covered *all southern New Jersey*. This is, unquestionably, wholly an erroneous view.[203] Reference to the map accompanying Mr. Lewis' paper shows, that the gravels that surround the Trenton gravels had emerged from the ocean level long previous to the deposition of the latest or Trenton gravels.

Personally, I can but express an opinion on the archæological significance of the traces of man found associated with these gravel deposits, and this is in nowise affected by the age and origin of the containing beds. Whatever age the geologists may assign to them, be it inter- or post-glacial, these traces of man must possess a very great antiquity.

It is evident that, at just such a locality as Trenton, where the river "widens out," traces of man, had he existed during the accumulation of the gravel, would be most likely to occur. This is true not only because there is here the greatest mass of the gravel, and the best opportunities for examining it in section; but the locality would be one most favorable for the existence of man, at the time. The higher ground in the immediate vicinity was sufficiently elevated to be free from the encroachments of both ice and water, and the climate,

[203] Since the above was in type, a late report by Professor Cook has been issued, in which he expresses views somewhat at variance with his earlier reports, and to a great extent in accordance with the views expressed in an article on these gravels, by Prof. H. C. Lewis, in the Proceedings of the Philadelphia Academy of Natural Sciences.

soil and fauna, all such as to make it possible for man to exist, at this time, in this locality. As Professor Cook, in this same report on page 14 has said: "It is difficult to get a clear conception of the condition of things which would have allowed the whole surface to become covered with thick ice, when that surface was nearly as it is now, and some points are not satisfactorily explained. It can be understood, however, that extreme cold was not necessary, for we see forests, pastures and grain fields about the lower ends of the present glaciers. There has been found in the terrace of modified drift at Trenton the tusk of a mastodon, which was evidently washed there when that mass of matter came down the valley of the Delaware with the torrents of water from the melting ice. It was about fourteen feet under the surface, and the gravel and stones were partially stratified over it. From these the inference seems plain that the climate at that time admitted of the growth of animals like the elephant in size and habits. Whatever theories or hypotheses may be adopted in regard to glaciers, the piles and other deposits of loose bowlders, gravel, sand and clay and the scratched rocks under them, which are found everywhere in northern New Jersey, are wonderfully like the deposits which are made by modern glaciers, and there is scarcely a feature in one but what can be paralleled in the other."

When we consider that not only the remains of the mastodon, but those of the bison have been found in this gravel, and that within a few yards of the spot where the tusk of the mastodon mentioned by Professor Cook, was found, palæolithic implements have been gathered, one at the same, and three at greater depths, it is apparent that we here have evidence of man's contemporaneity on the Atlantic coast, with the large mammals mentioned.

Certainly, it cannot be assumed that these mammalian bones were derived from the terminal moraine, where they were, at the time of their displacement, veritable fossils, and therefore their present association with human implements is merely accidental—that the pebbles and bowlders, being older than the relics of man, found with them, so also are the bones of those mammals which no longer belong to the fauna of this region. If the bones found in this gravel were petrifactions, and had all the characteristic marks of water-worn pebbles clearly traceable upon them, it might, then, be admitted that their present association did not show any evidence of equal antiquity. This, however, is not the case. The condition of these remains renders it clearly evident that the mighty currents, and floating icebergs of that ancient day, gathered at one and the same time, not only the bones of the mammals that roamed the neighboring forests, but likewise the weapons of the primitive people who preyed upon them.

It is desirable, at this point, to call attention to mammalian remains which have occured in superficial gravel deposits in New Jersey; possibly not of an earlier date than the age of the Trenton gravels, but not directly connected with them.

In referring to the discovery of walrus bones, in New Jersey, Mr. J. A. Allen, in his Monograph of North American Pinnipeds, p. 59 (Miscellaneous Publications, No. 12, U. S. Geol. and Geog. Survey of Territories), says: "In view of the now well-known former extension of the habitat of the Moose, Caribou, Reindeer, Musk Ox, and other northern mammals, southward to Kentucky, the * * * * hypothesis (referring to Leidy's statement that the New Jersey walrus bones may be the remains of the same species—as the living—which probably during the glacial period extended its habitation very far south of the latitude in which it has been found in the historic period) seems * * * * probable, and that the species in glacial times inhabited the eastern coast of the United States southward to Virginia, if not even beyond this point." Meagre, therefore, as are the traces of mammals in the Trenton gravels, it is evident that patient search will ultimately satisfactorily increase the list beyond that of the mastodon and bison. In the Geology of New Jersey, edition of 1868, page 740, Prof. E. D. Cope has mentioned the reindeer as among the most extinct mammalia of New Jersey. He says, "the Greenland Reindeer was a resident of New Jersey, when the walrus was on its shores, and when the climate resembled that of its present home. Antlers have been found in the gravel that covers the older formations everywhere." In the gravel that forms the eastern bank of the Delaware river, below Trenton, N. J., the late Prof. T. A. Conrad found a tooth, which he informed me, soon after, had been identified as that of the Caribou.

In this connection, also, it may be mentioned that remains of the woodland reindeer (*Rangifer caribou*) and of the bison have been found in an ordinary rock-shelter near Stroudsburg, Penn. The latter is believed to have been a remnant from a human feast, judging from marks of fire thereon.

In another cave, in Bucks Co., Pennsylvania, bones of the bison and moose are also reported.

In the first instance, stone and bone implements were associated with the remains of the animals mentioned, as well as those of the present fauna. These superficial "finds" have an important bearing upon the question of the discovery of bones of extra-limital and extinct mammals in the Trenton gravels. (Proceedings of the Philadelphia Academy of Sciences, p. 346, 1880.)

The commingling of relics of man and the bones of extinct animals is by no means confined to the Atlantic seaboard of this continent. The researches of Professor Whitney[204] demonstrate that this is true of the Pacific coast; while in the interior, evidence of the same association of man and the mastodon has been discovered by Professor Aughey,[205] in the Missouri valley. From the loess deposits in Nebraska, Professor Aughey has taken rude arrow or spearpoints, one at a depth of fifteen, and the other of twenty, feet below the surface; the latter beneath a vertebra of an elephant. In reference to these finds, Professor Aughey remarks, "It appears, then, that some old races lived around the shores of this ancient lake and paddled their canoes over its waters, and accidentally dropped their arrows in its waters or let them fly at a passing water-fowl. * * * * Thirteen inches above the point where the last-named arrow was found, and

---

204 Whitney. The Auriferous gravels of California. Cambridge, Mass., 1879.

205 Aughey. U. S. Geol. Survey of Colorado, etc. Hayden, An. Rep. 1874, p. 255. Washington, D. C., 1876.

within three inches of being on a line with it, in undisturbed loess, there was a lumbar vertebra of an elephant (*Elephas Americanus*). * * * * * It appears clear from this conjunction of a human relic and proboscidian remains that man here, as well as in Europe, was the contemporary of the elephant in at least a portion of the Missouri valley."

When such indefinite objects, as many of the palæolithic implements are, are offered as an indication of man's former presence, the question naturally arises: how are we to know that these rudely chipped pebbles are of artificial origin? This subject will now be considered.

The chance occurrence of single specimens of the ordinary forms of Indian relics, at depths somewhat greater than they have usually reached, even, in constantly cultivated soils, induced the author several years since, to examine carefully the underlying gravels, to determine if the common surface-found stone implements of Indian origin were ever found therein; except, in such manner as might easily be explained, as in the case of deep burials, by the uprooting of large trees, whereby an implement lying on the surface or immediately below it, might fall into the gravel beneath and subsequently become buried several feet in depth; and lastly, by the action of water, as where a stream, swollen by spring freshets, cuts for itself a new channel, and carrying away a large body of earth, leaves its larger pebbles and possibly stone implements of late origin upon the gravel of the new bed of the stream.

It was found that by all such means the most elaborately wrought Indian relics have occasionally been buried at considerable depths. It was also found, however, that there did occur in these underlying gravels, certain rudely shaped specimens of chipped stone, which have all the appearances of the stone implements of palæolithic times.

We find, also, on comparing a specimen of these chipped stones with an accidentally fractured pebble, that the chipped surfaces of the former all tend towards the production of a cutting edge, and there is no portion of the stone detached which does not add to the availability of the supposed implement as such; while in the case of a pebble that has been accidentally broken, there is necessarily all absence of design in the fracturing. Although the portions detached from these supposed stone implements are chipped with reference to the natural cleavage of the mineral, these larger surfaces, the result of single cleavages, are always supplemented by minor chippings along the edges, thus giving every indication of the original zigzag edge having been made comparatively straight by more careful work subsequently put upon it. This, of course, does not appear on a naturally fractured pebble.

These characteristic chippings obtained in the large jasper hoes and hatchets of the Indians, and a comparison of these with the ruder forms found in the gravel, show that identical means have produced the two forms: the difference being due to the want of skill in flint chipping, and in some measure to the greater difficulty in shaping this material, which differs essentially in its constitution from true flint, or the jasper, chert and chalcedony found in eastern North America; although it possesses a conchoidal fracture. It, may be mentioned, also, that although these implements are but little above the ordinary refuse of a modern quarry, and are often closely reproduced by the hammer, when fracturing rock for road-bed, yet they are none the less of artificial origin; and further, it must be borne in mind that this absence of careful workmanship is not wanting in the more recent productions of the Indians. From ancient graves in Massachusetts, from the stone graves in Tennessee, as well as from surface "finds" in Missouri, are several specimens—now in the Museum at Cambridge—which are in all respects, except in the mineral used, identical with the more specialized examples from the Delaware river gravels.

There is, in all the specimens collected, a considerable amount of weathering of the surfaces, the degree of which varies somewhat in the whole series, except where mineral other than argillite occurs. In such specimens, the alteration of the surface is much less.

The degree of weathering is of much importance, in its bearing upon both the age and origin of these objects. Not until a large series had been obtained from various depths, was it practicable to determine what variation occurred, in this respect; but lately, it has been found, on comparing a large series from depths varying from three feet to forty, that those from the greater depths were less weathered than the specimens found near the surface, except in a very few instances, when a few examples, showing a great amount of weathering, were found at depths exceeding twenty feet. Those specimens of palæolithic implements found upon the surface, on the other hand, intimately associated, as many of them are, with ordinary Indian relics, are invariably more weathered than those found *in situ*, in the deep, underlying gravels. In the uniformity of the degree of weathering of all the surfaces, we have evidence that the several chips removed, to produce the implement, were flaked off at the same time, and this, at once, places these objects in the category of artificial productions.

We derive from these facts, several important inferences bearing directly upon the age of these implements. Those that are but slightly weathered, and occur in deep, undisturbed gravel, were doubtlessly dropped in the river, and falling upon the gravelly bottom of the stream, were quickly covered by the constantly increasing mass of material which the swift currents were bearing downward from the ice-bound valley. These unweathered implements were comparatively new when lost. The more eroded examples from the same deep gravels may be considered as lost or discarded specimens that, after long exposure, were brought by the floods from some distant point. Those found upon the surface are such as were in use at the time of the gradual accumulation of the gravel, and from that distant time until now, have been exposed to the corroding influences of alternate winter's frost and summer's heat.

If the few specimens with deeply eroded surfaces, that have been taken from depths exceeding twenty feet, were weathered by long exposure prior to their burial in the gravel, then they must have been lost long prior to the occurrence of the transporting floods, and so made and used during the accession of glacial conditions, the second time (?) in the river valley. How long prior to that event, who shall say?

It has been suggested that these chipped surfaces might have been produced by frost action, and that these supposed implements therefore, were natural productions. Given a single fractured surface, which might be produced by the ordinary action of frost, and no other productive agency is required; but when we consider that instead of one, there are twenty or forty planes of cleavage, all equally weathered, and that an implement has been produced similar in form and finish to those neolithic specimens about whose origin there is no question, and we fail to see how nature, by any known or imaginable force, could so fashion either an oval pebble or an angular fragment of rock.

In this connection, attention may properly be called to many specimens of "chipped pebbles" which cannot be considered as implements, inasmuch as there is no trace of design in their present shapes. They are, indeed, artificially chipped over the greater portion of their surface, but they have no well defined point nor cutting edges. These irregular masses, usually smaller than the finished implements, bear no evidence of being crushed, although glacial action probably exposes fragments of rock or ice-encased pebbles more to such crushing force, than to any other, except perhaps that process of rubbing against denser mineral, which results in deeply incised striæ,—the so-called glacial scratches. The lithological character of argillite is such, that a given mass of this mineral, if exposed to a crushing force, will not fracture in such a way, as to resemble in any degree, the chipped pebbles, here referred to. When associated with the finished forms, and the same general character of weathering and of chipping is noticed on both, we cannot but consider them as identical in origin, and need have no hesitation in classing such designless forms, as broken specimens, as "failures," or, in some instances, as refuse chips, as they are found to have, in all respects, the same forms that are characteristic of the localities where neolithic implements of chert and jasper have been made.

One feature of them all, and especially of those from the deeper gravels, needs to be briefly referred to; this is the worn condition of the edges of the several surfaces produced by the detachment of the flakes. There are, especially in fig. 413, no well defined outlines of a single facet, although each separate flake can be traced on the surface of the implement. This partial wearing away, of these lines of separation of the several chips that have been removed, does not occur, to any marked degree, in jasper specimens that approach fig. 413 in shape, size and chipping. Whether this is the result of use previous to being lost or discarded, or of wear by long exposure to the shifting movements of sand and gravel, cannot now be determined; but of itself, it seems to connect closely these partly worn, yet clearly artificial forms, with rolled pebbles, which in outline only suggest the possibility of having once been chipped implements. While, naturally, broken pebbles may often approach in shape any of these forms of stone implements, it may at once be seen that it is, in every case, but an accidental resemblance.[206] The outline is obtained, but not that subsequent chipping that gives the finish that makes the implement desirable for use. The gravel bed, in which these "turtle back" celts and their modification have been found, contains a small percentage of angular pebbles, that have not lost all traces of recent fracture, and therefore are not as smooth and uniformly polished as an ordinary pebble. These specimens, when bearing marked resemblance to those clearly of artificial origin, may in fact have been fashioned by man, and have only partially lost, by the polishing action of water and sand, those indications of artificially produced fractures, which characterize the specimens here figured; but, as a rule, the angular pebbles are of natural formation, and their imperfectly ground and polished surfaces give evidence of the possibility, that, under favorable circumstances, a

---

[206] The relative abundance of these implements is perhaps a matter of some importance, in its bearing on the question of their origin. Were they natural forms, the peculiar force that operated to produce them, so marvellously like ordinary Indian relics as many of them are, would scarcely have been limited to so few pebbles as in this case: unless future exploration shall discover at some distant point a locality where only chipped pebbles occur. An effort has been made to estimate the comparative abundance of these palæolithic implements in the gravel deposit forming the bluff on the eastern bank of the Delaware river, and as near as can be determined, it is about one ten-thousandth of one per cent., or one in every million of pebbles. A sufficient number of these implements have certainly not as yet been gathered, to affect materially this calculation.

chipped implement might be associated with this gravel, even from the time of its deposition, and yet escape the wearing action by which its artificial features are obliterated. The deposit may be described as largely made up of ordinary smooth, water-worn pebbles varying in size from half an inch in diameter, to bowlders estimated to weigh from one to twenty tons.

Convinced that the so-called "turtle back" celts, which are the most primitive form of the chipped implements of the gravels, really are of artificial origin, many of them being identical in shape with the ordinary forms of European drift implements, and furthermore, since among the specimens found, are several spearhead-like implements, there can be but little doubt that, unassociated as they are with the common forms of surface-found relics, they are remains of an earlier and a ruder people, who occupied the eastern shores of this continent prior to the advent of the Indians, or are their immediate ancestors, as the case may be.

Mon. Mortillet[207] has, in a late communication to the Society of Anthropology of Paris, given an interesting account of the resemblance between the implements found in the valley of the Delaware, and those occurring in various localities in France. He says of a series of these implements collected by the author of this volume, and forwarded to him; "These stones do not chip so well as those of the Somme, because the latter are of *silex*, and silex is easily chipped. On the Delavarde [Delaware] River, there is no silex; men were there obliged to use a different stone—the trap, a sort of volcanic rock, slightly argillaceous, and very hard and difficult to chip. For this reason the axes that you see here are not as perfectly made as those from St. Acheul.

"In many parts of France rocks other than silex are employed and they are no better than those brought us. * * * * M. de Semallé has presented us with axes in quartzite which came from Bretagne,

[207] Mortillet. Bulletin de la Société d'Anthropologie de Paris, Tome Deuxième (III Série), p. 439. Paris, 1879.

and although our colleague assured us that he had carefully chosen them from among many that were not so marked, it is certain that they are chipped rudely enough, because in fact quartzite is exceedingly difficult to break as you wish it. Here now, I present you with axes in quartzite from the valley of the Garonne, and you can see that they are rudely chipped.

"It is interesting to see that the same epoch has produced similar industries in such different countries. This makes it more probable, that there was formerly a great bridge between America and Europe. The similarity of a great number of animals and of plants common to the two countries shows the existence of this communication. Still it may be supposed that certain kinds of birds could have crossed the ocean; seeds of plants may have been transported by the wind; but this explanation is insufficient for insects and is altogether inadmissible for terrestrial shells. How can it be imagined that snails and slugs, unable to live in water, could have passed from one continent to the other if there had been no [means of] communication between the worlds? Perhaps this communication may have taken place in the northern part of the ocean, in the latitude of Newfoundland."

Fig. 414 represents a carefully wrought stone implement, which may be considered as a typical example of this class of objects.

In its shape and size, as well as the circumstances under which it was found, it is a repetition, in America, of the hundreds of such finds, that are so common in many of the river valleys, both of France and England. Almond-shaped celts of this size and finish are not abundant when compared with the whole number of specimens found. They furnish evidence of the highest skill in chipping stone, and were probably among the most formidable weapons possessed by man at that time.

Fig. 414 was found by the writer, while watching the progress of an extensive excavation in Centre street, Trenton, N. J. It was nearly seven feet from the surface, surrounded by a mass of very large cobblestones and bowlders, one of the latter overlying it. It is moderately weathered; but the boundary ridges, of the various flakes that have been detached, are nearly obliterated, apparently by long exposure to shifting sands and running water.

Fig. 415 represents an average specimen of the flat-bottomed, peak-

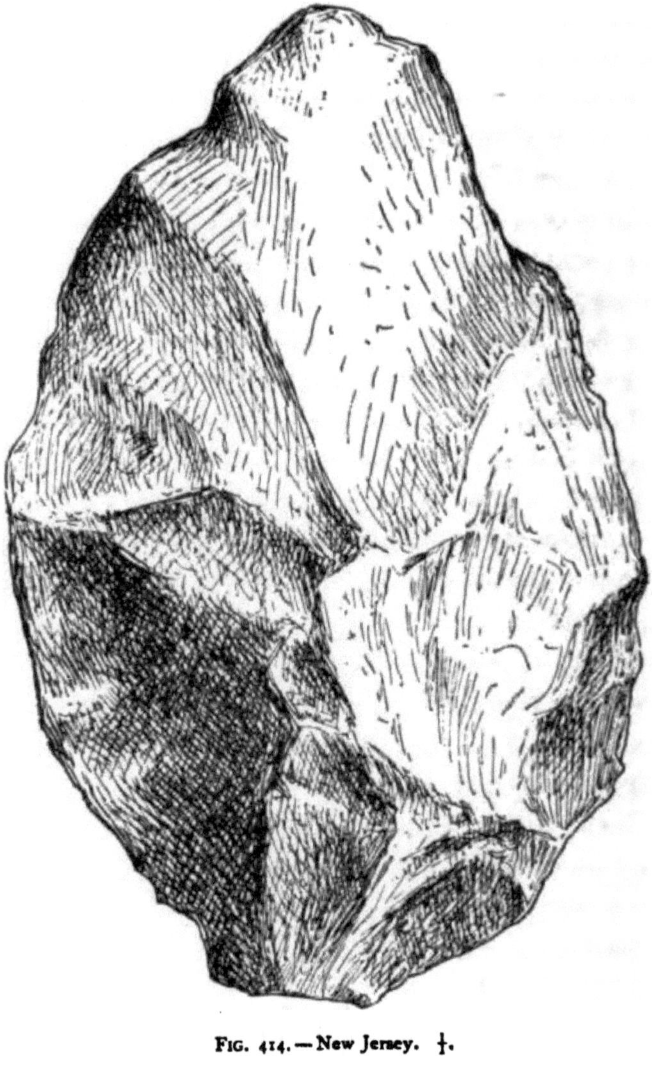

FIG. 414.—New Jersey. ½.

backed stones, known in some localities as "turtle-backs," a name that admirably describes their general appearance. These implements (for their artificial origin can scarcely be questioned) are very uniform in the character of the chipping, in the material, argillite, of which they are made, and the erosion of their surfaces, though in this latter respect they vary more than in any other.

The specimen here figured measures four inches in length and

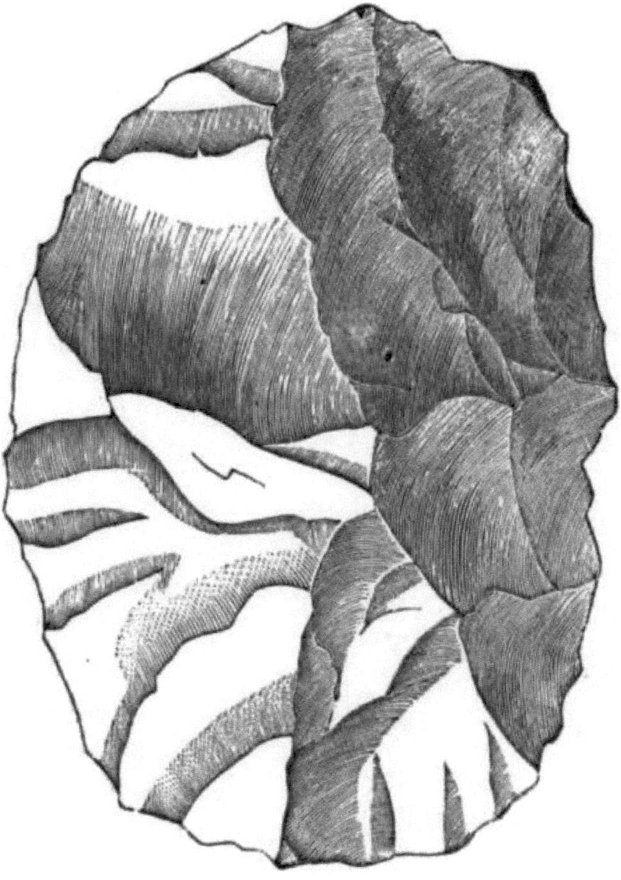

FIG. 415.—New Jersey. ½.

two and one-half inches in width. The bottom is nearly a perfect plane, and shows, by the slight indentations and scratch-like markings, that it has been chipped into its present shape, and not accidentally broken. Its greatest thickness is one and one-eighth inches; the "peak," or highest point of the back, being in the middle of the specimen, measured lengthwise, but rather nearer one side than the other. The broader side of the back does not appear to be any better adapted for cutting, than is the narrower or more abruptly descending side.

Although this stone, from long exposure, has become porous upon the surface, two edges still remain sharp and regular, and exhibit an amount of skill in "flint-chipping" about equal to that of the ordinary slate hoes, shovels, spears and arrowpoints. Close examination shows that the back has been worked into its present shape by a series of powerful blows, or by pressure. Usually large surfaces are left, in several instances planes being formed by the detachment of a single fragment of the rock, which extends from the peak to the edge of the implement.

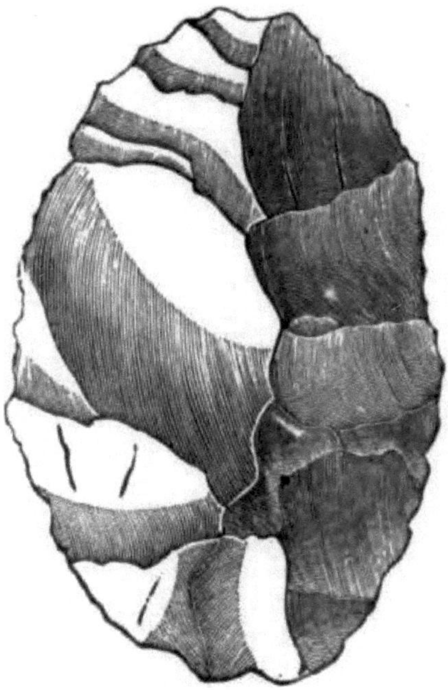

FIG. 416.—New Jersey. ½.

This specimen was found on the face of the gravel bluff that forms the eastern bank of the Delaware river, below Trenton, N. J. The original position, as to depth, was not noted at the time, as these objects were not then distinguished from the ordinary relics of the Indians. As many specimens, however, of the same pattern have been collected at this place, most of which were from undisturbed gravel, and at determined depths, it is quite evident that all these implements are from the same general locality, and while differing much in age, are not in any case to be referred to the Indians of a comparatively recent time.

Fig. 416 represents a somewhat smaller example of these "turtle backs," taken from the same locality. As in the preceding instance, this specimen is made of argillite, and has been considerably eroded by long exposure. This is believed to have occurred prior to its being deeply buried in the gravel bed, from which it was taken. The chipping, as in fig. 415, is coarse, but well designed, and has produced a moderately sharp cutting edge around the entire margin. As is nearly always the case with these "turtle backs," the under side is perfectly flat, and hence it is that these specimens, when resting upon their flat under surfaces, resemble the shells of our common box tortoise (*Cistudo*). To this fact they owe their popular name.

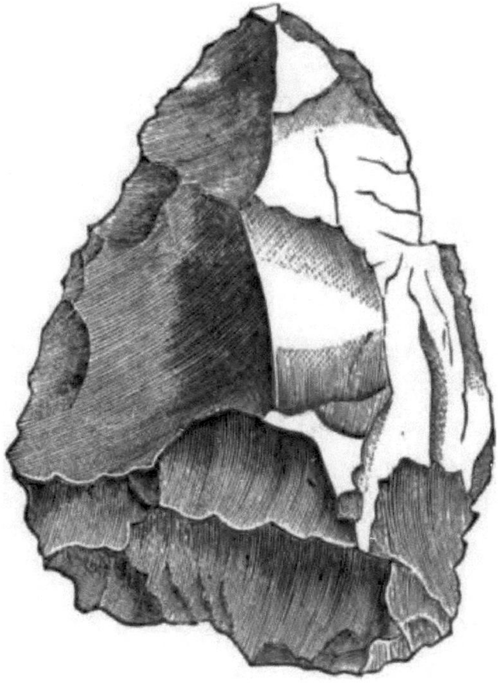

FIG. 417.—New Jersey. ½.

Fig. 417 represents a third example of the so-called "turtle backs," but varies in being pointed, or spear-shaped in its outline. Like the preceding oval specimens, this implement is flat upon its under side, but this side is not the natural surface of the pebble, worn smooth by water action, but the plane left by detaching a single large flake. The specimen has therefore the appearance of being the half of an ordinary palæolithic implement, which has been evenly split in two. This specimen, now preserved in the museum of the Academy of Science, at Salem, Mass., was one of the first found on the gravelly bluff facing the river at Trenton, N. J., although its age and origin were not recognized at the time.

The preceding specimens, excepting fig. 414, bear such a resemblance to a class of jasper implements, which are supposed to have been cutting tools, that we are warranted in assuming, that they were used for much the same purposes. The flat under surface does not give to these "turtle-backs" as good cutting edges as are found on those specimens that are chipped upon both sides; but, notwithstanding this objectionable feature, the edges are sufficiently sharp to be available for cutting flesh and similar yielding substances. For some such purpose, they were probably used.

Fig. 418 is an example of more elaborately wrought form, and is of interest from its resemblance to the European patterns of palæolithic implements, and from the fact that it is an excellent connecting link between the ruder forms, that have been figured, and those that are of higher design and better finish. This spear-shaped, or pointed implement is carefully shaped from an argillite pebble, and has well defined sharp, if not cutting, edges. The base is rounded, and preserves the natural surface of the pebble. The point is quite acute, and the sides have been produced by chipping, so that a comparatively uniform surface has resulted. The degree of weathering is uniform, and so far as this can be trusted as a guide, the specimen has had each flake removed at practically the same time.

This implement measures six inches in length, by from three to three and one-fourth inches in width, until near the point where it suddenly narrows.

It was found at the bluff at Trenton, in a narrow gorge, caused by running water which, however, had not displaced the material forming the sides of the little chasm. It was nine feet from the surface, and overtopped by a large bowlder. It bears considerable resemblance to certain chipped implements of jasper, porphyry and sandstone, which have been occasionally found on the surface associated with ordinary Indian relics; and which the writer supposes were largely used as "heads" for war-clubs. However this may be, an implement, like the one here described, might readily be mounted in a handle, or, having a blunt base, be held in the hand and wielded with terrible effect. Other examples of this form, mostly of argillite, have been collected from the same locality.

Fig. 419, *a b*, represents two views of a quite carefully wrought

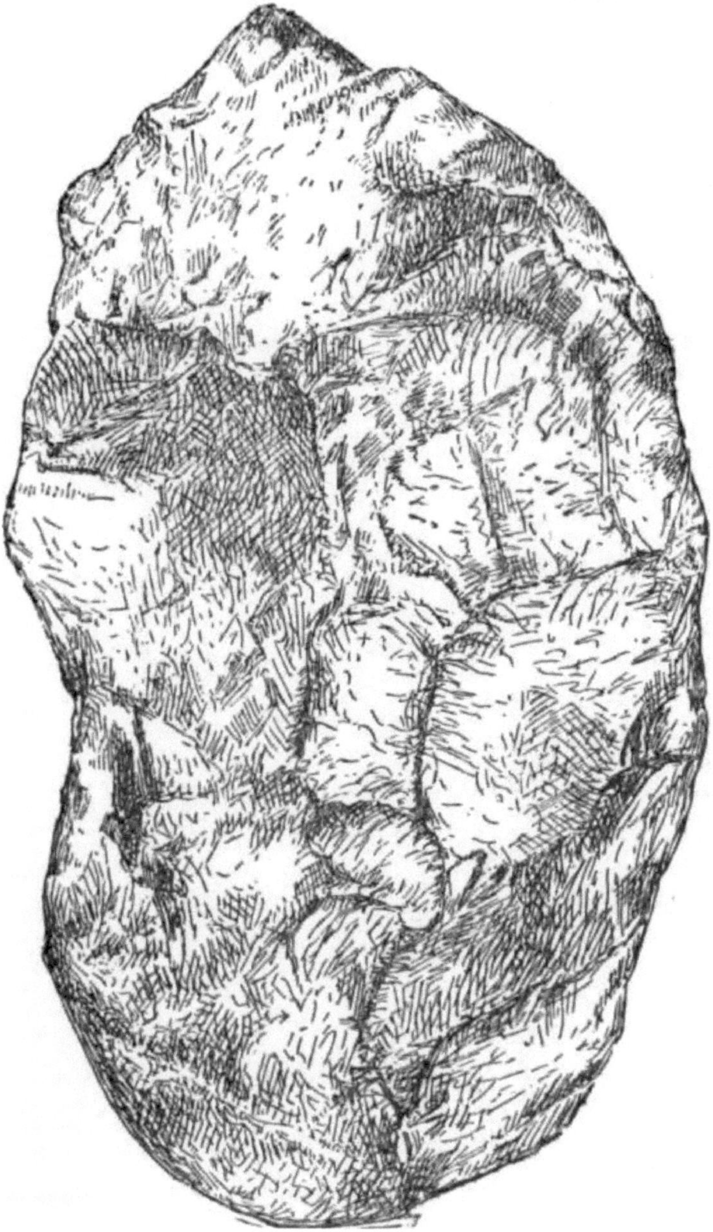

FIG. 418. — New Jersey. ½.

specimen of these rude implements, measuring nearly five inches in length, by two and one-half inches in average breadth, and less than two inches in greatest thickness. It is an excellent example of that form previously referred to, as a "turtle-back" celt. Of this specimen

FIG. 419 *a*.—New Jersey. ½.

Dr. M. E. Wadsworth says, "as far as can be told from examining its external surface, without any fresh fracture, I should consider it to be made of very compact argillite. It shows weathering, and also a more recent fracture, which has weathered to some extent. I should consider it very doubtful if this could be formed naturally." This specimen, like the preceding, came from the bluff facing the river. It was taken from a depth of three feet in from the face of the bluff, which was itself evidently the undisturbed gravel.

FIG. 419*b*.—New Jersey. ½.

It is desirable to state, in this connection, that the upper surface of this bluff was removed many years ago, when grading the street that now passes along the slowly crumbling escarpment formed by the wearing action of the river flowing at its foot. All the superficial soil and several feet of gravel were removed, and all specimens now found in the talus are necessarily from the gravel itself, and could not have been derived from the original surface.

A series of the earlier found of these argillite implements, derived from the gravel beds, were submitted to Dr. M. E. Wadsworth, of Cambridge, the eminent lithologist, to determine their mineralogical character, with especial reference to the possibility of the fractures that might arise from natural causes.

Of the specimen, fig. 420, Dr. Wadsworth says, "It is an argillite. It is highly indurated, with a conchoidal fracture, without cleavage, and fuses to a yellowish green or white glass which is feebly magnetic. The weathering which it shows could hardly have taken place except before it was covered with soil; it might possibly, but I think not probably, in a loose, open gravel. It is not at all likely to be of natural formation." The specimen is figured of natural size, and requires no further description, other than to remark that it was found in the *undisturbed* gravel of the bluff facing the Delaware, at a depth of six feet from the surface.

The word "undisturbed" is purposely emphasized, inasmuch as it

FIG. 420.—New Jersey. 1/1

is necessary, as Professor Pumpelly has pointed out, to determine whether the undisturbed specimens occur in the gravel as it exists when first exposed, or in a talus that may have been formed at the base of the bluff, and which, in some cases, may extend upwards, nearly to the top; as in the latter event it is possible that an implement might have very recently rolled down from the surface, and be, now, buried several feet from the face of the bluff. This possible occurrence has been duly considered in every instance, and no such displacement evidently had taken place, in the instance of the specimen here figured, or in that of others found both before and since.

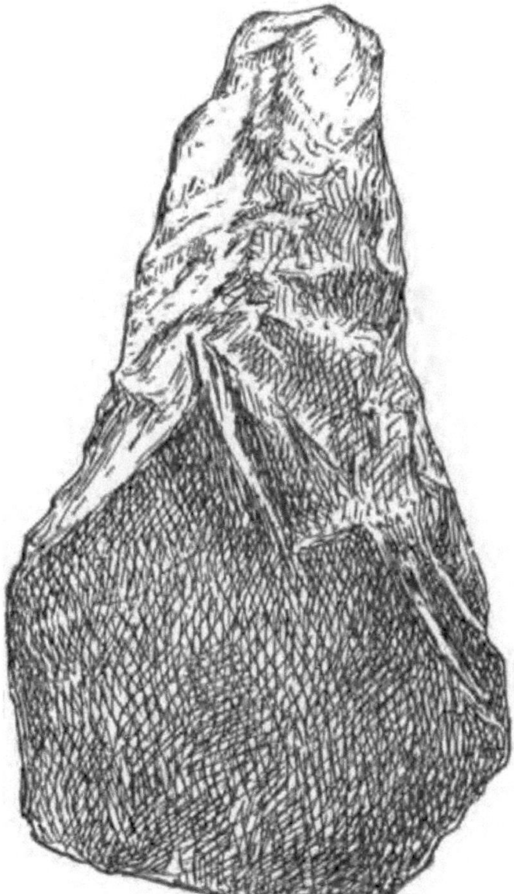

FIG. 421.—New Jersey. ½.

Fig. 421 represents a specimen formed of an argillite pebble, with a portion of the water-worn or weathered surface constituting the greater part of the base, on one side of the implement. The corresponding side is a uniform surface, but is less smooth, and exhibits every indication of being much less weathered, although it is greatly altered from a freshly fractured surface.

This specimen measures four and one-fourth inches scant in length. The base is, in width, a little less than one-half the length. The chipped portion decreases uniformly in width from the base, the flakes having been detached from both sides, and the edges. The specimen terminates in quite a blunt point, and does not appear to have been more acutely finished, than it now is. In general outline, fig. 421 closely resembles many of the European flint implements from the river valleys, and bears far more resemblance to many neolithic forms than do the majority of the chipped flints from tertiary deposits lately described in detail by M. Robiero.[208]

This specimen was taken from the gravel, at the bluff forming the eastern bank of the Delaware river at Trenton, at a depth of seven feet from the surface.

Fig. 422 represents a carefully shaped and creditably designed argillite implement, that is much like many of those found in Europe.

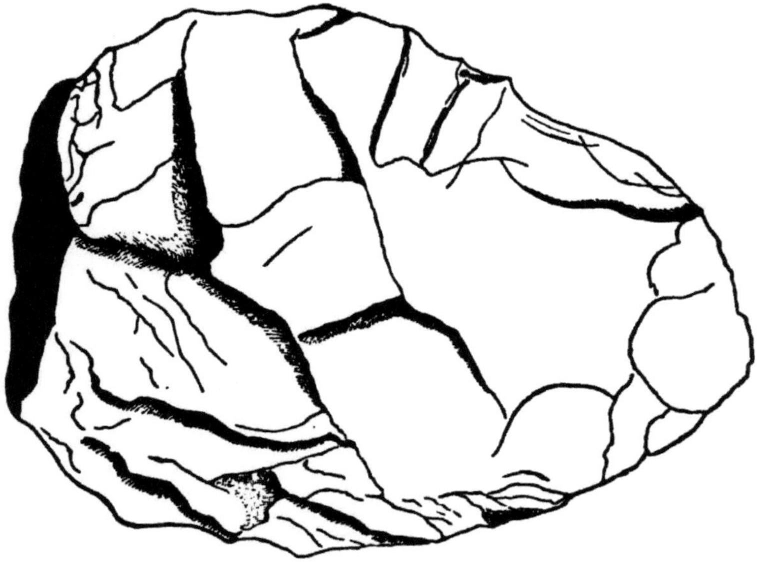

FIG. 422.—New Jersey. ½.

From the base, which is the unchipped natural surface of the argillite pebble, this implement is flaked equally on both sides, and brought to an edge along each margin, and at the same time it tapers to an obtuse point, sufficiently marked, however, to indicate that it was intended to pierce as well as to cut. The broad base, which is sufficiently wide to allow the specimen to stand upright without support,

---

208 Descrip. de Alguus. Silex E Quart. Lascados en contrados nos camados dos terrenos: Tertiar. e Quaternario. M. Carlos Robiero, Lisboa, 1877.

could never have been attached to a long shaft, and so it could never have been used as a spearhead.

Held in the hand, it would seem to be an awkward instrument for most purposes, but the broad base would serve to protect the hand, were it held in this way, and used for striking sudden blows. Hafted in some manner, an excellent weapon is obtained and one that would prove not only valuable, in close combats, but as a hunting implement, whenever an opportunity was given to strike a sudden blow.

Neolithic implements of this pattern, but usually longer, were probably used for grubbing roots, but implements of this pattern, when found in the river gravels, are more likely to have been ice-picks. If we are warranted in supposing that the people who made and used these palæolithic implements lived here during the prevalence of an arctic climate, then they must have had some means for cutting holes in the ice, and for such a purpose, the broad-based implements, like fig. 422, were admirably adapted. If used in this manner, they would be peculiarly liable to be lost through the holes cut in the ice.

This specimen was found, several years ago, in the bluff or gravel bed that formerly faced upon the east side of Cooper street, Trenton, and was in the loose gravel, which was then being removed. It was near the centre of a mass of small pebbles and sand, which was detached bodily, and from the foot of the bluff, and which, when it crumbled, exposed this specimen. Above the mass, in which this specimen was embedded, extended a stratum of sand of considerable thickness, overlying which was the surface soil. The specimen was at least at a depth of eight feet.

It was at this locality, that some time before, the writer found, *in situ*, three chipped masses of stone, supposed to have had an artificial origin. That two of them were palæolithic implements, subsequent discoveries have shown to be true. Of these, as early as 1873, the opinion was expressed,[209] that as they occurred so deeply in the earth, and in gravel and sand that showed no evidence of recent disturbance,

---

209 Abbott. American Naturalist, vol. vii, p. 207, figs. 36–38. Salem, Mass., 1873.

"they were fashioned and used by a people far antedating the race that subsequently occupied this same territory."

Fig. 423, *a*, *b*, represents the spearhead-like implement previously referred to. The illustration shows, at a glance, the artificial origin of the specimen. It is made of flint, and is the only instance of the occurrence of a drift implement of this mineral. This specimen was taken from the gravel, at a depth of six feet from the surface, on the site of the Lutheran Church, Broad street, Trenton, N. J. It was

FIG. 423.—New Jersey. ½.

found lying *in situ*, in a shallow stratum of coarse pebbles, and clearly showed by its surroundings that it had not become buried at this depth, subsequently to the deposition of the containing layer of pebbles.

The fact that this specimen was found at a depth, at which under exceptional circumstance ordinary Indian relics might be buried, and being of a different mineral from the characteristic forms of the gravel, might lead to the belief that this more artistically chipped flint spear-shaped implement, was an "intrusive" relic of Indian origin. The general character of this gravel-bed, even at this shallow depth—six feet from the surface—where this flint specimen occurred, was such as to convince any one, had they examined the locality at the time, that the specimen had not reached there subsequently to the deposition of the gravel itself. Fortunately, at the time, an exceptionally good opportunity of examining the locality was offered, and it was evident that the implement-bearing gravels, that can be readily distinguished from the later beds, here came to the surface. Bowlders of large size were upon the surface, and the sides of the excavation, from which was taken this specimen (fig. 423), showed by the close packing of the material constituting the mass, that it had not been recently disturbed, and that it had been deposited by the same forces that formed the gravel-bluff, now constituting the eastern bank of the river, nearly two miles distant. Immediately above it, *i. e.*, on the same horizon, but not directly over it, and continuously to the surface were numbers of large stones, several of them containing from six to ten cubic feet. In such a mass, and at such a depth, it is scarcely possible a spearpoint of the later Indians could have reached. The fact that the specimen is flint, and not argillite, has no bearing on the question of its being other than a palæolithic implement, inasmuch as in all well known localities in Europe, where palæolithic flints occur, there have been found occasional specimens made of other minerals.

Fig. 424 represents a very carefully chipped argillite implement that bears a marked resemblance to many of the European specimens of palæolithic implements. The specimen measures four and one-half inches in length, and a little less than two and one-half inches in its greatest width. In the chipping, this specimen varies somewhat from a typical turtle-back, in that the under, or flatter, side is somewhat chipped, especially along the edges, which throughout their entire length, exhibit traces of secondary chipping, whereby the edges were made more nearly straight. The general outline is that of a spear or lance-head, rather than an indefinitely shaped "chipped implement," as many of them are. There is in this instance a well defined point, and a broad, straight base, giving a general contour quite similar to certain jasper and slate "hoe blades," as this pattern of neolithic implements is sometimes called.

This specimen, fig. 424, was taken from the bluff facing the river, but two miles farther south than the exposure near Trenton, from which most of the specimens have been gathered. It was discovered in a *perpendicular exposure* of the bluff, immediately after the detachment of a large mass of material, and in a surface that had but the day before been exposed and had not yet begun to crumble. The specimen was twenty-one feet from the surface of the ground, and within a foot of the triassic clays that are here exposed. Directly over it, and in contact, was a bowlder of large size, probably weighing one hundred pounds; while at a distance of five feet above was a second much larger bowlder. The character of the mass, which was that of the bluff on the bank of the river near Trenton, was such as to render it impossible that this specimen could have reached this position subsequently to the deposition of the containing bed.

FIG. 424.—New Jersey. ⅔.

Fig. 425 represents a very artificial looking, and yet quite unique, form of chipped stone implement. It certainly bears no resemblance to any common form of neolithic weapon or domestic implement. In

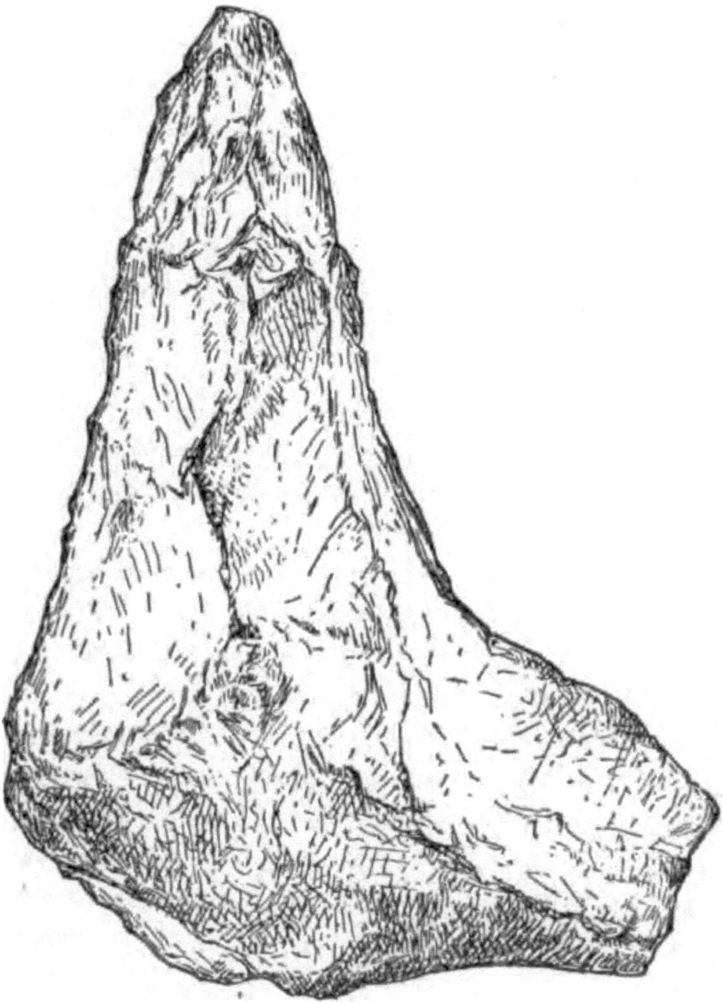

FIG. 425.—New Jersey. ½.

general, its appearance is that of a rude spear, such as not unfrequently occurs upon the surface, made of jasper and quartz; but the handle-like projection, which may or may not have been pointed originally, renders it difficult to determine the probable use of the implement; but that the specimen is artificial, and designed for some definite purpose, there can be no doubt.

This specimen measures four and five-eighths inches in length, and two inches in maximum width, exclusive of the projecting point or "handle" at one side. This projection is one and one-fourth inches in length. The chipping on this implement is quite well defined along the edges; and this, of itself, gives evidence of its artificial origin; for we do not find traces of secondary chipping, whereby zigzag lines are straightened, occurring among crushed or frost-fractured pebbles.

This unique form was found on the same gravelly bluff from which the preceding was taken, but at a point two miles distant, down the river. The specimen was exposed after a land-slide which occurred on Aug. 24th, immediately after a violent storm. A large mass of gravel was detached bodily, leaving a fresh surface of the bluff, from which this specimen projected. The depth from the surface was considerable, though it could not be accurately determined at the time.

Fig. 426 represents a specimen more like a weapon than any of the rude implements that have as yet been collected. Its shape seems to suggest its use, and, considering the rough workmanship that has been expended upon it, it is admirably adapted to the supposed use to which it was put. It foreshadows the tomahawk of more modern times. There appear to have been no fractures since the implement was made. The whole surface presents the same weather-worn appearance, and it is doubtful if even the rude edges were more regular in design or sharper than at present. Very nearly eight inches in length, the specimen may, for purposes of description, be divided into two sections—the front or blade of the weapon, and the hammer-head or back. The blade or front portion is four inches in length, forming nearly a continuous line with the top of the back; the elevation of the outline or margin being less than half an inch at the angle of the back and edge. Below, the line of the back and that of the blade form an obtuse angle; the blade being beneath, an inch and three-eighths wider than the narrower portion or hammer-head. The entire margin of this specimen has been chipped into its present shape and condition, giving it a rudely-rounded appearance at the top, edge, bottom, and extremity of the back. This chipping has not been done by an ordinary hammer-stone, pecking off the small fragments and producing the peculiar dotted appearance common to the ordinary grooved cobble-stone axes; but the stone has been *flaked* off in larger pieces, although the appearance varies somewhat from the shelly fracture of jasper. As a large portion of the side of this specimen is smooth, it is probable that the mass, as originally detached from

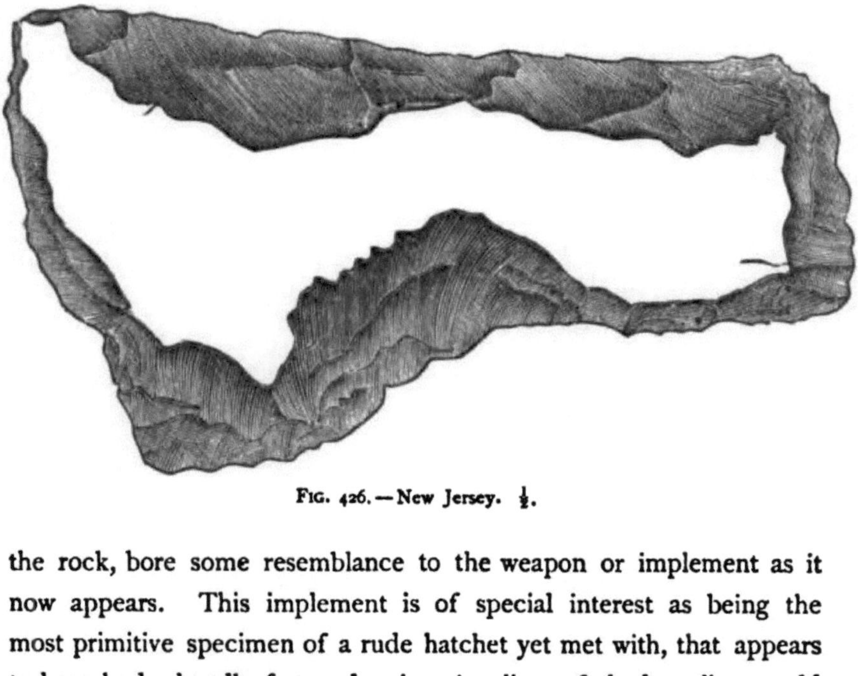

FIG. 426.—New Jersey. ½.

the rock, bore some resemblance to the weapon or implement as it now appears. This implement is of special interest as being the most primitive specimen of a rude hatchet yet met with, that appears to have had a handle fastened to it. A split or forked sapling could have been as readily attached to an axe of this shape as to any of the grooved forms. The shallow notch beneath, at the junction of the back and blade, was apparently so chipped to make the handle more secure.

This axe-like specimen was found upon the surface of a gravelly field, under circumstances that have no bearing upon the question of its age. The fact that it is of argillite, together with its weathered condition and the similarity of the chipping to those implements found in the gravel, renders it probable that it is not the handiwork of the Indian. It is described, in this connection, because its condition is such as to suggest that it had as early an origin as the term "palæolithic" implies.

Its palæolithic age is further rendered probable from the fact, that very recently, during an examination of the gravel bank facing the river, near Trenton, N. J., by Prof. W. Boyd Dawkins, Prof. Henry W. Haynes and others, a very characteristic specimen of a pointed palæolithic implement was found, that had a rude, but quite distinct encircling groove, and was therefore evidently intended to be hafted. Still another specimen, found in deep undisturbed gravel by the writer, was of such shape that it could scarcely have been used, unless attached to a handle.

Finally, it is desirable to add, in this connection, that these rude implements have been frequently found *in situ* by others, as well as myself, and thus evidence is not wanting to show that my own impressions as to their antiquity, based upon the circumstances of their position, when found, were not unwarranted.

In the Annual Report of the Museum at Cambridge, Mass., for 1880, the Curator has recently referred to this subject, in connection with the series of specimens found by me, and presented to that institution. Professor Putnam remarks that "others, including myself, have found implements in place in the gravel, and at a meeting of the Boston Society of Natural History, held Jan. 19, 1880, the subject was carefully discussed; and the evidence, supposed to be wanting by some as to the actual finding of specimens *in situ* in the gravel, was given in detail."

Having spoken at length of the deposits of gravel from which the characteristic implements here described have been taken, and the position of these gravels in the geological series having been shown, that we may appropriately determine the antiquity of these relics, it is necessary now to point out the relationship of the relics of an ancient people found in them, to the containing beds.

What seems to be a most conclusive argument in favor of the view that these relics of men are as old as, and in some cases it may be older than, the gravels that now surround them, is the fact that while the palæolithic implements are characteristic of the gravel, and neolithic implements of the surface, it is quite natural to find the former, as we find its containing bed, frequently cropping out upon the surface; while we never find this same soil at great depths, nor do the relics of the Indian, that now dot its surface, ever occur in such inexplicable positions. We can easily imagine an earthquake creating a deep chasm or crack in the surface, and inhuming a comparatively modern implement; but there are no traces of such cataclysmic action here, and if such an event had occurred, there would be other evidences than the commingling of objects from the surface with the underlying deposits; but such are wanting. Besides, if these rude forms were of identical origin with common Indian relics, then rude and elaborate alike, jasper, quartz, porphyry and slate together, axes, spears, pottery and ornaments, all of which are found upon the surface, should occur at these depths. Any disturbance that would bury one, would bury the others. Such, however, is not the case; and this one fact is, I think, of itself sufficient to show that there is a distinction to be drawn between these roughly chipped implements and the skilfully wrought productions of the Indians.

Furthermore, in considering the relationship of these rudely fashioned stone implements to the beds containing them, and the place of the latter in the geological history of the globe, it must first be borne in mind, that the many changes which have occurred in the past were periods of long duration, and that the changes of climate and the modifications of the dry land were all gradual occurrences. None were of such violence as to render the globe uninhabitable by man. The severity of the glacial climate itself, it is known, but partially destroyed, though it largely displaced, animal and vegetable life; and if the displacement of mammals is a clearly ascertained fact, it is quite safe to include man, if he also existed here, as I have endeavored to show was the case.

In the foregoing sketch of the indications of a primitive people that occupied the northern Atlantic seaboard of America, prior to the generally supposed recent advent of the North American Indian, nothing has been adduced to indicate the racial belongings of this preoccupying people. In the traces themselves of this supposed race that have been preserved, there is nothing that helps to solve the question of their pre-Indian or Indian origin; and we must consider them simply as rude productions of the Indians, who, as a people in the palæolithic stage of culture, reached our coast possibly as far back in time as the glacial epoch, unless other evidence can be brought to show that an earlier and ruder race once possessed this country. Have we any evidence of this?

A careful study of the relationship of the implements characteristic of the gravel, to the better known traces of the Indian — ordinary arrowheads, celts and axes — of itself, suggests the probability of the Indians being comparatively recent occupants of Eastern America; although, if measured by years, the duration of their occupancy extends far indeed, into the distant past.

The conclusion that the Indians were preceded by another people is based upon the fact that it is not practicable to trace any connection between the characteristic chipped palæolithic implements and the polished, pecked and finely wrought objects of Indian origin; the one form certainly not having any necessary connection with the other. The wide gap that exists between a full series of each of the two forms is readily recognized, when the two are brought together, and no one will hesitate to acknowledge it; but mere verbal descriptions of distinctive characteristics, prominent as they are to the eye, convey but little meaning.

The gap that exists between one of the more highly specialized palæolithic implements and a rude agricultural tool of Indian manufacture, is slight indeed, and the maker of the one might readily have made the other; but when we consider that a difference of material also, to a great extent, is characteristic of the two forms, and the rude argillite objects on the one hand are as marked a feature of the gravel beds, as the rude flint and slate hoes and shovels are common to the surface, it becomes evident to those who carefully examine the several forms from the two positions, that they are as widely separated as the fossils of different geological formations. Indeed, were the gravel beds, that contain these implements of palæolithic age and origin, by any means converted into solid rocks, these artificially shaped stones would become veritable fossils, and as clearly indicative of a well defined species of man, as the casts of *Cuculæa*, so abundant in our green sand marls, mark a well-known form of molluscan life once common in the Cretaceous seas.

As to the ordinary stone implements, it may be mentioned that those found upon the surface are all in accordance with what we know of the Indians, who, while occupants of the Atlantic coast of North America, were dwellers in a densely wooded country, with the distribution of land and water as it now is; but are not these palæolithic implements wholly out of place in like positions? One can scarcely conceive of any use for a "turtle-back celt," or for some of its modifications such as are seen in the limited range of patterns of the older forms, unless, perhaps as a rude weapon. If the environment of a race to any important extent determines the patterns of its weapons of war, and of the chase, as well as its domestic implements, then it is clearly evident that the well known forms of Indian stone implements most readily met their requirements; and with this people, depending so largely upon their skill with the bow for their subsistence, and armed with hatchets of another pattern, these palæolithic implements are unquestionably out of place. They are so far different, and more primitive in construction, as to suggest their use at a time long prior to the discovery of the bow, when another, and less wary fauna must needs be hunted, and they also indicate a lower degree of culture than that of the Indian at the time of the discovery of the continent by Columbus.

It has been shown that a variation in the physical condition of this country both as to distribution of land and water, and climate, with concomitant differences of fauna and flora, obtained during the glacial epoch; and to this period, or to a subsequent one of shorter durationl when a second glaciation of these valleys occurred, it would be more reasonable to ascribe these rudely fashioned implements, even if found only on the surface, but which, by their presence in the drift gravels, give us a faint glimpse of the primal race that occupied our shores.

When also, we consider that the several conditions of glacial times were largely those of Greenland and arctic America, and that there is unbroken land-communication between the desolate regions of the latter and our own more favored land, and more important than all, that there now dwells in this ice-clad country a race which, not only in the distant past, but until recently, used stone implements of rude patterns; it is natural to infer that the traces of a people found here, under circumstances that demonstrate a like condition of the country during their occupancy, are really traces of the same people.

This opinion, however, has been frequently questioned, and grave doubts expressed as to the relationship of American palæolithic man and the modern Eskimo. The publication of the recent volume of Prof. Wm. Boyd Dawkins, entitled "Early Man in Britain,"[210] renders it desirable, to pursue this subject somewhat further, although the marked similarity of the European and American palæolithic implements, and of the circumstances under which they occur, have been dwelt upon, and the recent remarks of M. Mortillet, on this subject, quoted in full. The conclusions of Prof. Dawkins are of unusual interest, inasmuch as his impression is that the older palæolithic man became totally extinct and the later palæolithic or "Cave-man" was of a different origin. This opinion is certainly at variance with the traces of post-glacial palæolithic man, as found on the Atlantic coast of America. The conclusions of Mr. Dawkins may be briefly stated, as follows:

Palæolithic man is separately considered as the River-drift man and the Cave-man; the former believed to be much the older people and known by the series of simplest patterns of stone implements, "found

[210] Early Man in Britain, by W. Boyd Dawkins, chaps. vi and vii, p. 124 et seq. London, MacMillan and Co., 1880.

in the late Pleistocene river beds." This River-drift man wandered over the greater part of Europe, and Asia, leading "a wandering feral life under feral conditions * * * * a hunter of a very low order, but not lower than the modern Australian." The Cave-man, on the contrary, was greatly restricted in his range, which of itself is held to be indicative of different age and race, and was far in advance of the River-drift man, in the variety and workmanship of his implements and weapons. If not two distinct peoples, these River-drift and Cave-men are certainly sections of the same race, which found their way into Europe at widely different times, the River-drift man being of far higher antiquity, and lived countless ages before the arrival of the Cave-man. "We are without a clue," writes our author "to the ethnology of the River-drift man, who most probably is as completely extinct at the present time as the woolly rhinoceros or the cave-bear; but the discoveries of the last twenty years have tended to confirm the identification of the Cave-man with the Eskimo."

In the earlier chapters of this volume, reference has been frequently made to the occurrence of rude arrowpoints, scrapers, and other forms of stone implements made of argillite which suggest an origin anterior to the ordinary jasper and quartz implements of the Indians; and in the description of the open air workshop sites of the Indians, where arrowheads were chipped in great numbers, the absence of argillite flakes in the accumulated *débris* was noted.

For these reasons and others more particularly referred to, in preceding chapters, it is believed that these more specialized argillite implements, although found, to a large extent, upon the surface, and associated with objects of Indian origin, really bear a closer relationship to the rude implements made by the American River-drift man, than to the Indian handiwork of more recent times.[211] If we are war-

[211] One marked result of the deforesting of the country, and its constant cultivation has been to remove, in great part, the many inequalities of the surface, and to dry up many of the smaller brooks. The hillocks have been worn down, the valleys filled up, and this, of course, has resulted in bringing to the surface, on the higher ground, the argillite implements which were at considerable depths, and in burying, in the valleys, the more recent jasper and quartz implements of Indian

ranted, as suggested by M. de Mortillet, in considering the Acheuléen hatchets of France, the flint implements of the River-drift man of England, and the argillite implements found in the valley of the Delaware, as the handiwork of the same people; then, tracing the parallelism further, it may be asked whether we have, in America, any evidence that this earliest or River-drift folk became extinct here, as Mr. Dawkins believed was the case in Europe. If we have no evidence, but on the contrary we are able to point out a united and continuous series of indications of palæolithic man's presence from so remote a period as the deposition of the Trenton gravel, until the comparatively recent date when man had become as advanced in culture, as the Cave-man of Europe, then the parallelism of the two continents is so far broken, that the culture in the one country is confined to one race, and in the other it is indicative of two.

Along the Atlantic coast of this continent there do not occur geological formations of a character favorable for the development of that particular phase of human culture known as Cave-life, as represented in Europe's prehistoric annals; but the same improvement in the patterns and finish of stone implements, and the use of bone, as well as stone, and all the distinguishing features of Cave-life, except that of the artistic representation of men and animals, are all readily traced as the gradually acquired improvements of the descendants of the ancient palæolithic man, who in nowise differed from the River-drift hunter of Europe.

This absence of geological formations calculated to preserve the earliest traces of man in unmistakable condition, and the scarcity of fossil remains in the gravel deposits renders the problem of determining when and by whom this continent was originally populated far more difficult to solve, than the like problem concerning Europe; but not-

origin, that were left upon the soil, when lost or discarded by the red-man. In the remnants of forests still remaining, where no such disturbance of the soil has occurred, the relative depths at which argillite and jasper respectively occur, indicate the greater age of the former. This is also shown by the position, in "meadow mud" of the so-called fish-spears, to which attention has already been called.

withstanding these disadvantages, there is sufficient evidence remaining, to warrant the assertion that the palæolithic man on the one hand, and the makers of the argillite spearpoints on the other, stadn in the relationship of ancestor and descendant, and if the latter, as is probable, is in turn the ancestor of the modern Eskimo, then does it not follow that the River-drift and Cave-man of Europe, supposing the relationship of the latter to the Eskimo to be correct, bear the same close relationship to each other, as do the American representatives of these earliest of people?

If this view be correct, it shows that the sequence of events, and advance of culture, have been practically synchronous in the two continents; and the parallelism in the archæology of America and Europe becomes something more than a "mere fancy." Nor is it improbable that future discoveries in Europe will bring to light the missing links, which, by their absence, seem to separate the River-drift hunter from the later Cave-man.

The subjoined table shows the essentially similar characteristic features of the prehistoric archæology of the two continents.

AMERICA.	EUROPE.
Palæolithic implements in Trenton gravel.	Flint implements in Pleistocene River-drift.
Argillite implements of more specialized patterns in alluvial deposits and surface.	More specialized flint implements of the caves.
Jasper and Quartz implements of North American Indian. Polished stone.	Neolithic stone implements of highest grade. Polished stone.
	Bronze.

The evidence of the presence of the Eskimo far southward of his present range rests on many and varied facts, some of which, it must be admitted, are of doubtful value.

It has been urgently claimed that no traces of man have been found along our seaboard that cannot, without violence to known facts, be referred to the Indians of the past few centuries, and therefore we are without warrant, in introducing a hypothetical race to explain away so simple a matter as the commingling of rude argillite and elaborate jasper arrowheads.

The manifold indications of prehistoric man on the Atlantic coast of North America cannot, however, be reduced to such a meagre array that a mere reference to the Indian is all that is necessary. The commonplace occurrence of gathering rude arrowpoints in the bottom lands, and of finding others of more artistic finish in an upland field, may, to a superficial observer, have no further meaning than that these objects are relics of the Indians who once roamed over this country; but to him who gathers tens of thousands of these relics, and notes, with conscientious care, the position in the earth of each; who rambles, not simply over ploughed fields, but scrutinizes every exposure of the upturned soil; digs deeply in the peaty accumulations of the meadow lands and scans every foot of the steep muddy banks of tide-water creeks; to him, these abundant relics of the vanished races tell a far different story, and hint, in no uncertain way, of diverse origins. If then, all are not Indian, to whom shall we refer the others except to the Eskimo?

It is not within the scope of this volume to refer, except briefly, to the historical evidences of the presence of the Eskimo, south of the St. Lawrence. Dr. Brinton [212] says, "there is reason to believe that at one time they possessed the Atlantic coast considerably to the south. The Northmen, in the year 1000, found the natives of Vinland, probably near Rhode Island, of the same race as they were familiar with in Labrador. They contemptuously call them *Skralingar*, chips, and describe them as numerous and short of stature (Eric Rothens Saga, in Mueller, *Sagænbibliothek*, p. 214). It is curious that the traditions of the Tuscaroras, who placed their arrival on the Virginian coast about 1300, spoke of the race they found there (called Tacci or Dogi) as eaters of raw flesh [213] and ignorant of maize (Lederer, *Account of North America*, in Harris, Voyages)."

If we could with full confidence refer the older shell heaps and all the more ancient traces of prehistoric man, down to so late a date as

---

[212] Brinton. Myths of the New World, 2d ed. p. 24. New York, 1876.

[213] "The name Eskimo is from the Algonkin word *Eskimantick*, eaters of raw flesh." Brinton.

A. D. 1300, to the Eskimo, the archæology of the Atlantic coast would be shorn of all obscurity, and every relic would possess a readily deciphered history. It may be added, that in view of the results of recent archæological and historical research, it is within the range of possibility, that this will be the crowning result of future investigation.

In his excellent article on the Tribes of the Extreme Northwest, Mr. Dall[214] remarks, "my own impression agrees with that of Dr. Rink, that the Innuit were once inhabitants of the interior of America; that they were forced to the west and north by the pressure of tribes of Indians from the south," and again, "there are many facts in American ethnology which tend to show that originally, the Innuit of the east coast had much the same distribution as the walrus, namely, as far south as New Jersey." The conclusion reached by Dr. Rink,[215] to which Mr. Dall refers, is, that the "Eskimo appear to have been the last wave of an aboriginal American race, which has spread over the continent from more genial regions, following principally the rivers and water-courses, and continually yielding to the pressure of the tribes behind them, until they have, at last, peopled the seacoast."

In a subsequent publication[216] Dr. Rink has repeated these conclusions. He remarks, "Quant à l'événement qui les a fait émigrer de leur pays primitif et se diriger vers le Nord, je suis disposé à croire que cela a été *une guerre*, mais je pense que cette guerre n'a fait que les *mettre en mouvement*, que la marche vers l'embouchure a été très-lente, et que la durée du séjour en ce lieu s'est prolongée.

"D'autre part, j'ai cherché à démontrer que c'est dans la partie nord-ouest de l'Amérique du Nord, dans la région du Mackenzie et de l'Athna, qu'il faut chercher le fleuve ou les fleuves à l'embouchure desquels les Esquimaux ont développé leur civilisation, et que leurs ancêtres sont issus des régions attenant aux cours supérieurs de ces fleuves.

"En effet, la grande majorité des Esquimaux se trouve en Amérique,

---

214 Contributions to N. A. Ethnology (U. S. Survey of Rocky Mt. Region), vol. i, p. 102.

215 Tales of the Eskimo, London, 1875.

216 Compte-Rendu de la Congrès international des Américanistes: L'habitat primitif des Esquimaux; M. H. Rink, p. 331. Luxembourg, 1878.

et un petit nombre seulement habite l'Asie. Cela paraît indiquer que ce peuple est originaire du continent américain. Cependant, tout en soutenant cette thèse, je ne me dissimule point que cette seule considération ne suffit pas pour que la question soit vidée.

"Mais, dans mes études comparatives sur les mœurs, la langue, la religion et les traditions des différentes tribus esquimaudes, j'ai déjà trouvé bien des choses qui confirment la thèse de *l'origine américaine des Esquimaux*, tandis qu'au contraire je n'ai trouvé que très-peu de faits favorables à la thèse de leur origine asiatique."

If we accept the conclusions of Dr. Rink, that the Eskimo is of American origin, and necessarily occupied, originally, a more southern portion of the continent, than that to which he is now confined, every difficulty in the solution of the problem of the antiquity of man on the Atlantic seaboard of America seems to vanish.

It has been shown that we have traces of early man that are apparently not of Indian origin. The important differences that distinguish them from Indian handiwork have been carefully pointed out, and further, it has been shown that they are a characteristic feature of a geological formation, that indicates that they are of greater antiquity than any known traces of the Indian.

The meagre evidence, on the other hand, that we have of the advent of the Indian on the North American continent, is strongly suggestive of his Asiatic origin, and possibly, of his derivation, in part, from a submerged continent, of which we have a trace, in certain islands in the Pacific ocean. Their arrival in America probably does not date from a period sufficiently remote, to afford geological evidences of antiquity, other than that offered by certain of the shellheaps.

If we admit the Asiatic origin of the Indian, and the American origin of the Eskimo, the greater antiquity of the latter is evident, and in the palæolithic implements of the river drift, and in the neolithic implements of the surface, we have the remaining traces of the handiwork of these two early peoples, who, throughout the unnumbered centuries of prehistoric times, and until less than three centuries ago, were the sole possessors of this continent.

THE

ANTIQUITY AND ORIGIN

OF THE

TRENTON GRAVEL.

BY

PROF. HENRY CARVILL LEWIS,

OF THE SECOND GEOLOGICAL SURVEY OF PENNSYLVANIA.

CHAPTER XXXIII.

THE AGE OF THE TRENTON GRAVEL.

THE discovery of palæolithic implements in a gravel at Trenton, and the important relation which this holds toward the question of the antiquity of man in eastern America, make a careful determination of its age a matter of much interest.

While in the present state of our knowledge, we can fix no exact date for this gravel, it is possible nevertheless to ascertain approximately the relative geological time in which it was deposited. To solve this problem we must determine in the first place what relation this gravel holds toward all the other surface deposits of the Delaware valley, and in the second place what connection each or any of these deposits has with the great glacier which once covered a large portion of northern America. The writer, having been for several years engaged in a special study of the more recent geological deposits of southeastern Pennsylvania, has divided them into a series of distinct formations; of which the oldest is a clay of wealden or sub-cretaceous age, and the newest, a modern mud which is now forming on the banks of the Delaware and other streams. Of the five clays and four gravels which he has distinguished for convenience of study, but of which several may hereafter prove to be of closely related age, it will be necessary here to refer only to those which bear directly upon the subject under consideration.

THE YELLOW GRAVEL.

Nearly the whole of southern New Jersey and a small adjoining portion of Pennsylvania are covered by a deposit of yellow gravel which has been variously known as quaternary, southern drift, etc. It extends southward all along the Atlantic coast in the region of tide water, rising some two hundred feet above the level of the ocean. As it caps the watershed between the Atlantic and the Delaware (elevation 190 feet), the writer has, in a former paper, [217] named it, for convenience, after a town in this watershed, where, in a railroad cut, it is well exposed,—calling it the "Glassboro gravel;" but in the present discussion, it will be sufficient to call it the YELLOW GRAVEL.

It is characterized by small waterworn pebbles, somewhat eggshaped in form, seldom above an inch in length, usually less, and composed of quartz or quartzite rocks. There are also occasional pebbles of flint, and of fossiliferous hornstone and chert. It contains no large boulders and has no pebbles of soft or readily decomposable rocks, and its pebbles have nearly all a weatherworn eaten appearance. Still other circumstances, such as the great amount of erosion it has suffered, and the decomposed state of the beds upon which it lies, point to the conclusion that it is an ancient deposit of aqueous origin, made during a submergence in preglacial times. Professor Cook, of New Jersey, states[218] that the glacial drift overlies and is more recent than the yellow gravel. This gravel, of newer tertiary age, is bounded on the northeast by a line of rocky hills which extends all along the southern Atlantic coast parallel to the ocean, and which we have called the UPLAND TERRACE.[219] This Upland terrace crosses the Delaware a few miles above Trenton, trending towards Princeton, and the yellow gravel is not found above this point.

THE PHILADELPHIA RED GRAVEL.

A more recent gravel, the PHILADELPHIA RED GRAVEL, is confined to the immediate valley of the Delaware. This gravel is a mixture of the

---

[217] The Trenton gravel and its relation to the Antiquity of Man. Proc. Ac. Nat. Sciences, Phila. 1880, p. 296.

[218] Report on Clays, p. 17.

[219] The Surface Geology of Philadelphia and vicinity. Proc. Acad. Nat. Sciences, Phila., 1880, p. 258.

yellow gravel with more recent pebbles brought down the river valley. It contains numerous pebbles and boulders of soft triassic shale and of other rocks of the upper Delaware, it holds waterworn boulders of sometimes two feet or more in length, and it is distinctly stratified in horizontal or undulating layers. This red gravel, colored by peroxide of iron, is more clayey than the yellow gravel and lies at a lower level within a channel cut through the other gravel. The writer has recognized the representatives of both of these gravels in the same relative positions, on the Potomac near Washington.

The red gravel has been apparently deposited by an ancient flood of the river of great volume, at a time when it rose one hundred or more feet higher than at present. The presence of flow and plunge motion and of alternate sandy layers indicates a rapidly flowing current. While its stratified character, its smooth waterworn pebbles, and the soft decomposed rock upon which it rests, all show that it was not transported by ice, yet the presence of boulders which can be traced to the northern valley of the river, the absence of all traces of former life in the gravel, and the altitude above the present river which it attains, point to the melting of a great glacier as the origin of the flood which formed it. It is more than probable that this gravel belongs to the CHAMPLAIN EPOCH, the epoch of the melting of the great glacier whose southern terminus in the Delaware valley was near Belvidere, sixty-five miles above Trenton.

THE PHILADELPHIA BRICK CLAY.

Resting unconformably upon the Philadelphia red gravel is the next deposit in order of time—the PHILADELPHIA BRICK CLAY. This clay, of a yellow color, and of varying depth and purity, is here confined to the valley of the Delaware and its tributaries, and is characterized by the presence of numerous boulders which become more frequent as the river is ascended. The writer has traced the boundaries of this boulder bearing clay up to the glaciated region and finds that it uniformly rises to a fixed limit of one hundred and fifty to one hundred and eighty feet above the river. Where the valley is wide, as at Philadelphia and Trenton, the clay is pure and fit for brick-making, but in narrow or steep portions of the valley the current has been too swift for the deposition of clay and it is represented by occasional, stranded, waterworn boulders. This clay rests against the upland terrace from Trenton to Philadelphia, at an elevation of one hundred and fifty feet. On the Lehigh river, a tributary of the upper Delaware, where[220] the bed of the river is more than two hundred feet higher than at Philadelphia, the clay rises one hundred and eighty feet above the river.

Whenever both clay and gravel are present, the clay lies unconformably upon the latter. Generally it lies in a series of crests and hollows upon the gravel, the clay occupying the hollows between the crests of gravel. The following section, observed in Philadelphia, shows six well marked waves of gravel and clay.

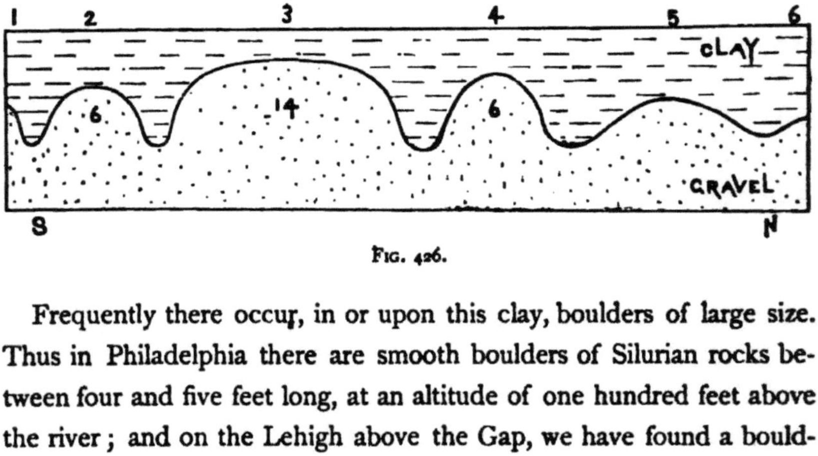

FIG. 426.

Frequently there occur, in or upon this clay, boulders of large size. Thus in Philadelphia there are smooth boulders of Silurian rocks between four and five feet long, at an altitude of one hundred feet above the river; and on the Lehigh above the Gap, we have found a boulder six feet long, elevated one hundred and fifty feet above the river at that place. In the vicinity of Bethlehem, thirty miles below the terminal moraine, the boulders in the clay sometimes show glacial striæ. It hardly admits of doubt that these boulders were borne by large cakes of floating ice derived from the base of the melting glacier.

220 We designate as "upper Delaware," the steep narrow portion of the river above tide water, and as "lower Delaware," the tidal portion of the river, or from Trenton, southward to Delaware bay.

That this was an epoch of submergence is indicated by the elevation of the deposit. While the underlying gravel was deposited by a rushing flood, it was not until quieter conditions had prevailed that clay could be formed. It is probable that this clay may be assigned to a period when the land stood one hundred and fifty feet or more below its present level, and when the cold waters from the melting glacier bore ice rafts which dropped their boulders. No shells or other signs of life have as yet been found in the brick clay, and it is inferred both that the water was fresh and that it had a temperature too low to support life.

In the consideration of the age of this clay, the amount of erosion it has suffered is an important feature. Unlike the modern alluvial clays of dark color, it does not appear on the immediate banks of streams, and has disappeared, wherever eroding agencies have been most active.

Finally, it is of interest to find that the clay which cements the unstratified "TILL," the "ground moraine" which covers the glaciated region to the north, is of a character so similar to the Philadelphia brick clay, that there is a strong probability that the latter was derived, in great part, directly from the grinding base of the glacier. The Philadelphia brickclay becomes more and more stony as we proceed northward, until in valleys at the base of the terminal moraine of the glacier its stones are almost as numerous as those of the true glacial till. Deposits of this boulder-bearing brickclay have more than once been confounded with glacial moraines. The latter, however, as is well known, may be distinguished by the abundance of angular and ice-scratched boulders and by the absence of stratification.

The relation of the Philadelphia brickclay to the till will be further discussed after the moraine and the other products of ice action have been described.

THE TRENTON GRAVEL.

The last and newest of all the gravels—a formation which when first studied at Philadelphia seemed of slight importance, and was called by the writer the "river gravel and sand," but which from its great development at Trenton is now appropriately known as the "TRENTON GRAVEL"—forms the subject of the present paper. At Philadelphia it lies close along the river, within all the older gravels, and rises but a few feet above the water. It is in this alluvial gravel, the latest, except the recent mud flats, of all the surface formations, and in this gravel only, that traces of man have been found.

The Trenton gravel at Philadelphia is composed principally of a sharp, micaceous sand, which below water-level is a quicksand, overlying a clean, dark gray gravel, whose pebbles are made exclusively of the rocks forming the upper valley of the river. The pebbles of this gravel, unlike those of the older gravels, are generally flat—a shape characteristic of true river gravels. Quartz pebbles are much less numerous than in the other gravels. Irregular strata of "bar-sand" frequently alternate with the gravel. The islands in the river and its banks are made of this gravel, and from data obtained from artesian wells, it appears that in the middle of the river it is about one hundred feet deep, lying upon rock. It therefore fills up an ancient channel of the once larger river, and the river now flows upon it. Occasional large boulders lie upon the sand.

It is to be especially noted that the Trenton gravel is bounded by a continuous hill of older red or yellow gravel, and that it therefore lies in a channel previously excavated through those gravels, down to the underlying rock. On tracing the Trenton gravel up the river, it is found gradually to extend farther from its banks and to rise to a greater elevation above it, until in the vicinity of Trenton, thirty miles above Philadelphia and at the head of tide-water, this formation extends several miles back from the river and rises between thirty and forty feet above it.

A few miles above Trenton the valley of the river narrows, and from here up, the river flows upon a rocky bottom, and the Trenton gravel is shallow and confined to the immediate vicinity of the river. It forms a low terrace, seldom over ten feet high, and extends as a "flat" on either side of the river. It continues up into glaciated regions, where it appears to form the lowest and most recent terrace. The writer has observed similar gravels on the Susquehanna and Allegheny rivers, and it is probable that they occur on all rivers rising in the glaciated region.

THE GEOLOGY OF TRENTON.

The great development of this formation at the city of Trenton, and the archæological interest attached to it at this place, call for a yet more detailed description.

Trenton lies at the junction of three great formations, the Azoic or Gneissic, the Triassic, and the Cretaceous. A narrow belt of steeply-inclined gneissic rocks, which in Pennsylvania are extensively developed, passes through the centre of the city, and is laid bare in several places. Resting unconformably upon the northern edge of the gneiss is a series of red sandstones and shales which have a gentle dip to the north. These belong to the Triassic formation, which extends for forty miles up the Delaware, and which is intersected by frequent trap dykes. Immediately south of Trenton is a plastic clay of lower cretaceous or Wealden age, which rests upon the southern edge of the gneiss, and dips very gently toward the sea. The more recent deposits of sand and gravel lie in horizontal strata upon these three formations, and often cover them so completely as to hide them from sight.[221]

The Trenton gravel, which is here coarser than at Philadelphia, extends in the shape of a horse-shoe eastward and southward of the city of Trenton. It forms a bay, which extends fully four miles back from the river, and which has one extremity in Trenton, at "Five Points," and the other at a distance of two miles below the city. It is a level plain, which is bounded throughout by a hill, on which appear the older yellow or red gravels and the brickclays. Since it was depos-

[221] For full descriptions of these formations, see Geology of New Jersey. 8vo. Newark, N. J. 1868. Geo. H. Cook, State Geologist.

ited, the river has cut down through it to the gneiss below, forming a bluff. South of the city the river is bordered by marshy meadows formed of recent alluvial mud, beyond which rises the bluff.

The writer has prepared a map (page 531) showing the extent of the Trenton gravel in this vicinity. That portion of the map which is unruled represents territory covered by red or yellow gravel and the brickclays. The upland terrace bounding these older deposits is outside of the limits of the map. The recent alluvial mud is shown bordering the river in places below the limit of tide-water, but is absent above that point. An interesting ancient island of red gravel is shown in Pennsylvania. It will be noticed that the Trenton gravel, confined to the immediate vicinity of the river above Trenton, suddenly spreads out like a bay at that city. Localities where palæolithic implements have been found below the surface are marked by a small cross.

The Trenton gravel, as exposed on the river bluff and in the numerous and long railroad cuts in the city, is seen to consist of a dark gray stratified gravel, overlaid by a gray sand. The gravel contains no clay, but has frequent sandy layers in which "flow and plunge" structure may be observed. Its pebbles are smooth, and generally of a flat oval shape. They are composed entirely of the rocks of the upper Delaware valley. This gravel has been well described by Professors Cook[222] and Shaler.[223] The sand overlying the gravel varies in depth from three to eight feet, and has all the characters of a true river sand. Waterworn boulders frequently lie in or upon this sand, and are rarely eight feet in length. It is difficult to account for the presence of such large boulders, except by assuming that they have been dropped from ice-rafts which floated down the once greatly enlarged river. They were dropped at a time immediately subsequent to that in which the gravel was deposited and when the violence of the flood had diminished. The depth of the

---

[222] Annual Report, 1877, p. 21.

[223] Annual Report of Peabody Museum, 1877, p. 44.

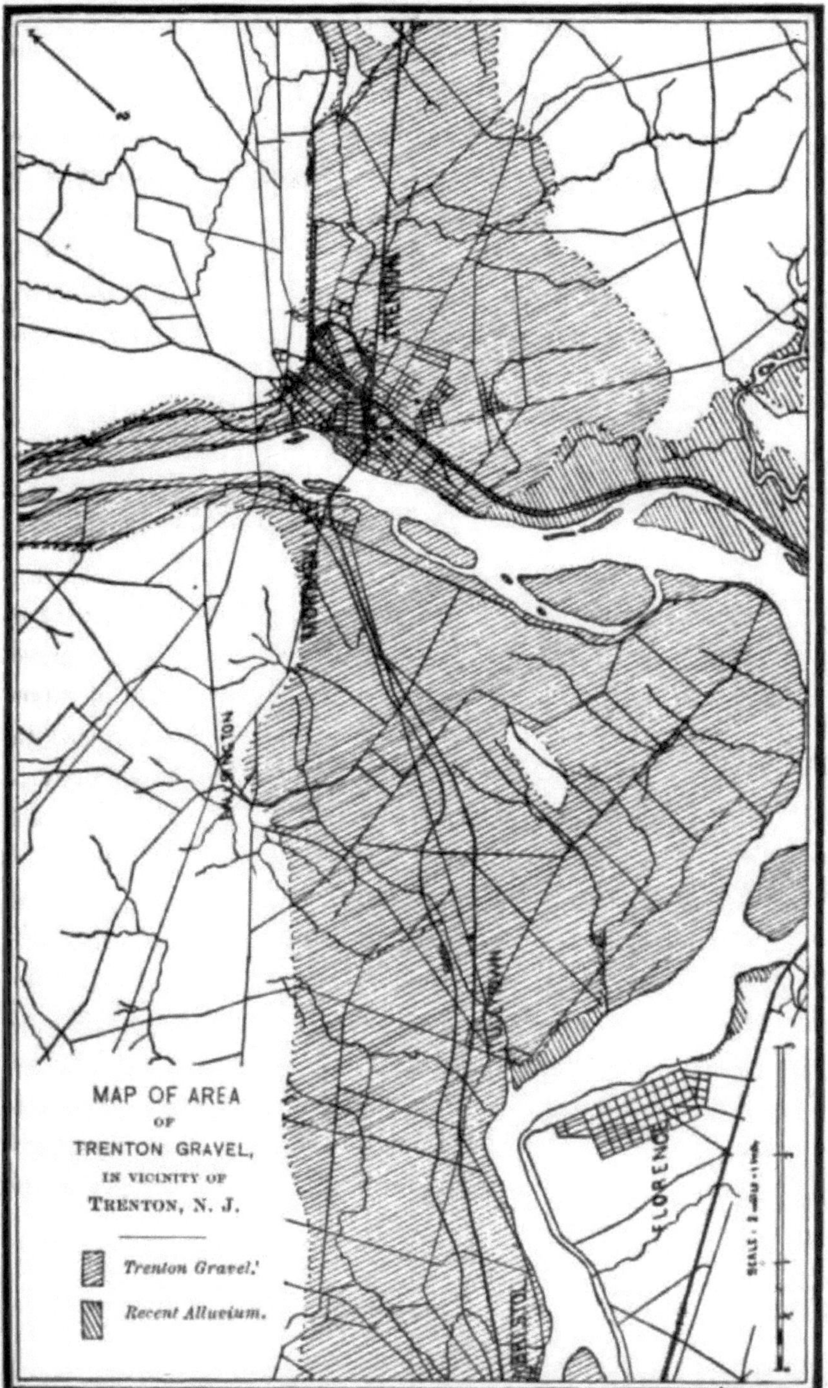
TRENTON
FLORENCE
MAP OF AREA
OF
TRENTON GRAVEL,
IN VICINITY OF
TRENTON, N. J.
Trenton Gravel.
Recent Alluvium.

Trenton gravel varies from perhaps forty feet in the centre of the "horse-shoe," near the river, to six or eight feet near its edges. Near the house of Dr. Abbott, not far from the extremity of the "horse-shoe," the Trenton gravel and sand is about fourteen feet thick and overlies the series of older strata which here compose the greater part of the bluff. The following section is seen about two miles south of Trenton.

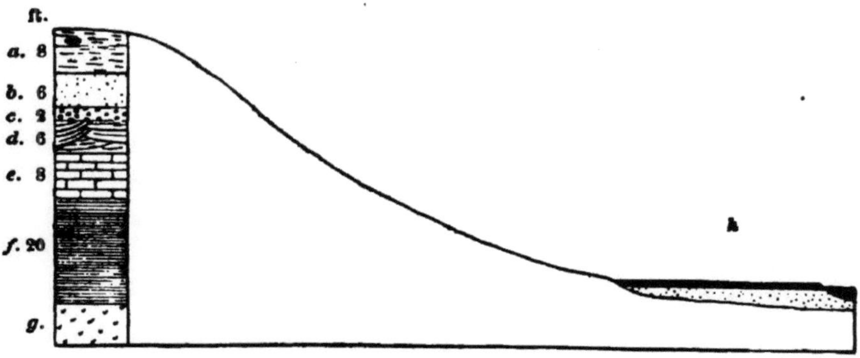

FIG. 427.—Section of bluff two miles south of Trenton, New Jersey. *a*, *b*, TRENTON GRAVEL; Implements—*a*, fine gray sand (boulder); *b*, coarse sandy gravel; c, red gravel; *d*, yellow gravel (preglacial); *e*, plastic clay (Wealden); *f*, fine yellow sand (Hastings?); *g*, gneiss; *h*, alluvial mud; *i*, Delaware river.

THE TRENTON GRAVEL A TRUE RIVER GRAVEL.

The presence of large boulders in the bluff at Trenton, and the extent and depth of the gravel at this place, have led to the supposition that there was here the extremity of a glacial moraine. Yet the absence of "till" and of scratched boulders, the absence of glacial striæ upon the rocks of the valley, and the stratified character of the gravel, all point to water action alone as the agent of deposition. The depth of the gravel and the presence of the bluff at this point, are explained by the peculiar position that Trenton occupies relatively to the river.

Trenton is in a position where naturally the largest amount of a river gravel would be deposited, and where its best exposures would be exhibited. It is at the point where a long, narrow valley, with precipitous banks and continuous downward slope, opens out into a wide, alluvial plain at a lower level. It is here that the rocky floor of the river suddenly descends to ocean level, and even sinks below it, forming the limit of tide-water. Thus any drift material which the flooded river swept down its channel would here, upon meeting open ground, be in great part deposited. Boulders which had been rolled down the inclined floor of the upper valley would here stop in their course and all be heaped up with the coarser gravel in the more slowly flowing water, except such as cakes of floating ice could carry oceanward. On the other hand, the finer gravel and sand would be deposited farther down the river. Thus it is that the material, which at Philadelphia is generally fine, grows coarser as the river is ascended.

We have seen that the gravel which at Philadelphia forms the bed of the river and rises only slightly above it, at Trenton forms a cliff nearly fifty feet high. The river has cut through the gravel at Trenton, but still flows upon it at Philadelphia. The fact follows as a natural consequence of the position of Trenton. Having heaped up a mass of detritus in the old river channel as an obstruction at the mouth of the gorge, the river, so soon as its volume diminished, would immediately begin wearing away a new channel for itself down to ocean level. This would be readily accomplished through the loose material, and would be stopped only when rock was reached. On the other hand, that gravel which had been deposited at places farther down the river where its bottom was below ocean level, would remain uneroded, or nearly so. When the river had attained the level of the ocean there would be no occasion to cut a deep channel, and it would therefore flow on top of the gravel which it had deposited.

It is necessary that this point should be understood, since it has been thought that to account for the high bank at Trenton, an elevation of the land must have occurred. It will be seen that the present explanation requires no change of level from that at present existing. An increase in the volume of the river will explain all the facts. The accompanying diagram will render this more clear.

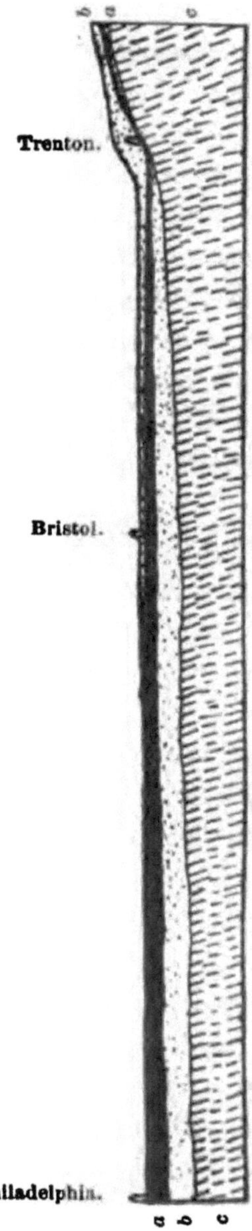

FIG. 428.—Longitudinal section of Valley of Delaware River, from Trenton to Philadelphia, showing position of the river gravel, with reference to the river. *a*, Delaware river; *b*, Trenton gravel; *c*, gneiss.

The fact of the river having cut through the gravel at Trenton, while at Philadelphia it flows upon it, is due to the configuration of the rock floor of the river, which at Trenton rises above ocean level, and at Philadelphia lies nearly 100 feet below it.

A few miles north of Trenton, all the older oceanic gravels disappear and two formations alone remain. These are the Philadelphia boulder-bearing brickclay and the Trenton gravel. Both are confined to the valley, and until we reach the region once covered by the glacier, no drift of any kind occurs above the limit of the brick-clay.

The Trenton gravel, now confined to the sandy, flat borders of the river, corresponding probably to the "intervale" of New England rivers, lies within a channel cut through the brick clays. That it is much more recent than the brickclay is shown both by the fresh appearance of its pebbles, and by the less amount of erosion it has suffered. Unlike the land covered by older surface formations, that covered by the Trenton gravel is remarkably level and free from hillocks or ravines. The change in topography may be well seen in the neighborhood of Trenton, and can be noticed almost anywhere along the valley. This difference is much more marked when comparison is made with the oceanic gravels. The Trenton gravel exhibits a topography peculiar to a true river gravel. Frequently, instead of forming a flat plain, it forms higher ground close to the present river channel than it does near its ancient bank. Moreover, not only does the ground thus slope downward on retreating from the river, but the boulders become smaller and less abundant. Both of these facts are in accordance with the laws of river deposits. In a time of flood the rapidly flowing water in the main channel, bearing detritus, is checked by the more quiet waters at the side of the river, and is forced to deposit its gravel and boulders as a kind of bank.

The section across the Delaware river above Trenton shows this topography and the relation of the Trenton gravel to the brickclay.

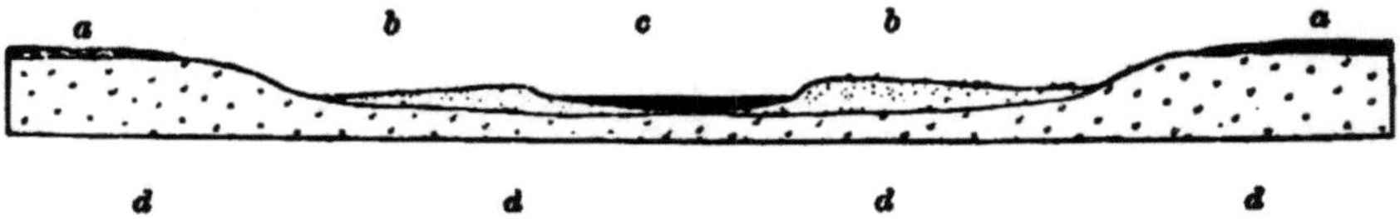

FIG. 429.—Section across the river at North Trenton. *a*, Philadelphia brickclay with boulders; *b*, Trenton gravel; *c*, Delaware river; *d*, gneiss.

Having now shown that the Trenton gravel is a true river gravel of comparatively recent age, it remains to point out the relation it bears to the glacial epoch.

THE TERMINAL MORAINE.

At or near the southern limit of the great ice-sheet is an accumulation of drift hills of characteristic rounded shape, forming a true terminal moraine. These hills, generally covered by large transported boulders, are connected together laterally, in an irregular manner to form a ridge at right angles to the ice motion. They are composed entirely of drift and form a remarkable accumulation which, rising sometimes over two hundred feet in height, can be traced continuously across the country.

Across northern New Jersey, Prof. Cook[224] has carefully traced it from Staten Island, on the east, to Belvidere, on the west; and has shown that it winds over hills and across valleys in such a manner, that by no other known agency than a great glacier, could it have been produced. Mr. Warren Upham[225] has traced it from Staten Island, eastward, through Long Island to Block Island and Cape Cod. Its course throughout Pennsylvania will shortly be made known. Similar moraine accumulations have been followed through a number of the western states by Profs. Whittlesey, Winchell, Chamberlin, and others. In Wisconsin, Prof. Chamberlin,[226] and in Minnesota, Mr. Upham,[227] have made investigations which indicate that these accumulations mark the halting places of the glacier during the period of its grand retreat.

This great moraine, which can be traced across the continent, marks the termination of the ice-sheet throughout the principal portion of the last glacial epoch. There is some evidence that in an earlier period a glacier advanced south of that limit. The moraine, throughout its whole course, is composed in great part of an unstratified deposit of angular and rounded boulders and pebbles embedded at all angles in a stiff clay, and very frequently scratched and polished. Occasional stratified sandy beds also occur. Curious features of glaciated regions which may be seen from Maine to Minnesota are the "kettleholes" or bowl-shaped depressions with no outlet, each of which perhaps marks the place where a mass of ice was buried in the sand, afterwards to melt and form a hollow. North of the moraine, the glacier has left undoubted traces in the universal covering of unstratified boulder clay or *till*, in the smoothed and grooved rocks, the transported boulders, the frequency of lakes and swamps caused by unequal distribution of the drift, the long gravel ridges known as *kames*, and the terraces along the rivers. These phenomena are wanting south of glacial action.

---

[224] Annual Reports of Geological Survey of New Jersey, 1877 and 1878.

[225] Proc. Amer. Assoc. Adv. Sci., vol. xxviii, 1879.

[226] Geology of Wisconsin, ii, p. 205, et seq., 1877.

[227] Geological Report of Minnesota, p. 73, 1879.

The materials forming the moraine in eastern Pennsylvania are in great part derived, not from far distant localities, but from the immediate vicinity to the north. The same may be said of the till and especially of its lower portions. There is no evidence of iceberg action in either of these deposits. As already stated, the Philadelphia brickclay, on the other hand, holds boulders generally brought from longer distances.

The moraine on the Delaware, sixty miles above Trenton, is distinctly marked on both sides of the river. In the valley, its materials have been modified by water action and spread out near Belvidere as a plain of stratified gravel. The pebbles on the higher ground show glacial scratches, while those in the valley have been subsequently waterworn. In Pennsylvania the moraine trends from Belvidere in a northwest direction, crossing the Kittatinny Mountain, east of the Wind Gap, and being well shown in each valley that it crosses.

During the melting of the glacier, either the ice or the morainic material so blocked up the Water Gap, as to form temporary lakes north of that point, by damming back the water. Thus we find, in the vicinity of Stroudsburg, Pennsylvania, a series of beautiful, level-topped terraces, the highest rising seventy-five feet above Brodhead's creek. Here, too, is what appears to be a "kame,"—a long, steep ridge of stratified gravel formed probably by sub-glacial streams, and afterwards partially covered by the terrace material. These curious features of glaciated regions have been well described by Rev. Geo. F. Wright[228] in Massachusetts, by Mr. Warren Upham[229] in New Hampshire and Vermont, and by Prof. G. H. Stone[230] in Maine, and it is shown that while more recent than the "till," they are older than the stratified valley drift.

The whole drainage area of the Delaware, north of the Water Gap, shows undoubted evidences of glaciation. There are many facts

---

[228] Proc. Boston Soc. Nat. Hist., vol. xix, p. 47; vol. xx, p. 210.

[229] Amer. Journ. Science, Dec. 1877, p. 460; New Hampshire Geological Survey, vol. iii.

[230] Proc. Amer. Assoc. Adv. Science, vol. xxix, 1880.

which indicate that the ice even close to its lower terminus had a thickness of over one thousand feet, which increased northward. Penobscot Knob, 2,100 above the sea and only about twelve miles north of the limit of the glacier shows transported boulders and glacial scratches on its very summit; while in the Wyoming valley, immediately north, the presence of a glacier is shown by terraces and kames.

THE AGE OF THE PHILADELPHIA BRICK CLAY.

In discussing the origin of the Trenton gravel, it will be most important that the age of the Philadelphia brickclay—a formation directly connected with the melting of the glacier—should be considered. We have already seen that while both the brickclay and the Trenton gravel are confined to the same valley, the former is of much greater extent than the latter, and was deposited at an earlier age. North of the moraine it is not uncommon to find stratified sands alternately with beds of clay, but in no case has the Trenton gravel been observed either to contain beds of clay or to alternate with them, and the conclusion already arrived at, that the Trenton gravel lies within a channel which had been excavated through the brickclay, is confirmed by all the facts observed. If, therefore, it can be shown that the Philadelphia brickclay is of Champlain age, and subsequent to the formation of the "till," it will necessarily follow that the Trenton gravel belongs to the extreme end of glacial times, or is post-glacial.

At Bethlehem, Pennsylvania, some fifteen miles below the moraine, a most instructive section is exposed. Here, upon the summit of a hill rising one hundred and eighty feet above the Lehigh river, the brickclay, holding large smooth boulders, lies unconformably upon a deep deposit of a stratified gravel, intermediate in its characters between the "modified drift" of glaciated regions and the Philadelphia red gravel. The gravel, of which thirty feet in depth is exposed, is distinctly stratified, and is composed of waterworn pebbles with occasional coarse sandy layers, and with no boulders. Identical sections may be observed in the kames and terraces north of the Delaware Water Gap. It is a well established fact that these formations overlie and are more recent than the unmodified till, which was deposited both as the terminal moraine, or beneath the ice-sheet as a ground moraine, and that they represent later stages in the melting of the glacier. The Philadelphia brickclay, now shown to be still more recent, belongs, therefore, to a late portion of the Champlain period. It appears to have been formed at a period of submergence during the retreat of the glacier.

A study of the valley of the Lehigh river throws much light upon the age of the brickclay. In the narrow valley extending from glaciated regions down to the Lehigh Water Gap, the clay is represented by waterworn boulders, often of large size, stranded on the banks. These are most numerous near the river, and are very scarce at their extreme outer limit, one hundred and eighty feet above the water. From the Lehigh Gap to the Delaware the valley is broad and the clay finely developed. Its boulders are so very much more numerous than they are at Philadelphia, that a section through it nearly resembles a section through a moraine, and the two phenomena have been confounded. The action of an ice-bearing flood of immense magnitude is clearly shown all along the river. Some of the clays which border the great lakes, rising one hundred to two hundred feet above them, are perhaps of similar age. In a paper on the age of the Trenton gravels, Rev. G. F. Wright,[231] who has examined a number of localities with the writer, estimates the glaciated region drained by the upper Delaware to be about six thousand square miles, and shows that even supposing the ice to have been fifteen hundred feet deep over this area, it would be impossible to provide for a flood of sufficient magnitude to account for the whole deposit of Philadelphia brickclay, without assuming an extensive depression of the valley. Professor Dana, in an exhaustive study of the floods produced in southern New England during the melting of the glacier, shows that the

[231] Proc. Bost. Soc. Nat. Hist., Jan. 19, 1881.

Connecticut river rose one hundred and fifty to one hundred and eighty feet above its present level. Many authorities might be cited to show the universality of this flood.

Whether the stratified drift which forms the New Haven plain[233] belongs to the epoch of this same great flood, or whether, with the Trenton gravel, it was formed in more recent times, is a question, the discussion of which brings us back to the problem presented at the beginning of this chapter, namely, the geological position of the Trenton gravel.

THE AGE OF THE TRENTON GRAVEL.

From the facts already cited, it will be seen that two hypotheses only can apply to the Trenton gravel. It is either *post*-glacial (A) or it belongs to the very last portion of the glacial period (B).

The view held by the late Thos. Belt[234] can no longer be maintained. In his numerous papers in the Quarterly Journal of the Geological Society of London, and in the Quarterly Journal of Science, he endeavored to prove the pre-glacial age of the implement-bearing beds of England and elsewhere. In a paper "On the Discovery of Stone Implements in Glacial Drift in North America," he fails to recognize any distinction between the gravels, and holds that the Trenton gravel is older than the brickclay or "pre-diluvial,"—*i. e.*, pre-champlain. As we have seen, the Trenton gravel is truly post-glacial. It only remains to define more strictly the meaning cf that term. There is evidence to support each of these hypotheses now set forth, and it may be that, in considering them in order, both may be found to be true.

(A) That the Trenton gravel is a *post-glacial* river deposit, made at a time when the river was larger than at present, is a conclusion warranted by many facts. We have seen that it represents the very

---

232 On Southern New England during the melting of the Glacier. Amer. Jour. Science, vol. x, Sept. to Dec., 1875.

233 Dana, *loc. cit.*, p. 414.

234 Quarterly Journal of Science, London, January, 1878, p. 55.

last of the gravel deposits of the upper Delaware valley. It cannot be assigned to the glacial period, except by assuming that there have been no river gravels deposited since that time — an assumption difficult to maintain. River gravels which are truly post-glacial occur in glaciated regions along the bottom of valleys. These lie only a few yards above the water, and are bordered by terraces of stratified drift. They form a sandy plain, on top of which boulders frequently lie.

Where terraces occur, they form the lowest of these and appear to have been made up from the materials of the older deposits. This same sand and gravel can be traced down the Delaware, past the terminal moraine, into the non-glaciated regions. At Belvidere, we can observe, 1st, the unstratified moraine, some distance back from the river; 2nd, the stratified drift formed from the moraine and spread out as a plain at a lower level, and which is of the same age as the terraces above the Water Gap; 3rd, a sandy river gravel newer than either of these, which forms the low, sandy plain along the river and is of limited extent. Farther south, we find this last gravel all along the stream. It becomes deeper as we go down the valley. Frequently it may be observed to form a bank close to the stream and to slope gently downwards towards the side of the valley — a feature already described as belonging to this gravel.

Finally, on reaching Trenton, we find this same gravel, with the same characteristics, but of greater depth and extent, spread out to form the plain upon which part of that city stands. There has been no break in the sequence of facts observed, and the conclusion is forced upon us that the deposit at Trenton is the same as that which borders the upper parts of the river, and that both are post-glacial. If the Trenton gravel were the same as the upper terraces of the glaciated regions, there should be some traces of such terraces between the moraine and Trenton. Yet none such have been observed, and the only continuous terrace is the lowest one, which finally merges into the Trenton plain.

The Trenton gravel differs in several respects from the stratified drift of the New Haven plain — a deposit typical of New England rivers. During a recent examination of that locality, under the guidance of Prof. Dana, it was observed that it resembled the gravel of the terraces more than that of the Trenton plain. Unlike a true river gravel, the boulders did not lie on top of the deposit, but below it, as though the glacier, in its retreat, had first dropped the boulders and then covered them with sand and gravel carried along by the flood issuing from its base. The New Haven plain, unlike that of Trenton, is characterized by numerous kettle-holes,—the result probably of ice action,—and all of its features suggest that it was formed while the melting glacier was close at hand. The Trenton gravel, on the other hand, shows no evidences of ice action. That the boulders upon its surface were dropped from ice-cakes is however probable. The materials of the gravel are composed of a mixture of pebbles brought from the stratified moraine drift of Belvidere and northward with pebbles formed in the river bed farther south, both of which the flood has brought down and restratified.

A flood of sufficient extent to produce the deposit at Trenton need not necessarily be of very great magnitude. From the upper border of Pennsylvania to Trenton, the Delaware descends over nine hundred feet—an average fall of five feet per mile. Even from the Delaware Water Gap to Trenton the descent is about four feet to the mile. Since in a great part of its course the valley is a narrow one, it will be seen that a moderate increase of the volume of the river at its headwaters could produce all the effects observed at the point where the valley suddenly opens out. A similar post-glacial flood has been recognized in England and upon the continent. Mr. Tylor[235] calls the age of the flood the "Pluvial period," remarking,[236] that "the existence of a glacial period almost necessitates that of a pluvial period, commencing prior to the glacial and continuing after it, occupying a region south of that occupied by the ice and snow."

Although the Trenton gravel was subsequent to the great melting

---

[235] Quart. Journ. Geol. Soc., vol. xxii, p. 463; vol. xxiv, p. 103.

[236] *Loc. cit.*, vol. xxiv, p. 120.

which produced the brickclay, it is possible that it was immediately subsequent to the final disappearance of the last traces of the ice at the headwaters of the Delaware, and that it is post-glacial only in a local sense. It is more recent than the glacier at the time of its retreat from Belvidere, but there is no proof that the glacier did not linger considerably later in more northern regions.

(B) Thus the second hypothesis may be true, as well as the first. In considering the Trenton gravel as entirely post-glacial, there arises the difficulty of assigning a sufficient origin for the flood which formed it. No flood within the historical epoch has been known to at all approach in magnitude those which in time deposited the Trenton gravel. No boulders of the size found in and upon that gravel are now carried down the river by floating ice. At the time of the Trenton gravel floods, the lower part of the site of Philadelphia, the whole of that of Bristol and Tullytown, and nearly all of that of Trenton, were submerged. No rain-storms within the recollection of man, or mentioned in tradition, could have supplied such an amount of water, and no origin for such extraordinary rains is suggested, except under a very different climate or by evaporation from a melting glacier.

That the climate was then cold is further indicated not only by the suggestion that there was then probably very large masses of boulder-bearing ice floating in the enlarged river, but also from the fact that fossil remains of arctic animals, as the reindeer and walrus, have been found in post-glacial deposits in New Jersey and Pennsylvania, which indicate a continuance of a colder climate than now, after the disappearance of true glacial conditions. The frequent occurrence of boulders resting upon the sand overlying the gravel suggests the grounding of large ice-cakes derived from some mass of ice large enough to be called a glacier.

Since the present channel of the river at Trenton has been excavated after the deposition of the Trenton gravel at that place, and since such excavation would necessarily begin so soon as the river ceased to deposit any gravel, it follows that the river could have flowed on top of the deposit at Trenton only when, as a flood of great volume and rapidity, it bore along large masses of gravel. Although possible, it is difficult to separate completely such a flood from the melting of a glacier. Yet, if a glacier, it must have been very different, both in age and extent, from that whose melting caused the Philadelphia brickclay. Judging from comparative erosion alone, one might be induced to think that perhaps as much time elapsed between the deposition of the brickclay and that of the Trenton gravel as has elapsed from the latter period to the present day.

From the limited extent of the Trenton gravel, it is inferred that if caused by a glacial flood, such glacier must have been either a local one or at least have had its southern extremity confined to the Delaware valley. The melting of a local glacier in the Catskill Mountains would probably result at the headwaters of the Delaware in a continued flood of sufficient volume, if supplemented by the action of floating ice, to form the Trenton gravel. Whether such a glacier was a lingering remnant of the great glacier which had retired from Pennsylvania and New Jersey, and still existed farther north, or whether there was a separate and more recent glacier belonging to a second glacial epoch, is as yet an open question.

There are not wanting evidences of a second glacial epoch in America. Intercalated beds, which, according to their geographical position contain land plants or marine shells, have frequently been found with true "till" both above and below them. These offer "undeniable evidence that animals and plants occupied the land during temperate interglacial epochs, preceded and followed by an arctic climate and ice-sheets like those now covering the interior of Greenland and the Antarctic continent."[237] Prof. Chamberlin, of Wisconsin, in a recent letter to the writer, suggests that the Philadelphia red gravel and brickclay were formed at the time of the first and most extended glaciation, and a channel excavated through it during the interval of deglaciation; while the second advance of the glacier formed the New Jersey moraine, and *its* final retreat, the Trenton gravel.

---

[237] Geol. of Minnesota, Report for 1877, p. 37; Report for 1879, p. 115.

A second glacial period in Europe, known as the "Reindeer Period," has long been recognized. It appears to have followed that in which the clays were deposited and the terraces formed, and may therefore correspond with the period of the Trenton gravel. If there have been two glacial epochs in this country, the Trenton gravel cannot be earlier than the close of the later one. If there has been but one, traces of the glacier must have continued into comparatively recent times, or long after the period of submergence. The Trenton gravel, whether made by long continued floods which followed a first or second glacial epoch,—whether separated from all true glacial action or the result of the glacier's final melting,—is truly a post-glacial deposit, but still a phenomenon of essentially glacial times—times more nearly related to the Great Ice Age than to the present.

THE ANTIQUITY OF MAN.

Interesting as is the solution of any geological problem, it is doubly so when it involves the question of the antiquity of the human race. Archæology now joins with geology to make history. When we find that the Trenton gravel contains implements of human workmanship so placed with reference to it that it is evident that at or soon after the time of its deposition man had appeared on its borders, and when the question of the antiquity of man in America is thus before us, we are tempted to inquire still further into the age of the deposit under discussion.

It has been clearly shown by several competent archæologists that the implements that have been found are a constituent part of the gravel, and not intrusive objects. It was of peculiar interest to find that it has been only within the limits of the Trenton gravel, previously traced out by the writer, that Dr. Abbott, Prof. F. W. Putnam, Mr. Lucien Carr, and others, have discovered these implements, *in situ*. The map accompanying this chapter, on which each place is marked where implements have been found beneath the surface, illustrates this point. At the localities on the Pennsylvania Railroad, where extensive exposures of these gravels have been made, the deposit is undoubtedly undisturbed. No implement could have come into this gravel except at a time when the river flowed upon it and when they might have sunk through the loose and shifting material. All the evidence points to the conclusion that at the time of the Trenton gravel flood, Man in a rude state, with habits similar to those of the River-drift Hunter of Europe, and probably under a climate similar to that of more northern regions, lived upon the banks of the ancient Delaware, and lost his stone implements in the shifting sands and gravel of the bed of that stream. The term "Eskimo period" has been suggested[238] for that of the Trenton gravel, in accordance with the view that present boreal races are the descendants of the ancient palæolithic man.

The actual age of the Trenton gravel, and the consequent date to which the antiquity of man on the Delaware should be assigned, are questions which geological data alone are insufficient to solve. The only clew, and that a most unsatisfactory one, is afforded by calculations based upon the amount of erosion. This, like all geological considerations, is relative rather than absolute, yet several calculations have been made, which, based either upon the rate of erosion of river channels, or the rate of accumulation of sediment, have attempted to fix the date of the close of the glacial epoch. By assuming that the Trenton gravel was deposited immediately after the close of this epoch, an account of such calculations may be of interest. If the Trenton gravel is *post*-glacial in the widest acceptation of the term, a yet later date must be assigned to it.

When a student of surface geology, who has lived south of glacial action, examines for the first time the true glacial drift and sees the kame-like ridges and bowl-shaped depressions maintaining regular outlines and steep slopes, he cannot but be struck with the comparatively recent look of these deposits. He cannot but believe that if the great periods of time have elapsed since their deposition, which

---

[238] Lewis. Proc. Acad. Nat. Sci. Philadel., 1880, p. 308.

some geologists maintain, the gravel ridges would be rounded down and the kettle-holes filled up by the erosive action of frost, rain and wind. Recent investigations in glacial geology are bringing forward many evidences that the final disappearance of the glacier in eastern America was not far remote.

Prof. Chamberlin[239] remarks that "no sensible denudation has taken place in Wisconsin since the glacial times in either drift bearing or driftless areas. Mr. Upham,[240] speaking of the lakes which dot the surface of Minnesota, says, "the lapse of time since the ice age has been insufficient for rains and streams to fill these basins with sediment or to cut outlets low enough to drain them; though in many instances we can see such changes slowly going forward."

Rev. G. F. Wright,[241] in a paper entitled "*An attempt to calculate approximately the date of the Glacial Era in Eastern North America, from the depth of sediment in one of the bowl-shaped depressions abounding in the Moraines and Kames of New England,*" finds that the accumulation of peaty matter in a typical kettle-hole in Massachusetts, whether caused by growth of vegetation or by winds and rains, is equal to a level deposit of eight feet in thickness. At the rate of one inch in a century, which is probably less than the true rate, this would place the close of the glacial epoch at less than ten thousand years ago.

A still more recent estimate has been made by Dr. Andrews,[242] who, from calculations based upon the erosive action of the great lakes, concludes that the total lake deposits made since the glacial epoch, were formed within seventy-five hundred years.

Another source of calculation is the recession of the falls of a river since glacial times. The most notable calculation of this kind is that made upon the recession of the Falls of Niagara. A gorge seven miles in length has been cut from Lewiston to the present falls. Mr.

---

[239] Geology of Wisconsin, vol. ii, p. 632, 1877.

[240] Geology of Minnesota, Report for 1879, p. 72.

[241] Amer. Journal Science, vol. xxi, Feb. 1881, p. 120.

[242] Transactions Chicago Academy of Sciences, vol. ii.

Bakewell estimated the annual cutting backward of the falls to be about one yard a year, but Prof. James Hall[243] and Sir Charles Lyell[244] thought that one foot a year was a more probable amount. They showed that beds containing recent shells and mastodon teeth occurred in the banks above the gorge, at the whirlpool, three miles below the falls, and also on Goat Island above the falls, indicating that in the Champlain epoch the waters of Lake Erie extended up over the gorge and present falls, and that since that period a large portion of the gorge had been excavated. They found also at the whirlpool an ancient pre-glacial channel, which, having been filled with drift in glacial times, had forced the river to cut a new channel through the rock since that period.

There are here, therefore, data for calculating the close of the glacial epoch. If the whole gorge has been cut out since that epoch, at the rate of one foot per year, thirty-five thousand years would be required. It has been, however, more than once suggested that a portion of the gorge is pre-glacial. Prof. Dana[245] supposed about one mile of it to be pre-glacial, but Mr. Belt[246] after a personal investigation concludes that the gorge above the whirlpool was excavated nearly up to the present position of the falls in pre-glacial times. After giving the evidences upon which he founds his opinion, he says:[247] "If the conclusion at which I have arrived is correct, that the gorge from the whirlpool to the falls is pre-glacial, and that the present river has only cut through the softer beds between Queenstown and the whirlpool, and above the latter point merely cleared out the pre-glacial gorge in the harder rocks, twenty thousand years or even less is amply sufficient for the work done, and the occurrence of the glacial epoch, as so measured, will be brought within the shorter

---

243 Geology of New York, vol. iv, p. 383, et seq.

244 Travels in North America, 1841–'2, vol. i, p. 22, et seq. See, also, Proc. Geol. Society of London, vol. ii, p. 77, vol. iii, p. 595, vol. iv, p. 19.

245 Manual of Geology, p. 590.

246 Quarterly Journal of Science, April, 1875, p. 135.

247 *L. c.*, p. 154.

period that, from other considerations I have argued, has elapsed since it was at its height."

A calculation of a similar kind has been made by Prof. N. H. Winchell [248] upon the recession of the falls of St. Anthony, since the last glacial epoch. These falls, in the Mississippi river, were discovered in 1680, and a continuous record of their recession may be found since then. A narrow gorge, formed by their recession, extends from the falls to Fort Snelling, eight miles south. Below this point the valley widens, and shows indisputable evidence of having been excavated in pre-glacial times. From the falls to Fort Snelling, however, the drift, which lies above the rocky walls of the gorge, has been cut through so as to form a bluff on either side; this fact, when taken in connection with others pointing to the same conclusion, clearly shows the post-glacial age of this gorge. An ancient channel of the river, now filled with glacial-drift, is described and the evidence seems decisive that, since the glacial epoch, the river, having been forced out of its old channel, has cut out a new one eight miles long, through the rock. Unlike the rocks at Niagara, those at the Falls of St. Anthony are horizontal and of unvarying composition, and any conclusions made here will be of much greater accuracy. Prof. Winchell gives three separate measurements, which result in the following terms of years required for the total recession, viz.:—12,103 years; 6,276 years; and 8,202 years. He holds that an average of these rates—8,860 years—represents the time which has elapsed since the maximum cold of the last glacial epoch.

Thus we find, that if any reliance is to be placed upon such calculations, even if we assume that the Trenton gravel is of glacial age, it is not necessary to make it more than ten thousand years old. The time necessary for the Delaware to cut through the gravel down to the rock is by no means great. When it is noted that the gravel cliff at Trenton was made by a side wearing away as at a bank, and when it

---

248 Geol. of Minnesota, Annual Rep. for 1876, p. 156 et seq. See also, Quart. Journal Geol. Society of London, Nov. 1878, p. 886.

is remembered that the erosive power of the Delaware was formerly greater than at present, it will be conceded that the presence of the cliff at Trenton will not necessarily infer its high antiquity; nor in the character of the gravel is there any evidence that the time of its deposition need have been long. It may be that as investigations are carried further, it will result not so much in proving man of very great antiquity, as in showing how much more recent than usually supposed was the final disappearance of the glacier.

In all these subjects we are but at the threshold of understanding. We are entering a field where many sciences meet and where each must help the other. No single investigation is sufficient. In the present discussion, the aim has been to define the age of the Trenton gravel, and the consequent antiquity of Man in the Delaware valley solely with reference to geological events.

At the present stage of the discussion, theories are necessarily somewhat provisional; but as facts multiply and others enter this wide field of investigation, we may hope at no distant date to arrive at more definite and trustworthy conclusions.

INDEX.